The Antioxidant Vitamins C and E

Editors

Lester Packer
Department of Molecular Pharmacology
and Toxicology
University of Southern California
Los Angeles, California

Maret G. Traber
Linus Pauling Institute
Oregon State University
Corvallis, Oregon

Klaus Kraemer
BASF Aktiengesellschaft
Ludwigshafen, Germany

Balz Frei
Linus Pauling Institute
Oregon State University
Corvallis, Oregon

LONDON AND NEW YORK

First published 2002 by AOCS Press.

Published 2018 by Routledge
2 Park Square, Milton Park, Abingdon, Oxon OX14 4RN
52 Vanderbilt Avenue, New York, NY 10017

Routledge is an imprint of the Taylor & Francis Group, an informa business

Library of Congress Cataloging-in-Publication Data

The antioxidant vitamins C and E / editors, Lester Packer ... [et al.].
    p. cm.
Proceedings of a symposium on vitamins C and E, held March 6-9, 2002 at the 2002 World Congress of the Oxygen Club of California, Santa Barbara, California.
Includes index.
  ISBN 1-893997-29-4 (hardcover : alk. paper)
  1. Vitamin C--Congresses. 2. Vitamin E--Congresses. 3. Antioxidants--Congresses. I. Packer, Lester.

QP772.A8 A586 2002
613.2'86--dc21

2002152005

ISBN 13: 978-1-893997-29-5 (hbk)

Acknowledgment

The Oxygen Club of California (OCC) symposium on the antioxidant vitamins C and E and this book were made possible through support from BASF Aktiengesellschaft.

Preface

Early in the last century, vitamins C and E were identified as essential micronutrients for humans. Subsequent studies established the important roles of these vitamins as the body's major dietary antioxidants. Vitamin C (ascorbic acid) and the various forms of vitamin E react with many different free radicals and reactive oxygen and nitrogen species. When scavenging free radicals, vitamin E in cell membranes and lipoproteins and vitamin C in aqueous intra- and extracellular fluids form one- and two-electron oxidation products. Initially, demonstrations in chemical systems and later in biochemical systems showed that vitamin E radicals formed by oxidant or free radical exposure were reduced by ascorbic acid, resulting in regeneration of vitamin E. Vitamin E, which is usually present only in relatively low concentrations in the body, may thus be spared by this vitamin C action. Both vitamins appear to interact in a network of redox-active antioxidants including thiols, NADH- and NADPH-dependent enzymes, and bioflavonoids.

In addition to their antioxidant functions as electron or hydrogen donors in free radical reactions, vitamins C and E exhibit many other activities in biological systems. Ascorbic acid is a coenzyme (or more accurately, a cosubstrate) in hydroxylation reactions during collagen biosynthesis and also functions in various metabolic reactions. Most studies with the various members of the vitamin E family of molecules have focused on α-tocopherol, which is maintained in the circulation and tissues by the α-tocopherol transfer protein(s). Interestingly, other vitamin E forms, such as γ-tocopherol and α-tocotrienol, exhibit unique properties in various cell and in vitro systems and are the subject of emerging interest.

Both vitamins C and E by virtue of regulating oxidative processes through their antioxidant actions have important effects on regulating cell signaling and gene expression. Although oxidants and oxidative stress conditions affect many of the essential pathways involved in gene transcription and protein synthesis, these antioxidants may themselves be involved in redox regulation of cell signaling. Hence, vitamins C and E affect cell growth and proliferation and also participate in some of the cell death pathways involving apoptosis or necrosis. Importantly, α-tocopherol specifically inhibits certain cell signaling pathways dependent upon protein kinase C and interacts with a number of binding proteins.

Clearly, the areas of greatest interest for both the public and the scientific community are the roles of vitamins C and E in disease prevention, in retarding the progression of age-related chronic and degenerative disorders, and in promoting healthy aging. The introductory chapters (1 and 9) in this book provide a concise historical background of vitamins C and E and summarize many important discoveries concerning their biological functions and their roles in prevention of dietary deficiency, chronic disease and oxidative stress. Other chapters detail current research on the biological actions of vitamins C and E in biochemical, cell and physiological systems, and their roles in "oxidative stress-related" diseases and disorders. These latter chapters highlight the importance of vitamins C and E in metabolic and inflammatory disorders, in age-related diseases and in healthy aging.

Many areas of future study are apparent from the findings on vitamins C and E reported in this volume. The findings raise questions concerning the relative importance of free radical scavenging as opposed to other cellular regulatory mechanisms. In addition, determination of the optimal amounts of vitamins C and E required during aging for health through food sources as opposed to vitamin supplements, which are highly popular, warrants further clinical and epidemiologic investigations.

We thank the OCC (Oxygen Club of California) 2002 World congress co-sponsored by the Linus Pauling Institute for hosting a workshop on "The Antioxidant Vitamins C and E." The workshop, which was sponsored by BASF, was the basis for the invitations to contribute to this volume. We are especially grateful to all of the authors, who are leading scholars in vitamins C and E research, for their excellent articles.

Lester Packer
Maret G. Traber
Klaus Kraemer
Balz Frei

Contents

Chapter 1

Vitamin C: An Introduction

Jane V. Higdon and Balz Frei

Linus Pauling Institute, Oregon State University, Corvallis, OR 97331

Structure and Chemistry of Vitamin C (L-Ascorbic Acid)

The chemical name for L-ascorbic acid is 2,3-didehydro-L-threo-hexano-1,4-lactone. Carbon 5 of ascorbic acid (Fig. 1.1) is asymmetric, making two enantiomeric forms possible; L-ascorbic acid is the naturally occurring and biologically active form. L-Ascorbic acid is a water-soluble 6-carbon α-ketolactone with two enolic hydrogen atoms (pK_{a1} at carbon 3 = 4.17 and pK_{a2} at carbon 2 = 11.57; Fig. 1.1) (1,2). At physiologic pH, >99% of L-ascorbic acid is ionized to L-ascorbate, which can donate a hydrogen atom ($H^+ + e^-$) to produce the resonance-stabilized ascorbyl radical (Fig. 1.1). The ascorbyl radical can donate a second electron to form the 2-electron oxidation product of ascorbate, dehydroascorbic acid (DHA), or dismutate to form ascorbate and DHA (Fig. 1.1). Alternatively, the ascorbyl radical may be enzymatically reduced back to ascorbate by NADH-dependent semidehydroascorbate reductase or the NADPH-dependent selenoenzyme, thioredoxin reductase. DHA can be reduced back to ascorbate by the glutathione-dependent enzyme, glutaredoxin, or thioredoxin reductase (3). If not recycled to ascorbate, DHA is irreversibly hydrolyzed to 2,3-diketo-L-gulonic acid (DKG), which does not function as an antioxidant. Further degradation of DKG results in the formation of oxalic acid and L-threonic acid. Other catabolites include, among many others, L-xylonic acid, L-lyxonic acid, and L-xylose (4). The term vitamin C is generally used to describe all compounds that qualitatively exhibit the biological activity of ascorbate, including ascorbate and DHA (2).

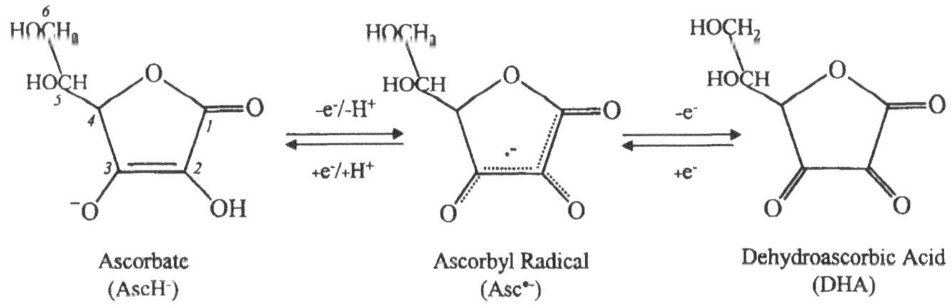

Fig. 1.1. Oxidation of ascorbate (AscH$^-$) by two successive one-electron oxidation steps to give the ascorbyl radical (Asc$^{\bullet-}$) and dehydroascorbic acid (DHA), respectively.

1

Biological Activities of Vitamin C

Electron Donor for Enzymes

The physiologic functions of vitamin C are related to its efficacy as a reducing agent or electron donor. Vitamin C is known to be a specific electron donor for eight human enzymes (1). Three of those enzymes (proline hydroxylase, lysine hydroxylase, and procollagen-proline 2-oxoglutarate 3-dioxygenase) participate in the post-translational hydroxylation of collagen, which is essential for the formation of stable collagen helices (5). Many of the manifestations of the vitamin C–deficiency disease, scurvy, are related to defective collagen synthesis, including blood vessel fragility (petechiae, ecchymoses, and inflamed bleeding gums), tooth loss, bone and connective tissue disorders, and impaired wound healing. Vitamin C is also necessary for the maximal activity of two dioxygenase enzymes (γ-butyrobetaine 2-oxoglutarate 4-dioxygenase and trimethyllysine 2-oxoglutarate dioxygenase) required for L-carnitine biosynthesis. L-Carnitine is required for the transport of activated fatty acids across the inner mitochondrial membrane and plays critical roles in modulating energy metabolism (6). Fatigue and lethargy, which are early signs of scurvy, likely are related to carnitine deficiency. Vitamin C is also a cosubstrate for dopamine-β-monooxygenase, the enzyme that catalyzes the conversion of the neurotransmitter dopamine to norepinephrine. Neuropsychiatric changes associated with scurvy, including depression, mood swings, and hypochondria, could be related to deficient dopamine hydroxylation (3). Two other enzymes are known to require vitamin C as a cosubstrate, although their connection to the pathology of scurvy has not been established. Peptidyl glycine α-amidating monooxygenase is required for peptide amidation and 4-hydroxylaphenylpyruvate dioxygenase is required for tyrosine metabolism (5).

Vitamin C may also play a role in the metabolism of cholesterol to bile acids and in steroid metabolism as a cosubstrate of the enzyme 7α-monooxygenase (7). The hydroxylation of xenobiotics and carcinogens by the cytochrome P450 family of enzymes is also enhanced by reducing agents, such as vitamin C (8). Vitamin C has been found to enhance the activity of endothelial nitric oxide synthase (eNOS) by maintaining its cofactor tetrahydrobiopterin in the reduced, and thus active, form (9). Synthesis of nitric oxide (NO) by eNOS plays a critical role in maintaining normal endothelial function (10). The finding that vitamin C enhances eNOS activity through its activity as a reducing agent supports the idea that sufficient vitamin C may contribute to the prevention of cardiovascular diseases (see below).

Recent research suggests that a family of oxygen-dependent prolyl hydroxylase enzymes plays an important role in the ability of cells to recognize and respond to hypoxia (11). Under normoxic conditions, hydroxylation of a conserved proline residue on the hypoxia-inducible transcription factor α-subunit (HIF-1α) results in rapid proteasome-mediated degradation. Hypoxia inhibits prolyl hydroxylation, allowing HIF-1α to accumulate and migrate to the nucleus where it activates hypoxia-responsive genes (12). Although ascorbate is known to increase the activity of these oxygen-dependent prolyl hydroxylase enzymes *in vitro* (11,13), its significance to this "oxygen sensing" pathway *in vivo* remains to be determined.

Antioxidant Activity

An antioxidant has been defined as "a substance that, when present at low concentrations compared with those of an oxidizable substrate, significantly delays or prevents oxidation of that substrate" (14). Several properties make vitamin C an ideal antioxidant in biological systems. First, the low one-electron reduction potentials of ascorbate and the ascorbyl radical enable these compounds to react with and reduce virtually all physiologically relevant reactive oxygen species (ROS) and reactive nitrogen species (RNS), including superoxide, hydroperoxyl radicals, aqueous peroxyl radicals, singlet oxygen, ozone, nitrogen dioxide, nitroxide radicals, and hypochlorous acid (15). Hydroxyl radicals also react rapidly, although not preferentially, with vitamin C; hydroxyl radicals are so reactive that they combine indiscriminately with any substrate at a diffusion-limited rate. Vitamin C also acts as a co-antioxidant by regenerating α-tocopherol from the α-tocopheroxyl radical. This may be an important function because *in vitro* experiments have found that α-tocopherol can act as a prooxidant in the absence of co-antioxidants such as vitamin C (16). Another property that makes vitamin C an ideal antioxidant is the low reactivity of the ascorbyl radical formed when ascorbate scavenges ROS or RNS. The ascorbyl radical is neither strongly oxidizing nor strongly reducing and it reacts poorly with oxygen. Thus, when a reactive radical interacts with ascorbate, a much less reactive radical is formed. The ascorbyl radical scavenges another radical or rapidly dismutates to form ascorbate and DHA. Alternatively, the ascorbyl radical and DHA can be reduced enzymatically or recycled back to ascorbate (see above).

Dietary Iron Absorption

The ability of ascorbate to maintain metals ions in a reduced state is critical to the function of the mono- and dioxygenases discussed above (7,8). Concomitant consumption of vitamin C from food or supplements enhances nonheme iron absorption from a single meal in a dose dependent manner (17), probably because the reduction of iron by ascorbate makes it less likely to form insoluble complexes with phytate and other ligands (1). Iron deficiency is the most common nutrient deficiency in the world and is associated with a number of adverse health effects (18). Consequently, the potential for increased vitamin C intake to improve iron nutritional status by increasing the bioavailability of dietary nonheme iron has received considerable attention. Despite consistent findings of enhanced iron absorption from a single meal in the presence of vitamin C, several intervention studies were not able to demonstrate that increasing vitamin C intake improved iron nutritional status (19,20). More recent research indicates that the enhancing effect of vitamin C on iron absorption from a complete diet, rather than a single meal, may be offset partially by dietary inhibitors of iron absorption (21). Further research is required to determine whether increasing vitamin C intake is an effective strategy for improving iron nutritional status.

Uptake, Distribution, and Metabolism of Vitamin C

Intracellular Transport

Ascorbate is actively transported into cells *via* Sodium-dependent Vitamin C Transporters known as SVCT1 and SVCT2 (22). SVCT1 is expressed mainly on the epithelial surfaces of the intestine and kidney, and in the liver, whereas SVCT2 expression has been found in a number of tissues, including neurons, endocrine tissue, and bone. This distribution suggests that SVCT1 is involved with bulk transport of ascorbate, whereas SVCT2 is involved in tissue-specific ascorbate uptake (23). DHA can be transported into cells by facilitated diffusion *via* the glucose transporters, GLUT1, GLUT3 and to some extent, GLUT4. Intracellularly, DHA is immediately reduced to ascorbate through chemical reduction by glutathione or enzymatic reduction (4).

Intestinal Absorption

Intestinal absorption of ascorbate appears to occur mainly through active transport *via* sodium-dependent ascorbate transporters on the apical side of enterocytes, whereas the transport mechanism responsible for basolateral efflux of ascorbate from enterocytes is not yet known (23). Intestinal absorption of DHA has not been well characterized in humans (4,5). Although DHA has antiscorbutic effects when administered to humans, at least one study suggests that the absorption of DHA is less than that of ascorbate (24). Glucose has been found to inhibit ascorbate and DHA uptake by human small intestinal brush border membrane vesicles (25) and by human neutrophils (26).

Bioavailability is defined in pharmacokinetic terms as the difference between the increase in plasma levels of a substance after a dose given intravenously and that after the same dose given orally. The only study examining the true bioavailability of ascorbate calculated that the bioavailability of a liquid solution of ascorbate given to fasting men at steady state was >80% for doses ≤100 mg, 78% for a 200-mg dose, 75% for a 500-mg dose, and 62% for a 1250-mg dose (27).

Plasma Concentrations

Plasma ascorbate concentrations in people without scurvy generally range from 11 to 90 μmol/L, whereas DHA is not generally detectable in plasma (<2% of total ascorbate). Plasma ascorbate concentrations <11 μmol/L indicate vitamin C deficiency, and concentrations between 11 and 28 μmol/L represent marginal vitamin C status (2). Data on plasma ascorbate levels collected from 1988 to 1994 during the third National Health and Nutrition Examination Survey (NHANES III) indicated that the prevalence of vitamin C deficiency in the United States ranges from 9% in women to 13%

in men and the prevalence of marginal vitamin C status ranges from 17% in women to 24% in men (28).

In two pharmacokinetic studies conducted in healthy young men and women, steady-state plasma ascorbate concentrations increased rapidly at vitamin C doses between 30 and 100 mg/d, suggesting that varying vitamin C intake within that range may result in significant differences in the availability of ascorbate to tissues (29,30). At doses of 200 mg/d, the rate of increase in steady-state plasma ascorbate concentrations decreased, and plasma ascorbate levels increased very little at doses >400 mg/d.

Tissue concentrations

Tissue ascorbate concentrations vary greatly depending on the tissue type, with the highest concentrations found in adrenal and pituitary glands and slightly lower concentrations found in liver, spleen, lens, pancreas, kidney, and brain (4). Intracellular ascorbate concentrations in lymphocytes, neutrophils, and monocytes obtained from healthy young men and women were saturated at vitamin C doses of 100 mg/d and reached concentrations of 1–4 mmol/L, i.e., at least one order of magnitude higher than plasma concentrations (29,30).

Excretion

Urine is the primary route of excretion for ascorbate and its metabolites in humans. In the kidney, ascorbate is filtered by the glomerulus and actively reabsorbed by sodium-dependent ascorbate transporters in the proximal tubules (4,23). Active reabsorption of ascorbate is saturable, and human plasma ascorbate concentrations appear limited by the capacity for renal reabsorption. In healthy men and women at steady-state conditions, the threshold for urinary ascorbate excretion was observed at vitamin C intakes between 60 and 100 mg/d, and most of the ascorbate from intakes ≥500 mg/d was excreted in the urine within 24 h (29,30). Because DHA cannot be detected in plasma, no information is available regarding renal excretion or reabsorption of DHA (4). Limited research in healthy men indicates that high doses of supplemental vitamin C ranging from 1000 to 10,000 mg/d increases urinary oxalate excretion, but not above normal reference ranges of 20–60 mg/d (28).

Sources and Intake Recommendations

Recommended Intake of Vitamin C

Unlike most mammals, humans and other primates obtain vitamin C exclusively from their diets because they lost the ability to synthesize vitamin C from glucose during the course of evolution. This defect is due to mutations in the gene encoding the final enzyme of the vitamin C biosynthetic pathway, L-gulonolactone oxidase (1). To prevent scurvy, an adult must consume ~10 mg/d of vitamin C, an amount easily obtained from as little as one serving/d of most fruits and vegetables. Although the amount of vitamin C required for the prevention of scurvy in humans has been well established, the optimal intake of vitamin C is likely to be considerably higher and to vary with life stage, gender, and disease state.

The current recommended dietary allowance (RDA) for vitamin C is 90 mg/d for men and 75 mg/d for women (31). The recommended intake for smokers is 35 mg/d higher than for nonsmokers, because ascorbate turnover is ~35 mg/d greater in smokers, presumably due to increased oxidative stress and other metabolic differences. The previous RDA of 60 mg/d for men and women was based in part on the prevention of scurvy with a 4-wk margin of safety (1). The current RDA is based on the vitamin C intake required for 80% neutrophil saturation with little urinary loss in healthy men. At the time the recommendation was made, similar data were not available for women, and the RDA was extrapolated on the basis of body weight. Recently, the results of a similar study in women were published along with a recommendation that the RDA for vitamin C be raised to 90 mg/d for women as well (30).

Although the current RDA is no longer based solely on the prevention of scurvy, it continues to be based primarily on the prevention of deficiency, rather than the prevention of chronic disease and the promotion of optimal health. Pharmacokinetic studies in healthy young men and women found that leukocytes generally became saturated at vitamin C intakes between 100 and 200 mg/d, and these intake levels have been associated epidemiologically with decreased risk of chronic disease, particularly cancer, heart disease, and stroke. Thus, for healthy individuals, a vitamin C intake of at least 200 mg/d should be considered on the basis of preventing chronic disease and promoting optimal health. The amounts of vitamin C required to maintain optimal body levels in special populations, such as children, pregnant women, and older adults, have not been established. Similarly, the amounts of vitamin C required to derive therapeutic benefits in individuals affected by chronic diseases are not known and most probably are higher than current recommendations for healthy individuals.

Dietary Sources of Vitamin C

A daily vitamin C intake of at least 200 mg is easily obtained by consuming 5 servings of fruits and vegetables. Increased fruit and vegetable intakes have been consistently associated with decreased risk of chronic diseases (32–34), and indeed most of the epidemiologic evidence associating increased vitamin C intake with decreased chronic disease risk is based on vitamin C consumption from fruits and vegetables. Therefore, fruits and vegetables should be considered the preferred vehicle for increasing one's vitamin C intake. Fresh vegetables, fruits, and juices are the richest sources of vitamin C. The vitamin C content in foods may be decreased by prolonged storage and some cooking practices. Boiling vegetables has been found to result in vitamin C losses from 50 to 80%. Steaming vegetables in minimal amounts of water or cooking them in a microwave oven substantially decreases the loss of vitamin C from cooking (35,36).

Supplements

Vitamin C supplements are a significant source of vitamin C intake in the United States. Data from NHANES III indicate that ~40% of the U.S. population surveyed had taken at least one nutritional supplement in the past month (37). The most common ingredient in those supplements was vitamin C, which was present in 45% of the supplements. In general, studies comparing the bioavailability of vitamin C from foods with that from supplements have found little difference (38). Studies comparing the bioavailability of vitamin C from different types of vitamin C supplements (ascorbic acid, mineral ascorbates, ascorbate plus vitamin C metabolites, and ascorbate with flavonoids) have not generally found differences (39). One exception is a study that found plasma ascorbate levels and 24-h urinary ascorbate excretion to be increased by 35% when ascorbate was taken with a citrus extract containing bioflavonoids (40). The significance of these results is unclear given recent findings that flavonoids may inhibit cellular uptake of ascorbate (41,42).

Milestones in Vitamin C Research

Recognition of the Biological Importance of Vitamin C in the Prevention of Scurvy and Associated Symptoms

Diseases likely to be scurvy have been reported throughout written history. Known as the "calamity of sailors," scurvy has also been recorded during famines, sieges, imprisonment, and long expeditions over land. James Lind reported the benefits of citrus fruits in treating scurvy in his "Treatise on Scurvy" in 1753, but it was not until 1795 that the British admiralty mandated a daily dose of citrus juice for British seamen, the origin of the term, "limey" (43). Although it was acknowledged that citrus fruits could prevent and cure scurvy, the concept that the disease was caused by the lack of an essential nutrient in the diet was not generally accepted at the beginning of the 20th century. Reports by Axel Holst and Theodor Frölich in 1907 that scurvy could be produced experimentally in guinea pigs by feeding a diet lacking fresh fruits or vegetables, and the proposal by Casmir Funk in 1912 that scurvy, pellagra, rickets, and beriberi were due to dietary deficiencies of factors he called "vitamines" led to 20 years of intensive efforts toward isolating the antiscorbutic factor (44). The complementary findings of two research groups, that of Charles King at the University of Pittsburgh in the United States and that of Albert Szent-Györgyi at the University of Szeged in Hungary, led to the discovery of vitamin C as the antiscorbutic factor in 1932 (45,46). In 1933, E.L. Hirst and W.N. Hayworth elucidated the structure of vitamin C, using material isolated by Szent-Györgyi, and vitamin C was first synthesized the same year (47). Szent-Györgyi was awarded the 1937 Nobel Prize for Physiology or Medicine, in part, for his work in isolating vitamin C as the antiscorbutic factor, and Hayworth was awarded the Nobel Prize for Chemistry the same year. The isolation, identification, and synthesis of ascorbic acid laid the foundation for research into the role of vitamin C in health and disease that continues today.

Recognition of the Role of Vitamin C as an Important Biological Antioxidant

In the 1950s, the free radical theory of aging hypothesized that free radicals arising from enzymatic and nonenzymatic reactions inside and outside cells contributed to the aging process (48). However, the presence of free radicals in biological systems was not generally considered likely until the discovery of superoxide dismutase in 1969 (49). Currently, there is a great deal of evidence that ROS and RNS play significant roles in aging and a number of chronic diseases. Ascorbate and the ascorbyl radical readily scavenge virtually all physiologically relevant ROS and RNS (see above). Ascorbate has also been found to be the most effective endogenous aqueous-phase antioxidant in human plasma under many different oxidizing conditions (50,51). Although other endogenous antioxidants are able to decrease the rate of detectable lipid peroxidation, only ascorbate is reactive enough to intercept oxidants before they can cause detectable oxidative damage to lipids. These experimental data are in agreement with a thermodynamic hierarchy or "pecking order," indicating that vitamin C is the terminal small molecule antioxidant (15).

Stimulation of Interest in the Use of Vitamin C to Prevent and Treat Diseases Other than Scurvy

During the late 1960s, Linus Pauling, the only person ever to receive two unshared Nobel Prizes (Chemistry in 1954 and Peace in 1962), became fascinated with the role of vitamin C in maintaining health. In 1970, he wrote the book, *Vitamin C and the Common Cold*, which generated a great deal of public interest as well as scientific controversy over the potential health benefits of supplemental vitamin C. Later, Dr. Pauling became increasingly interested in potential roles of vitamin C and other micronutrients to prevent and treat certain chronic and hereditary—as opposed to deficiency—diseases, a scientific discipline he termed "orthomolecular medicine." To that end, he published several more books and established the Linus Pauling Institute of Science and Medicine (52). After Linus Pauling's death in 1994, the Linus Pauling Institute moved to the campus of Oregon State University, his undergraduate alma mater, where scientists continue to conduct research into the roles of vitamins, micronutrients, and phytochemicals in disease prevention and treatment.

On the basis of cross-species comparisons, evolutionary arguments, and the amount of vitamin C likely consumed in a raw plant food diet, Dr. Pauling reasoned in 1970 that the optimum daily vitamin C intake would be at least 2300 mg/d for an adult with an energy requirement of 2500 kcal/d (53). Scientific evidence accumulated since then from epidemiologic, biochemical and clinical studies, many of which are reviewed in this book, has established that much more moderate vitamin C intakes of between 100 and 200 mg/d are associated with tissue saturation in healthy adults and reduced risk from chronic disease (1–3). Although several of Dr. Pauling's views on vitamin C and health have been proved incorrect, his pioneering efforts in stimulating scientific, medical, and popular interest in the roles of micronutrients, and vitamin C in particular, in promoting optimal health and preventing chronic disease have had a lasting effect, for better or for worse, and cannot be ignored.

Beneficial Effects of Dietary Vitamin C in Chronic Disease Prevention

Cardiovascular diseases. Numerous large epidemiologic studies have found a significant inverse relationship between vitamin C intake and cardiovascular or cerbrovascular disease risk (3). Those studies that did not find a relationship between vitamin C intake and cardiovascular disease risk often compared individuals who were already consuming close to 100 mg/d of vitamin C with those consuming higher amounts. Because human leukocytes in healthy young men and women are saturated at vitamin C doses of ~100 mg/d (29,30), it is possible that once tissue saturation has been achieved, additional protective effects of vitamin C against cardiovascular diseases are small and, therefore, difficult to detect. Consistent with this notion, numerous prospective studies have found low plasma ascorbate levels at baseline to be associated with a subsequent increase in the risk of heart disease or stroke (3,54,55).

Endothelium-derived NO is a critical molecule for maintaining healthy endothelial function (56). In addition to inducing vasodilation by stimulating vascular smooth muscle relaxation, NO inhibits other potentially atherogenic processes, such as smooth muscle proliferation, platelet aggregation, and leukocyte-endothelial cell interactions. Endothelium-dependent vasodilation of the brachial artery can be measured noninvasively in humans. Treatment with vitamin C has consistently resulted in improved endothelium-dependent vasodilation in individuals with coronary artery disease, angina pectoris, hypercholesterolemia, hypertension, or diabetes (3). One study found endothelium-dependent vasodilation in patients with coronary artery disease to improve by 40% after oral supplementation with 500 mg/d of vitamin C for 4 wk (57). As previously noted, ascorbate has been found to increase NO synthesis in endothelial cells by maintaining the NOS cofactor, tetrahydrobiopterin, in the reduced form, thereby enhancing eNOS activity (9).

Hypertension significantly increases the risk of cardiovascular and cerebrovascular diseases. Plasma ascorbate levels were inversely correlated with systolic and diastolic blood pressure in a cross-sectional study of >500 men and women in the U.K. (58). In healthy people who were fed a vitamin C–deficient diet for 30 d and a vitamin C–sufficient diet for another 30 d, plasma ascorbate levels were also inversely correlated with diastolic blood pressure (59). At least two intervention studies in patients with mild-to-moderate hypertension have found that 4 wk of oral supplementation with 500 mg/d of vitamin C resulted in significant decreases in systolic or diastolic blood pressure (60,61).

Diabetes mellitus. A number of observational studies have found that people with diabetes mellitus have plasma or serum ascorbate levels at least 30% lower than those without the disease (62). Plasma ascorbate levels were also significantly and inversely correlated with glycosylated hemoglobin (HbA1c) levels in a cross-sectional study of >6000 British men and women (63). However, it is not yet clear whether lower plasma ascorbate concentrations observed in diabetics are related to an increased requirement for vitamin C or other factors, such as diet and lifestyle. A large population-based study found that differences in serum ascorbate concentrations among >200 individuals with newly diagnosed diabetes and 1800 individuals without diabetes disappeared after adjustment for vitamin C consumption and cigarette smoking (64). Diabetes represents a condition of increased oxidative stress (65,66), which may be related to the fact that diabetic individuals are at more than twice the risk of death from cardiovascular and cerebrovascular diseases (67). Vitamin C supplementation has also been found to improve endothelium-dependent vasodilation (see above), which is commonly impaired in diabetic individuals. Thus, ensuring sufficient vitamin C intake may be beneficial in preventing some of the complications of diabetes mellitus.

Cancer.   A number of well-designed epidemiologic studies have suggested a protective role for dietary vitamin C, especially with respect to cancers of the lung and digestive tract. For the most part, dietary vitamin C intake, mainly from fruits and vegetables, rather than supplements, appeared to be the source of the protective effect (3,34). Such studies are the basis for dietary guidelines endorsed by the U.S. Department of Agriculture and the National Cancer Institute, which recommend at least five servings of fruits and vegetables per day. In general, prospective studies in which the lowest intake group consumed >86 mg/d of vitamin C did not observe significant differences in cancer risk, whereas those studies that observed significant cancer risk reductions found them in people consuming at least 80–110 mg/d of vitamin C (3). One prospective study that followed 870 men for 25 y found that those who consumed >83 mg/d of vitamin C had a striking 64% reduction in lung cancer compared with those who consumed <63 mg/d (68).

A number of observational studies have found increased dietary vitamin C intake to be associated with decreased risk of stomach cancer, and laboratory experiments have shown that vitamin C inhibits the formation of carcinogenic N-nitroso compounds in the stomach (69). Infection with the bacteria, *Helicobacter pylori*, is known to increase the risk of stomach cancer and also appears to lower the vitamin C content of stomach secretions (70). Although two intervention studies did not find a decrease in the occurrence of stomach cancer with vitamin C supplementation (31), more recent research suggests that vitamin C supplementation may be a useful addition to standard *H. pylori* eradication therapy in reducing the risk of gastric cancer (71).

Cataracts.   Cataracts are a leading cause of blindness in the United States, occurring more frequently and becoming more severe with age. Some, but not all, studies have observed increased dietary vitamin C intake (72,73) and increased blood levels of vitamin C (74) to be associated with a decreased risk of cataracts. Two intervention trials examined the effect of vitamin C supplementation in combination with other nutrients on cataract risk. Supplementation of >2000 men and women in Linxian, China with 120 mg/d of vitamin C and 30 µg/d of molybdenum for 5 y resulted in a nonsignificant 23% reduction in cataract risk, whereas in a separate but similar trial, multivitamin and mineral supplementation resulted in a significant 36% risk reduction (75). More recently, a 7-y intervention trial of a daily antioxidant supplement containing 500 mg of vitamin C, 400 ɪᴜ of vitamin E and 15 mg of β-carotene in >4000 men and women found no difference between the antioxidant combination and a placebo on the development and progression of age-related cataracts (76). Consequently, the relationship between vitamin C intake and cataract risk requires further clarification before specific recommendations can be made.

Recognition of a Vitamin C Deficiency Disease Other than Scurvy

The creation of mice deficient in the ortholog (Slc23a1) of the sodium-dependent ascorbic acid transporter, SVCT2, has recently led to the observation of a vitamin C deficiency syndrome other than scurvy. Despite maternal vitamin C supplementation, ascorbate was undetectable or markedly reduced in the tissues of newborn *Slc23a1⁻/⁻* mice, indicating that the Slc23a1 is required for ascorbate transport across the placenta and into many fetal tissues (77). Slc23a1 deficient mice died within minutes of birth due to respiratory failure and intraparenchymal brain hemorrhage. Newborn *Slc23a1⁻/⁻* mice did not exhibit the generalized vascular fragility that is characteristic of scurvy, and normal skin 4-hydroxyproline levels indicated that the post-translational processing of collagen was not affected, also suggesting they were not suffering from scurvy. The lethal effects of this ascorbate transporter deficiency in newborn mice provide evidence of previously unrecognized roles for vitamin C in perinatal development.

Conclusion

The consequences of insufficient vitamin C intake have been recorded throughout history, yet ascorbic acid was not isolated and recognized as the antiscorbutic factor until 1932. Although the need for small amounts of vitamin C to prevent scurvy is now widely recognized, optimal intakes of vitamin C for health promotion and chronic disease prevention and treatment remain controversial. Determining the optimal vitamin C intakes and understanding underlying biochemical mechanisms will require a plethora of scientific approaches, including biochemical, pharmacokinetic, toxicologic, epidemiologic, and intervention studies, some of which are detailed in the chapters that follow this introduction.

References

1. Levine, M., Rumsey, S.C., Daruwala, R., Park, J.B., and Wang, Y. (1999) Criteria and Recommendations for Vitamin C Intake, *J. Am. Med. Assoc. 281*, 1415–1423.
2. Jacob, R.A. (1999) Vitamin C, in *Nutrition in Health and Disease*, 9th edn. (Shils, M.E., Olson, J.A., Shike, M., and Ross, A.C., eds.) pp. 467–483, Williams & Wilkins, Baltimore.
3. Carr, A.C., and Frei, B. (1999) Toward a New Recommended Dietary Allowance for Vitamin C Based on Antioxidant and Health Effects in Humans, *Am. J. Clin. Nutr. 69*, 1086–1107.

4. Rumsey, S.C., and Levine, M. (1998) Absorption, Transport, and Disposition of Ascorbic Acid in Humans, *J. Nutr. Biochem. 9*, 116–130.

5. Levine, M., Rumsey, S., Wang, Y., Park, J., Kwon, O., Xu, W., and Amano, N. (1996) Vitamin C, in *Present Knowledge in Nutrition*, 7th edn. (Ziegler, E.E., and Filer, L.J., eds.) pp. 146–159, ILSI Press, Washington.

6. Rebouche, C.J. (1999) Carnitine, in *Nutrition in Health and Disease*, 9th edn. (Shils, M.E., Olson, J.A., Shike, M., and Ross, A.C., eds.) pp. 505–512, Williams & Wilkins, Baltimore.

7. Burri, B.J., and Jacob, R.A. (1997) Human Metabolism and the Requirement for Vitamin C, in *Vitamin C in Health and Disease* (Packer, L., and Fuchs, J., eds.) pp. 341–366, Marcel Dekker, New York.

8. Tsao, C.S. (1997) An Overview of Ascorbic Acid Chemistry and Biochemistry, in *Vitamin C in Health and Disease* (Packer, L., and Fuchs, J., eds.) pp. 25–58, Marcel Dekker, New York.

9. Huang, A., Vita, J.A., Venema, R.C., and Keaney, J.F., Jr. (2000) Ascorbic Acid Enhances Endothelial Nitric-Oxide Synthase Activity by Increasing Intracellular Tetrahydrobiopterin, *J. Biol. Chem. 275*, 17399–17406.

10. Gewaltig, M.T., and Kojda, G. (2002) Vasoprotection by Nitric Oxide: Mechanisms and Therapeutic Potential, *Cardiovasc. Res. 55*, 250–260.

11. Bruick, R.K., and McKnight, S.L. (2001) A Conserved Family of Prolyl-4-Hydroxylases That Modify HIF, *Science 294*, 1337–1340.

12. Bruick, R.K., and McKnight, S.L. (2002) Transcription. Oxygen Sensing Gets a Second Wind, *Science 295*, 807–808.

13. Jaakkola, P., Mole, D.R., Tian, Y.M., Wilson, M.I., Gielbert, J., Gaskell, S.J., Kriegsheim, A., Hebestreit, H.F., Mukherji, M., Schofield, C.J., Maxwell, P.H., Pugh, C.W., and Ratcliffe, P.J. (2001) Targeting of HIF-α to the von Hippel-Lindau Ubiquitylation Complex by O_2-Regulated Prolyl Hydroxylation, *Science 292*, 468–472.

14. Halliwell, B., and Gutteridge, J.M.C. (1999) *Free Radicals in Biology and Medicine*, 3rd edn., Oxford University Press, New York.

15. Buettner, G.R. (1993) The Pecking Order of Free Radicals and Antioxidants: Lipid Peroxidation, Alpha-Tocopherol, and Ascorbate, *Arch. Biochem. Biophys. 300*, 535–543.

16. Upston, J.M., Terentis, A.C., and Stocker, R. (1999) Tocopherol-Mediated Peroxidation of Lipoproteins: Implications for Vitamin E as a Potential Antiatherogenic Supplement, *FASEB J. 13*, 977–994.

17. Hallberg, L., Brune, M., and Rossander, L. (1986) Effect of Ascorbic Acid on Iron Absorption from Different Types of Meals. Studies with Ascorbic-Acid-Rich Foods and Synthetic Ascorbic Acid Given in Different Amounts with Different Meals, *Hum. Nutr. Appl. Nutr. 40*, 97–113.

18. Yip, R. (2001) Iron, in *Present Knowledge in Nutrition* (Bowman, B.A., and Russel, R.M., eds.) 8th edn., pp. 311–328, ILSI Press, Washington.

19. Hunt, J.R., Gallagher, S.K., and Johnson, L.K. (1994) Effect of Ascorbic Acid on Apparent Iron Absorption by Women with Low Iron Stores, *Am. J. Clin. Nutr. 59*, 1381–1385.

20. Cook, J.D., Watson, S.S., Simpson, K.M., Lipschitz, D.A., and Skikne, B.S. (1984) The Effect of High Ascorbic Acid Supplementation on Body Iron Stores, *Blood 64*, 721–726.

21. Cook, J.D., and Reddy, M.B. (2001) Effect of Ascorbic Acid Intake on Nonheme-Iron Absorption from a Complete Diet, *Am. J. Clin. Nutr. 73*, 93–98.

22. Tsukaguchi, H., Tokui, T., Mackenzie, B., Berger, U.V., Chen, X.Z., Wang, Y., Brubaker, R.F., and Hediger, M.A. (1999) A Family of Mammalian Na$^+$-Dependent L-Ascorbic Acid Transporters, *Nature 399*, 70–75.
23. Hediger, M.A. (2002) New View at C, *Nat. Med. 8*, 445–446.
24. Sabry, J.H., Fisher, K.H., and Dodds, M.L. (1958) Utilization of Dehydroascorbic Acid, *J. Nutr. 68*, 457–466.
25. Malo, C., and Wilson, J.X. (2000) Glucose Modulates Vitamin C Transport in Adult Human Small Intestinal Brush Border Membrane Vesicles, *J. Nutr. 130*, 63–69.
26. Washko, P., and Levine, M. (1992) Inhibition of Ascorbic Acid Transport in Human Neutrophils by Glucose, *J. Biol. Chem. 267*, 23568–23574.
27. Graumlich, J.F., Ludden, T.M., Conry-Cantilena, C., Cantilena, L.R., Jr., Wang, Y., and Levine, M. (1997) Pharmacokinetic Model of Ascorbic Acid in Healthy Male Volunteers During Depletion and Repletion, *Pharm. Res. 14*, 1133–1139.
28. Johnston, C.S. (2001) Vitamin C, in *Present Knowledge in Nutrition*, 8th edn. (Bowman, B.A., and Russel, R.M., eds.) pp. 175–183, ILSI Press, Washington.
29. Levine, M., Conry-Cantilena, C., Wang, Y., Welch, R.W., Washko, P.W., Dhariwal, K.R., Park, J.B., Lazarev, A., Graumlich, J.F., King, J., and Cantilena, L.R. (1996) Vitamin C Pharmacokinetics in Healthy Volunteers: Evidence for a Recommended Dietary Allowance, *Proc. Natl. Acad. Sci. USA 93*, 3704–3709.
30. Levine, M., Wang, Y., Padayatty, S.J., and Morrow, J. (2001) A New Recommended Dietary Allowance of Vitamin C for Healthy Young Women, *Proc. Natl. Acad. Sci. USA 98*, 9842–9846.
31. Food and Nutrition Board and Institute of Medicine (2000) Vitamin C, in *Dietary Reference Intakes for Vitamin C, Vitamin E, Selenium, and Carotenoids*, pp. 95–185, National Academy Press, Washington.
32. Bazzano, L.A., He, J., Ogden, L.G., Loria, C.M., Vupputuri, S., Myers, L., and Whelton, P.K. (2002) Fruit and Vegetable Intake and Risk of Cardiovascular Disease in US Adults: The First National Health and Nutrition Examination Survey Epidemiologic Follow-Up Study, *Am. J. Clin. Nutr. 76*, 93–99.
33. van't Veer, P., Jansen, M.C., Klerk, M., and Kok, F.J. (2000) Fruits and Vegetables in the Prevention of Cancer and Cardiovascular Disease, *Public Health Nutr. 3*, 103–107.
34. Steinmetz, K.A., and Potter, J.D. (1996) Vegetables, Fruit, and Cancer Prevention: A Review, *J. Am. Diet. Assoc. 96*, 1027–1039.
35. Vanderslice, J.T., and Higgs, D.J. (1991) Vitamin C Content of Foods: Sample Variability, *Am. J. Clin. Nutr. 54*, 1323S-1327S.
36. Rumm-Kreuter, D., and Demmel, I. (1990) Comparison of Vitamin Losses in Vegetables Due to Various Cooking Methods, *J. Nutr. Sci. Vitaminol. (Tokyo) 36 (Suppl. 1)*, S7–S14; discussion S14–S15.
37. Balluz, L.S., Kieszak, S.M., Philen, R.M., and Mulinare, J. (2000) Vitamin and Mineral Supplement Use in the United States. Results from the Third National Health and Nutrition Examination Survey, *Arch. Fam. Med. 9*, 258–262.
38. Gregory, J.F., III (1993) Ascorbic Acid Bioavailability in Foods and Supplements, *Nutr. Rev. 51*, 301–303.
39. Johnston, C.S., and Luo, B. (1994) Comparison of the Absorption and Excretion of Three Commercially Available Sources of Vitamin C, *J. Am. Diet. Assoc. 94*, 779–781.
40. Vinson, J.A., and Bose, P. (1988) Comparative Bioavailability to Humans of Ascorbic Acid Alone or in a Citrus Extract, *Am. J. Clin. Nutr. 48*, 601–604.

41. Park, J.B., and Levine, M. (2000) Intracellular Accumulation of Ascorbic Acid Is Inhibited by Flavonoids Via Blocking of Dehydroascorbic Acid and Ascorbic Acid Uptakes in HL-60, U937 and Jurkat Cells, *J. Nutr. 130*, 1297–1302.

42. Song, J., Kwon, O., Chen, S., Daruwala, R., Eck, P., Park, J.B., and Levine, M. (2002) Flavonoid Inhibition of Sodium-Dependent Vitamin C Transporter 1 (SVCT1) and Glucose Transporter Isoform 2 (GLUT2), Intestinal Transporters for Vitamin C and Glucose, *J. Biol. Chem. 277*, 15252–15260.

43. Carpenter, K.J. (1986) *The History of Scurvy and Vitamin C*, Cambridge University Press, Cambridge.

44. Sauberlich, H.E. (1997) A History of Scurvy and Vitamin C, in *Vitamin C in Health and Disease* (Packer, L., and Fuchs, J., eds.) pp. 1–24, Marcel Decker, New York.

45. Svirbely, J.L., and Szent-Györgi, A. (1932) Hexuronic Acid as the Antiscorbutic Factor, *Nature 129*, 576.

46. King, C.G., and Waugh, W. (1932) The Chemical Nature of Vitamin C, *Science 75*, 357–358.

47. Haworth, W.N., and Hirst, E.L. (1933) Synthesis of Ascorbic Acid, *J. Soc. Chem. Ind. 52*, 645–647.

48. Harman, D. (1956) Aging: A Theory Based on Free Radical and Radiation Chemistry, *J. Gerontol. 11*, 298–230.

49. Knight, J.A. (1998) Free Radicals: Their History and Current Status in Aging and Disease, *Ann. Clin. Lab. Sci. 28*, 331–346.

50. Frei, B., England, L., and Ames, B.N. (1989) Ascorbate Is an Outstanding Antioxidant in Human Blood Plasma, *Proc. Natl. Acad. Sci. USA 86*, 6377–6381.

51. Frei, B. (1991) Ascorbic Acid Protects Lipids in Human Plasma and Low-Density Lipoprotein Against Oxidative Damage, *Am. J. Clin. Nutr. 54*, 1113S-1118S.

52. Marinacci, B. (1994) Linus Pauling: Scientist for the Ages, in *Linus Pauling: In Memorium*, pp. 7–14, Linus Pauling Institute of Science and Medicine, Palo Alto.

53. Pauling, L. (1970) Evolution and the Need for Ascorbic Acid, *Proc. Natl. Acad. Sci. USA 67*, 1643–1648.

54. Khaw, K.T., Bingham, S., Welch, A., Luben, R., Wareham, N., Oakes, S., and Day, N. (2001) Relation Between Plasma Ascorbic Acid and Mortality in Men and Women in Epic-Norfolk Prospective Study: A Prospective Population Study. European Prospective Investigation into Cancer and Nutrition, *Lancet 357*, 657–663.

55. Yokoyama, T., Date, C., Kokubo, Y., Yoshiike, N., Matsumura, Y., and Tanaka, H. (2000) Serum Vitamin C Concentration Was Inversely Associated with Subsequent 20-Year Incidence of Stroke in a Japanese Rural Community. The Shibata Study, *Stroke 31*, 2287–2294.

56. Carr, A., and Frei, B. (2000) The Role of Natural Antioxidants in Preserving the Biological Activity of Endothelium-Derived Nitric Oxide, *Free Radic. Biol. Med. 28*, 1806–1814.

57. Gokce, N., Keaney, J.F., Jr., Frei, B., Holbrook, M., Olesiak, M., Zachariah, B.J., Leeuwenburgh, C., Heinecke, J.W., and Vita, J.A. (1999) Long-Term Ascorbic Acid Administration Reverses Endothelial Vasomotor Dysfunction in Patients with Coronary Artery Disease, *Circulation 99*, 3234–3240.

58. Bates, C.J., Walmsley, C.M., Prentice, A., and Finch, S. (1998) Does Vitamin C Reduce Blood Pressure? Results of a Large Study of People Aged 65 or Older, *J. Hypertens. 16*, 925–932.

59. Block, G. (2002) Ascorbic Acid, Blood Pressure, and the American Diet, *Ann. N.Y. Acad. Sci. 959*, 180–187.

60. Duffy, S.J., Gokce, N., Holbrook, M., Huang, A., Frei, B., Keaney, J.F., Jr., and Vita, J.A. (1999) Treatment of Hypertension with Ascorbic Acid, *Lancet 354*, 2048–2049.

61. Hajjar, I.M., George, V., Sasse, E.A., and Kochar, M.S. (2002) A Randomized, Double-Blind, Controlled Trial of Vitamin C in the Management of Hypertension and Lipids, *Am. J. Ther. 9*, 289–293.

62. Will, J.C., and Byers, T. (1996) Does Diabetes Mellitus Increase the Requirement for Vitamin C?, *Nutr. Rev. 54*, 193–202.

63. Sargeant, L.A., Wareham, N.J., Bingham, S., Day, N.E., Luben, R.N., Oakes, S., Welch, A., and Khaw, K.T. (2000) Vitamin C and Hyperglycemia in the European Prospective Investigation into Cancer—Norfolk (EPIC-Norfolk) Study: A Population-Based Study, *Diabetes Care 23*, 726–732.

64. Will, J.C., Ford, E.S., and Bowman, B.A. (1999) Serum Vitamin C Concentrations and Diabetes: Findings from the Third National Health and Nutrition Examination Survey, 1988–1994, *Am. J. Clin. Nutr. 70*, 49–52.

65. Sampson, M.J., Gopaul, N., Davies, I.R., Hughes, D.A., and Carrier, M.J. (2002) Plasma F_2 Isoprostanes: Direct Evidence of Increased Free Radical Damage During Acute Hyperglycemia in Type 2 Diabetes, *Diabetes Care 25*, 537–541.

66. Devaraj, S., Hirany, S.V., Burk, R.F., and Jialal, I. (2001) Divergence Between LDL Oxidative Susceptibility and Urinary F(2)-Isoprostanes as Measures of Oxidative Stress in Type 2 Diabetes, *Clin. Chem. 47*, 1974–1979.

67. American Diabetes Association (2000) Diabetes Facts and Figures, [Web page], http://www.diabetes.org/main/info/facts/facts2.jsp, accessed July 23, 2002.

68. Kromhout, D. (1987) Essential Micronutrients in Relation to Carcinogenesis, *Am. J. Clin. Nutr. 45*, 1361–1367.

69. Mirvish, S.S. (1994) Experimental Evidence for Inhibition of *N*-Nitroso Compound Formation as a Factor in the Negative Correlation Between Vitamin C Consumption and the Incidence of Certain Cancers, *Cancer Res. 54*, 1948S–1951S.

70. Jarosz, M., Dzieniszewski, J., Dabrowska-Ufniarz, E., Wartanowicz, M., Ziemlanski, S., and Reed, P.I. (1998) Effects of High Dose Vitamin C Treatment on *Helicobacter pylori* Infection and Total Vitamin C Concentration in Gastric Juice, *Eur. J. Cancer Prev. 7*, 449–454.

71. Feiz, H.R., and Mobarhan, S. (2002) Does Vitamin C Intake Slow the Progression of Gastric Cancer in *Helicobacter pylori*-Infected Populations? *Nutr. Rev. 60*, 34–36.

72. Taylor, A., Jacques, P.F., Chylack, L.T., Jr., Hankinson, S.E., Khu, P.M., Rogers, G., Friend, J., Tung, W., Wolfe, J.K., Padhye, N., and Willett, W.C. (2002) Long-Term Intake of Vitamins and Carotenoids and Odds of Early Age-Related Cortical and Posterior Subcapsular Lens Opacities, *Am. J. Clin. Nutr. 75*, 540–549.

73. Jacques, P.F., Chylack, L.T., Jr., Hankinson, S.E., Khu, P.M., Rogers, G., Friend, J., Tung, W., Wolfe, J.K., Padhye, N., Willett, W.C., and Taylor, A. (2001) Long-Term Nutrient Intake and Early Age-Related Nuclear Lens Opacities, *Arch. Ophthalmol. 119*, 1009–1019.

74. Simon, J.A., and Hudes, E.S. (1999) Serum Ascorbic Acid and Other Correlates of Self-Reported Cataract Among Older Americans, *J. Clin. Epidemiol. 52*, 1207–1211.

75. Sperduto, R.D., Hu, T.S., Milton, R.C., Zhao, J.L., Everett, D.F., Cheng, Q.F., Blot, W.J., Bing, L., Taylor, P.R., Li, J.Y., (1993) The Linxian Cataract Studies. Two Nutrition Intervention Trials, *Arch. Ophthalmol. 111*, 1246–1253.

76. Eye Disease Case Control Study Group (2001) A Randomized, Placebo-Controlled, Clinical Trial of High-Dose Supplementation with Vitamins C and E and Beta Carotene for Age-Related Cataract and Vision Loss: AREDS Report No. 9, *Arch. Ophthalmol. 119*, 1439–1452.

77. Sotiriou, S., Gispert, S., Cheng, J., Wang, Y., Chen, A., Hoogstraten-Miller, S., Miller, G.F., Kwon, O., Levine, M., Guttentag, S.H., and Nussbaum, R.L. (2002) Ascorbic-Acid Transporter Slc23a1 Is Essential for Vitamin C Transport into the Brain and for Perinatal Survival, *Nat. Med. 8*, 514–517.

Chapter 2

Vitamin C Pharmacokinetics in Healthy Men and Women

Mark Levine, Yaohui Wang, and Sebastian J. Padayatty

Molecular and Clinical Nutrition Section, Digestive Diseases Branch, National Institute of Diabetes and Digestive and Kidney Diseases, National Institutes of Health, Bethesda, MD 20892–1372

Introduction

Recommended Dietary Allowances (RDA) for vitamin C were increased in 2000 by the Food and Nutrition Board of the U.S. National Academy of Sciences as part of new Dietary Reference Intake guidelines (1,2). Independent of these guidelines, recommended vitamin C intake was also increased in many countries (3–5). These increases were based in part on new pharmacokinetics data in healthy subjects. Here we describe why these pharmacokinetics studies were undertaken and what they showed.

Before 2000, the RDA for vitamin C were based on prevention of deficiency with a margin of safety (6,7). Recommendations were not based on biochemical function. We recognized that preventing deficiency might not be equivalent to ideal nutrient intake (8). Therefore, we proposed in 1986 and 1987 that the RDA for vitamin C and other water-soluble vitamins should be based on vitamin function in relation to concentration (8,9). Several data sets must be obtained to achieve this goal (10,11). It is necessary to know vitamin C concentrations in humans in relation to dose across a wide dose range. Once relevant concentrations in humans have been measured, several functional outcomes must be characterized in relation to these concentrations. Functional outcomes must be determined in cells and tissues rather than simply for isolated vitamin C–dependent reactions *in vitro*. Findings must then be extended to humans, with targeted biochemical and clinical outcomes measures. Such data represent specific concentration-function relationships, the foundation of recommendations for ideal nutrient intake.

A key part of this proposal is the relationship between ingested vitamin and the resulting concentrations achieved in plasma and tissues. Information was available about vitamin C concentrations in humans (12–22). However, it was difficult to interpret the overall data because of flaws in study design, execution, or analyses. The flaws included the following: use of vitamin C assays that were not specific or sensitive at low concentrations; narrow dose range of administered vitamin C; use of a diet deficient in other vitamins and minerals; use of radiolabeled vitamin C without verification of radiolabel metabolism *in vivo*; lack of verification of steady state for vitamin

C dose; and outpatient or uncertain dietary control of vitamin C ingestion. To obtain information about a variety of doses, it became necessary to combine data from different studies in which there was variability in analytical techniques and dietary control, making comparisons unwieldy. Only limited and incomplete data were available from studies in which there was dietary control of vitamin C ingestion in inpatients (14). The problems in data interpretation using such comparisons are evident in a recent meta-analysis (22). Thus, the goal of consistent dose-concentration data for vitamin C in humans was elusive.

Therefore, in 1991, we enrolled men in an inpatient study to determine vitamin C plasma and tissue concentrations as a function of doses over a wide dose range. Vitamin C was measured using a sensitive high-performance liquid chromatography (HPLC) coulometric electrochemical detection assay that was validated using human blood samples under clinical sampling conditions (23,24). Healthy men ($n = 7$) completed this study (25), with hospitalization of each subject for ~5 mo. In 1995, the study was extended to women. Women ($n = 15$) completed the study (26), with hospitalization of each subject for ~6 mo. The results of these studies are described here.

Subjects and Methods

The study was approved by the Institutional Review Board, National Institute of Diabetes and Digestive and Kidney Diseases, National Institutes of Health. Written informed consent was obtained from all enrolled subjects. Men ($n = 7$) ages 20–26 y and women ($n = 15$) ages 19–27 y enrolled and completed the study. Complete study details are described elsewhere (25–28).

Subjects were initially screened by written questionnaire and telephone interview by physicians. Potentially acceptable subjects were then screened in person. Subjects underwent complete history, full physical examination, laboratory testing, psychological testing, and interviews by several staff members. Selected subjects were in good health, were nonsmokers, used no medications or supplements, and did not use alcohol or illicit drugs.

Subjects were hospitalized on an endocrinology-metabolism ward to control nutrient intake. Men were hospitalized 146 ± 23 d and women were hospitalized 186 ± 28 d. For the entire hospitalization, subjects consumed a vitamin C–deficient diet that contained <5 mg of vitamin C daily and utilized a computerized 14-d cycle selective menu design with >300 menu choices. The diet was sufficient in energy, protein, fat, and saturated fat. Other vitamin and mineral deficiencies were prevented by supplement administration so that only vitamin C intake was restricted (27). Pure vitamin C, when administered, was given twice daily in water (pH adjusted to 6.5) to subjects who had fasted for at least 2 h before breakfast and dinner.

Upon hospital admission, subjects began the depletion phase of the study. Consumption of the study diet caused vitamin C depletion. When plasma concentrations declined to 7–8 μmol/L, subjects were depleted without clinical scurvy. Neutrophils were isolated to measure vitamin C content, 36-h bioavailability sampling for vitamin C (15 mg) was performed, and 24-h urine samples were collected to measure excreted vitamin C, creatinine, and other metabolites.

Subjects then entered the repletion phase of the study. Vitamin C (15 mg) twice daily was administered until subjects achieved steady state for this dose (30 mg daily). Bioavailability sampling (36-h) for the daily dose was performed, and subjects underwent apheresis (cell separation) to obtain monocytes, lymphocytes, and platelets for analyses of vitamin C content. After 24-h urine samples were collected and neutrophils were isolated, the vitamin C dose was increased and the sequence repeated at the new dose. Using this study design, subjects received in succession, daily vitamin C doses of 30, 60, 100, 200, 400, 1000, and 2500 mg, with bioavailability sampling for vitamin C doses of 15, 30, 50, 100, 200, 500, and 1250 mg. All measurements for vitamin C were analyzed by HPLC with coulometric electrochemical detection. All vitamin C samples were analyzed in triplicate with SD < 5% of the mean. No vitamin C degradation occurred under processing and storage conditions. Dehydroascorbic acid was <2% of plasma vitamin C and could not be distinguished from 0. Plasma data are predose values from morning samples from subjects who were fasting.

All experimental results are displayed as mean ± SD. When not displayed, the SD was less than symbol size. Data were not available for all subjects at all doses for the following reasons: inability of some subjects to remain hospitalized for the entire study; intravenous access limitations; and inadvertent nursing errors or sample loss. Numbers of subjects for whom data were available are indicated in figure legends.

Results

Fasting plasma vitamin C concentrations as a function of vitamin C dose and study day are shown in Figure 2.1A and B for men and women, respectively. Plasma vitamin C concentrations are shown at all doses. Vitamin C plasma concentrations were determined at least 2–3 times/wk in all subjects. Subjects required different amounts of time to achieve steady state at doses <100 mg/d. Because some subjects remained at doses longer than others, gaps are displayed between doses.

Steady state was attained when plasma vitamin C concentrations reached equilibrium for a given dose. Steady-state plasma concentration was defined as the mean of at least 5 plasma samples over at least 7 d with <10% SD, and the first sample included in the steady-state calculation was ≥90% of the mean. The steady-state value for the highest dose in one male subject was based on four samples. For men, 86% of steady-state values were based on ≥6 samples. For women, all steady-state values were based on ≥6 samples, with 85% of steady-state values based on ≥7 samples. Each subject achieved steady state for a dose before the next dose was given. Steady state was evident from visual inspection of data and was always confirmed using the calculations described. An example of steady state at the 60-mg dose is shown in Figure 2.2.

Steady-state plasma values were calculated for every subject at every dose and displayed as a function of dose (Fig. 2.3A and B for men and women, respectively). There was a sigmoid relationship between dose and steady-state plasma concentra-

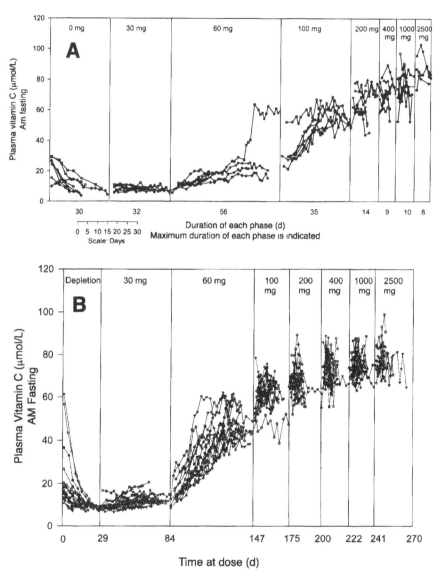

Fig. 2.1. Vitamin C plasma concentration as a function of days at dose. Doses are indicated at the top of each panel. Each symbol represents a different subject. There is a 1-d gap between all doses for bioavailability sampling. See text for details. (A) Plasma concentration as a function of days at dose in men. Doses through 400 mg/d were received by 7 subjects, through 1000 mg/d by 6 subjects, and through 2500 mg/d by 3 subjects. *Source*: Ref. 25. (B) Plasma concentration as a function of days at dose in women. Doses through 200 mg/d were received by 15 subjects, through 1000 mg/d by 13 subjects, and through 2500 mg/d by 10 subjects. *Source*: Ref. 26.

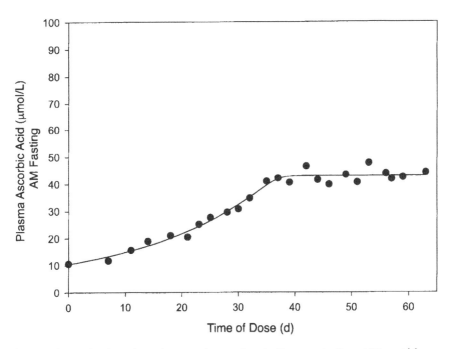

Fig. 2.2. Determination of steady-state plasma vitamin C concentration at 60 mg/d for one female subject. *Source*: Ref. 26. See Ref. 25 for male subject data.

tion at doses <400 mg/d for both men and women. However, the curve for women was shifted to the left compared with that for men (Fig. 2.3C). By repeated-measures analysis of variance, plasma concentrations for women at doses of 30–100 mg daily were higher than for men ($P2 = 0.01$), but differences disappeared at doses >100 mg/d. The first dose beyond the steep (linear) portion of the sigmoid curve for both sexes was 200 mg/d. This dose produced a plasma concentration of ~70 μmol/L, similar to the concentration at which the sodium vitamin C transporter (SVCT)2 approaches V_{max} (29,30). Several factors could be responsible for the sigmoid shape of the relationship between dose and plasma concentration, i.e., vitamin C uptake, absorption, and excretion. These were investigated in turn as described below.

Tissue uptake of vitamin C was determined by measuring vitamin C concentrations in cells that could be obtained without harming subjects. Circulating cells and platelets found in blood were ideal indicators of vitamin C distribution in tissues. Vitamin C concentrations were determined at steady state over the dose range in neutrophils, monocytes, and lymphocytes for both sexes, and additionally in platelets of women (Fig. 2.4A, B). Most cells saturated at doses of 100–200 mg/d.

Bioavailability, or the fraction of the dose absorbed, was determined from oral and intravenous vitamin C administration when steady state for a dose was reached. Data have been calculated for men, and analysis is ongoing for data from women.

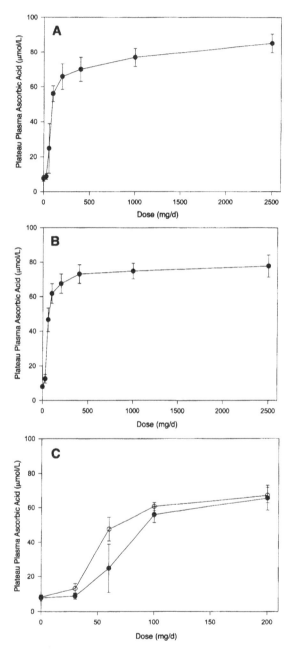

Fig. 2.3. Steady-state plasma vitamin C concentrations as a function of dose for all doses for all men (A) and for all women (B), and comparison of values for men (A) and women (B) at doses of 0–200 mg/d (C). See Figure 2.1 and text for details. *Sources*: Ref. 25,26.

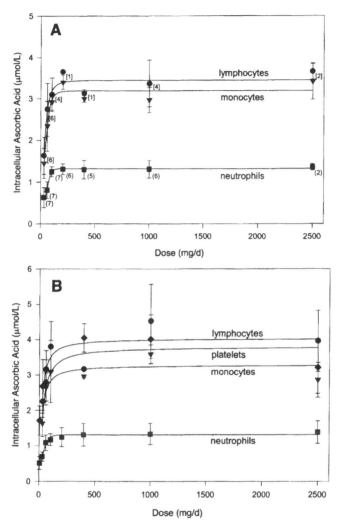

Fig. 2.4. Intracellular vitamin C concentrations (mmol/L) in circulating cells as a function of dose. Cells were isolated when steady state was achieved for each dose. (A) Cells from male subjects. Numbers in parentheses at each dose indicate the number of men from whom neutrophils were obtained. Numbers in brackets at each dose indicate the number of men from whom lymphocytes and monocytes were obtained. See Ref. 25 for details. (B) Cells and platelets from female subjects. For neutrophils, samples were available from 13 women at doses of 0–200 mg/d; from 11 women at doses of 400 and 100 mg/d; and from 10 women at 2500 mg/d. For lymphocytes, monocytes, and platelets, samples were available from 13 women at 30 mg/d; from 12 women at 60 mg/d; from 6 women at 100 mg/d; from 2 women at 400 and 1000 mg/d; from 9 women at 2500 mg/d. See Ref. 26 for details.

Examples of data obtained for bioavailability determinations are shown in Figure 2.5. As a first approach, area under curve (AUC) calculations (linear trapezoidal analyses) were used to calculate bioavailability at doses of 200, 500, and 1250 mg (Table 2.1). At each of these three doses, bioavailability was calculated as the ratio of the area under the oral dose divided by the area under the intravenous dose. By simple visual inspection, it can be seen that the AUC was similar at the 200-mg dose orally and intravenously, but was much less for the oral dose compared with the intravenous

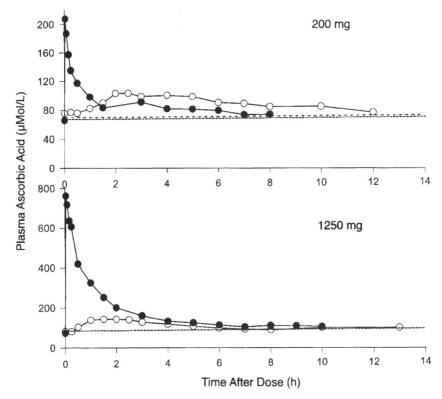

Fig. 2.5. Vitamin C bioavailability in plasma. Upper panel: bioavailability in one male subject for 200 mg. Lower panel: bioavailability in one male subject for 1250 mg. For each dose, vitamin C was administered at 8:00 AM orally; sample values (○) are shown for the times indicated. Baseline is indicated by the dashed line with larger spaces. After 24 h, the same dose was given intravenously and samples were taken for the times indicated (●). Baseline is indicated by the dashed line with smaller spaces. For oral doses, samples before zero time and between 13 and 24 h are not shown for clarity. Bioavailability was calculated using the linear trapezoidal method. Bioavailability was the ratio of the area of the oral dose divided by the area of the intravenous dose. The area after the curve returned to baseline was assumed to equal zero. See Ref. 25 for details.

TABLE 2.1
Vitamin C Bioavailability in Men at Doses of 200, 500, and 1250 mg Determined by the Linear Trapezoidal Method (Area Under the Curve)[a]

Dose (mg)	Bioavailability (% ± SD)
200	112 ± 25
500	73 ± 27
1250	49 ± 25

[a]*Source:* Ref. 25.

dose at 1250 mg (Fig. 2.5). AUC calculations are based on pharmacokinetics assumptions that volume of distribution and clearance are constant. However, these assumptions are not valid at vitamin C doses <200 mg. A more sophisticated pharmacokinetics model was developed and is the first to include the three parameters of nonlinear absorption, nonlinear elimination, and nonlinear tissue distribution (28). This model was used to calculate bioavailability based on plasma values when vitamin C was administered orally and intravenously for all doses at steady state (Table 2.2). Data analyzed using the model indicated that as vitamin C dose increased, bioavailability (absorption) decreased.

Plasma was nearly saturated with vitamin C at an oral dose of 200 mg/d (Fig. 2.3). However, bioavailability did not decrease proportionally at higher doses (Table 2.1). One potential explanation was urinary excretion of absorbed vitamin C. During bioavailability sampling when vitamin C was administered orally and intravenously, urine was collected from men and women and vitamin C excretion was measured (Fig. 2.6A, B, and insets). The threshold dose for urinary excretion of vitamin C was between 60 and 100 mg for both sexes. With intravenous administration of 500 and 1250 mg, nearly the entire dose was excreted in urine for both sexes. With oral administration of these doses, urine excretion was less than that of intravenous administration, likely because bioavailability was less at higher doses compared with lower ones (Table 2.1) (25). Considered together, the data show that at doses ≥500 mg/d, when plasma is saturated, most of absorbed vitamin C is excreted in urine.

TABLE 2.2
Vitamin C Bioavailability in Men at All Doses Calculated Using a Pharmacokinetics Model with Nonlinear Absorption, Elimination, and Tissue Distribution[a]

Dose (mg)	Bioavailability (% ± SD)
15	85 ± 20
30	85 ± 20
50	84 ± 20
100	82 ± 20
200	78 ± 22
500	75 ± 24
1250	62 ± 34

[a]*Source:* Ref. 28.

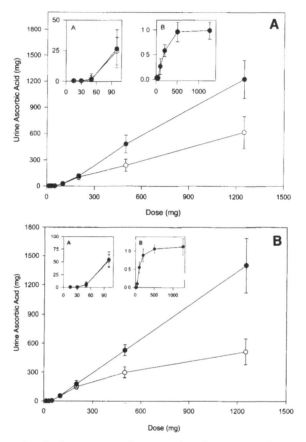

Fig. 2.6. Urinary vitamin C excretion as function of single vitamin C doses at steady state in men (A) and women (B). Urine was collected during determination of vitamin C bioavailability for each dose. Vitamin C excretion was determined after administration of vitamin C given either orally (○) or intravenously (●). (A) Urine was collected from oral sampling for 24 h, and then from intravenous sampling for 9–10 h. Data from oral and intravenous administration were available from 7 subjects at doses of 15, 30, 50, and 100 mg; from 6 subjects at 200 mg; from 6 subjects at 500 mg; from 3 subjects at 1250 mg. See Ref. 25 for details. (B) Urine was collected from oral and intravenous sampling for 24 h. Data from oral and intravenous administration were available from 11 subjects at doses of 15, 30, 50, and 200 mg; from 10 subjects at 100 mg; from 8 subjects at 500 mg; and from 9 subjects at 1250 mg. See Ref. 26 for details. For both panels: Inset A shows vitamin C excretion for single oral (○) or intravenous (●) doses of 15–100 mg. The x-axis indicates dose and the y-axis indicates amount (mg) excreted in urine. Inset B shows the fractional excretion (the fraction of the dose excreted) after intravenous administration of single doses of vitamin C. The x-axis indicates dose and the y-axis indicates fractional excretion (vitamin C excreted in urine in milligrams divided by the vitamin C dose in milligrams). The minimum amount of ascorbate excreted was <0.4 mg.

Discussion

The data presented here describe vitamin C pharmacokinetics in healthy young men and women, and define the range of physiologic vitamin C concentrations at which functional outcomes are possible. In both sexes, there was a steep sigmoid relationship between dose and steady-state plasma concentration at doses between 30 and 100 mg/d. By 400 mg/d, plasma at steady state was saturated, with little effect of higher doses. Cells saturated at lower doses. Plasma saturation was due to several factors, including cell and tissue saturation at daily doses >100–200 mg, decreased bioavailability at doses >200 mg, and increased renal excretion at doses >100–200 mg (28). Some of these findings have already provided a new basis for recommended dietary allowances of vitamin C (2–5).

Several aspects of the data were surprising. One striking finding was the observed tight control of plasma and tissue concentrations as a function of dose. Healthy humans apparently strive to reach plasma concentrations of 70–80 μmol/L; once this concentration range is achieved, it is not exceeded despite large increases in oral ingestion. Ascorbate doses at ≥400 mg/d had virtually no effect on increasing steady-state plasma and tissue concentrations, and 200 mg/d of vitamin C resulted in nearly complete saturation of plasma and tissues.

Why is there such tight control of vitamin C concentrations, particularly in plasma? There are two general explanations, i.e., either tight control is necessary to avoid harm or it is beneficial directly or indirectly. For the first possibility, tight control of plasma concentrations may be to avoid harm arising perhaps *via* potential prooxidant toxicity. Theoretically, ascorbic acid could act as a prooxidant under certain conditions especially at higher concentrations. For the second possibility, tight control of extracellular concentrations may allow higher local concentrations to occur under certain conditions, for example, by release of ascorbate from cells into tissue. These concentrations might act in a paracrine or autocrine fashion, perhaps participating in signal transduction. Such local ascorbate release has been described in cell systems and animals with respect to the adrenal gland (31,32) and brain (33). If there were strong advantages to tight control for these or other reasons, it would be predicted that vitamin C transporters would have less genetic variation than normally predicted, for example, as determined by single nucleotide polymorphisms in the known vitamin C transporters SVCT1 and SVCT2. Such genomic analyses are underway.

Another surprising finding was that tight control of vitamin C concentrations could be bypassed for several hours when vitamin C was given intravenously. Obviously, intravenous administration is not relevant physiologically, but may have unexpected implications pharmacologically (34,35). Vitamin C given intravenously results in vasodilatation, although differences between oral physiologic and intravenous pharmacologic administration have sometimes been misunderstood (36,37). Nevertheless, independent of recommendations for physiological benefit, vitamin C given pharmacologically achieves far higher concentrations, and these perhaps might have therapeutic importance. By analogy, oral penicillin may be ineffective in treatment of life-threatening meningitis, whereas intravenous penicillin is effective because far higher concentrations are achieved. There are theoretical possibilities for therapeutic use of vitamin C intravenously, and these possibilities should be explored rigorously.

An additional surprising finding was that vitamin C dose-concentration relationships in plasma were shifted to the left for women compared with men. In hindsight, these differences might be due to differences in muscle mass. Muscle mass is greater in men than women. Although mRNA for vitamin C transporters have not been reported in muscle, muscle contains higher ascorbic acid concentrations than plasma. Although ascorbate concentrations in muscle are lower than in many other tissues, muscle mass contributes significantly to total body weight, and therefore muscle is predicted to contain a substantial total amount of ascorbate. Because women have less muscle mass than men, women should achieve ascorbate saturation in muscles at lower doses than men. This in turn would permit ascorbate plasma concentrations in women to rise higher compared with men at low doses, which was what was observed. In future studies, lean body mass will be measured to verify these possibilities.

There are limitations to these pharmacokinetics data. Prolonged hospitalization was necessary; thus, large numbers of subjects could not be studied. The subjects were healthy adult nonsmokers <28 y old. Pharmacokinetics are not known in older subjects, smokers, diabetics, patients with cardiovascular disease, patients with end-stage kidney disease on dialysis, and patients with other conditions. Prolonged hospitalization is not practical or possible for many of these relevant populations. It is possible that under some conditions or in some patients, vitamin C pharmacokinetics curves are shifted to the right, or that saturation values differ from healthy subjects. Methods should be developed to study at least some of the relevant patient populations, for example by decreasing hospitalization time.

Another limitation of the pharmacokinetics data is that pure ascorbic acid was used to determine steady-state plasma concentrations and bioavailability. However, recommended dietary allowances provide guidelines for ingesting vitamin C in the diet from foods. It is possible that other substances in foods rich in vitamin C could decrease absorption. For example, vitamin C is found in high amounts in many fruits and vegetables. These foods also contain many other compounds, with flavonoids as one example. Recent evidence shows that flavonoids inhibit the intestinal vitamin C transporter SVCT1 (38). Flavonoid inhibition occurred when this transporter was expressed in expression systems, when cells were transfected to overexpress the transporter, and when bioavailability was determined in animals given vitamin C and flavonoids. It is unknown whether flavonoids or other compounds in foods inhibit vitamin C absorption in humans. If such inhibition of absorption occurred, vitamin C dose concentration curves would be shifted to the right.

There is substantial evidence that fruits and vegetables are beneficial to human health. Five or more servings of fruits and vegetables daily may be protective in heart disease, stroke, cancer, cataract development, and longevity (39,40). Five or more daily servings of fruits and vegetables will provide 210–300 mg of vitamin C. This amount of vitamin C should produce near-saturation of cells and plasma, even if flavonoids decreased absorption, without causing harm. In certain circumstances, fruits and vegetables may be ill-advised or contraindicated, for example, in patients with end-stage renal disease on dialysis or patients with inflammatory bowel disease. Otherwise, for most adults, the best advice is to eat five varied servings of fresh fruits and vegetables daily to maximize health and to obtain vitamin C from these foods.

References

1. Yates, A.A., Schlicker, S.A., and Suitor, C.W. (1998) Dietary Reference Intakes: The New Basis for Recommendations for Calcium and Related Nutrients, B Vitamins, and Choline, *J. Am. Diet. Assoc. 98*, 699–706.
2. Food and Nutrition Board, Panel on Dietary Antioxidants and Related Compounds (2000) Vitamin C, in *Dietary Reference Intakes for Vitamin C, Vitamin E, Selenium, and Carotenoids*, pp. 95–185, National Academy Press, Washington.
3. Deutsche Gesellschaft für Ernährung (German Society of Nutrition), Österreichische Gesellschaft für Ernährung (Austrian Society of Nutrition), Schweizerische Gesellschaft für Ernährungsforschung (Swiss Society of Nutrition Research), Schweizerische Vereinigung für Ernährung (Swiss Union of Nutrition) (2000) *Empfohlene Referenzwerte für die Nährstoffzufuhr (Recommended Values for Nutrient Intake)*, pp. 137–144. Umschau-Braus, Frankfurt.
4. Health and Nutrition Information Society (2000) *Recommended Dietary Allowances (Dietary Reference Intakes)* 6th edn., Daiichi Shuppan, Tokyo.
5. Birlouez-Aragon L., Fieux, B., Potier de Courcy, G., Hercberg, C. (2000) Vitamin C, in *Apports Nutritionnels Conseilles pour la Population Francaise* (Martin, A., ed.) 3rd edn., pp. 215–230, Lavoisier Tec and Doc, London.
6. Food and Nutrition Board (1980) Vitamin C (Ascorbic Acid), in *Recommended Dietary Allowances* (Committee on Dietary Allowances, ed.) pp. 72–81, National Academy of Sciences, Washington.
7. Food and Nutrition Board (1989) *Recommended Dietary Allowances*, 10th edn., National Academy Press, Washington.
8. Levine, M. (1986) New Concepts in the Biology and Biochemistry of Ascorbic Acid, *N. Engl. J. Med. 314*, 892–902.
9. Levine, M., Hartzell, W. (1987) Ascorbic Acid: The Concept of Optimum Requirements, *Ann. N.Y. Acad. Sci. 498*, 424–444.
10. Levine, M., Dhariwal, K.R., Washko, P.W., Butler, J.D., Welch, R.W., Wang,Y., and Bergsten, P. (1991) Ascorbic Acid and In Situ Kinetics: A New Approach to Vitamin Requirements, *Am. J. Clin. Nutr. 54*, 1157S–1162S.
11. Levine, M., Cantilena, C.C., Dhariwal, K.R. (1993) In Situ Kinetics and Ascorbic Acid Requirements, *World Rev. Nutr. Diet. 72*, 114–127.
12. Baker, E.M., Saari, J.C., and Tolbert, B.M. (1966) Ascorbic Acid Metabolism in Man, *Am. J. Clin. Nutr. 19*, 371–378.
13. Baker, E.M., Hodges, R.E., Hood, J., Sauberlich, H.E., and March, S.C. (1969) Metabolism of Ascorbic-1-^{14}C Acid in Experimental Human Scurvy, *Am. J. Clin. Nutr. 22*, 549–558.
14. Baker, E.M., Hodges, R.E., Hood, J., Sauberlich, H.E., March, S.C., and Canham, J.E.

(1971) Metabolism of [14]C- and ^{3}H-Labeled L-Ascorbic Acid in Human Scurvy, *Am. J. Clin. Nutr. 24*, 444–454.

15. Kallner, A., Hartmann, D., and Hornig, D. (1979) Steady-State Turnover and Body Pool of Ascorbic Acid in Man, *Am. J. Clin. Nutr. 32*, 530–539.

16. Garry, P.J., Goodwin, J.S., Hunt, W.C., and Gilbert, B.A. (1982) Nutritional Status in a Healthy Elderly Population: Vitamin C, *Am. J. Clin. Nutr. 36*, 332–339.

17. Omaye, S.T., Skala, J.H., and Jacob, R.A. (1986) Plasma Ascorbic Acid in Adult Males: Effects of Depletion and Supplementation, *Am. J. Clin. Nutr. 44*, 257–264.

18. Jacob, R.A., Skala, J.H., and Omaye, S.T. (1987) Biochemical Indices of Human Vitamin C Status, *Am. J. Clin. Nutr. 46*, 818–826.

19. Jacob, R.A., Pianalto, F.S., and Agee, R.E. (1992) Cellular Ascorbate Depletion in Healthy Men, *J. Nutr. 122*, 1111–1118.

20. VanderJagt, D.J., Garry, P.J., and Bhagavan, H.N. (1987) Ascorbic Acid Intake and Plasma Levels in Healthy Elderly People, *Am. J. Clin. Nutr. 46*, 290–294.

21. Blanchard, J., Conrad, K.A., Watson, R.R., Garry, P.J., and Crawley, J.D. (1989) Comparison of Plasma, Mononuclear and Polymorphonuclear Leucocyte Vitamin C Levels in Young and Elderly Women During Depletion and Supplementation, *Eur. J. Clin. Nutr. 43*, 97–106.

22. Brubacher, D., Moser, U., and Jordan, P. (2000) Vitamin C Concentrations in Plasma as a Function of Intake: a Meta-Analysis, *Int. J. Vitam. Nutr. Res. 70*, 226–237.

23. Washko, P.W., Hartzell, W.O., and Levine, M. (1989) Ascorbic Acid Analysis Using High-Performance Liquid Chromatography with Coulometric Electrochemical Detection, *Anal. Biochem. 181*, 276–282.

24. Dhariwal, K.R., Hartzell, W.O., and Levine, M. (1991) Ascorbic Acid and Dehydroascorbic Acid Measurements in Human Plasma and Serum, *Am. J. Clin. Nutr. 54*, 712–716.

25. Levine, M., Conry-Cantilena, C., Wang, Y., Welch, R.W., Washko, P.W., Dhariwal, K.R., Park, J.B., Lazarev, A., Granulich, J.F., King, J., and Cantilena, L.R. (1996) Vitamin C Pharmacokinetics in Healthy Volunteers: Evidence for a Recommended Dietary Allowance, *Proc. Natl. Acad. Sci. USA 93*, 3704–3709.

26. Levine, M., Wang, Y., Padayatty, S.J., and Morrow, J. (2001) A New Recommended Dietary Allowance of Vitamin C for Healthy Young Women, *Proc. Natl. Acad. Sci. USA 98*, 9842–9846.

27. King, J., Wang, Y., Welch, R.W., Dhariwal, K.R., Conry-Cantilena, C., and Levine, M. (1997) Use of a New Vitamin C-Deficient Diet in a Depletion/Repletion Clinical Trial, *Am. J. Clin. Nutr. 65*, 1434–1440.

28. Graumlich, J.F., Ludden, T.M., Conry-Cantilena, C., Cantilena, L.R., Jr., Wang, Y., and Levine, M. (1997) Pharmacokinetic Model of Ascorbic Acid in Healthy Male Volunteers During Depletion and Repletion, *Pharm. Res. 14*, 1133–1139.

29. Tsukaguchi, H., Tokui, T., Mackenzie, B., Berger, U.V., Chen, X.Z., Wang, Y.X., Brubaker, R.F., and Hediger, M.A. (1999) A Family of Mammalian Sodium-Dependent L-Ascorbic Acid Transporters, *Nature 399*, 70–75.

30. Daruwala, R., Song, J., Koh, W.S., Rumsey, S.C., and Levine, M. (1999) Cloning and Functional Characterization of the Human Sodium-Dependent Vitamin C Transporters hSVCT1 and hSVCT2, *FEBS Lett. 460*, 480–484.

31. Levine, M., Asher, A., Pollard, H, and Zinder, O. (1983) Ascorbic Acid and Catecholamine Secretion from Cultured Chromaffin Cells, *J. Biol. Chem. 258*, 13111–13115.

32. Levine, M., and Morita, K. (1985) Ascorbic Acid in Endocrine Systems, *Vitam. Horm. 42*, 1–64.

33. Rebec, G.V., and Pierce, R.C. (1994) A Vitamin as Neuromodulator: Ascorbate Release into the Extracellular Fluid of the Brain Regulates Dopaminergic and Glutamatergic Transmission, *Prog. Neurobiol. 43*, 537–565.

34. Padayatty, S.J., and Levine, M. (2000) Reevaluation of Ascorbate in Cancer Treatment: Emerging Evidence, Open Minds and Serendipity, *J. Am. Coll. Nutr. 19*, 423–425.

35. Padayatty, S.J., and Levine, M. (2001) New Insights into the Physiology and Pharmacology of Vitamin C, *Can. Med. Assoc. J. 164*, 353–355.

36. Kaufmann, P.A., Gnecchi-Ruscone, T., di Terlizzi, M., Schafers, K.P., Luscher, T.F., and Camici, P.G. (2000) Coronary Heart Disease in Smokers: Vitamin C Restores Coronary Microcirculatory Function, *Circulation 102*, 1233–1238.

37. Padayatty, S.J., and Levine, M. (2001) Vitamin C and Coronary Microcirculation [Letter], *Circulation 103*, E117.

38. Song, J., Kwon, O., Chen, S., Daruwala, R., Eck, P., Park, J.B., and Levine, M. (2002) Flavonoid Inhibition of SVCT1 and GLUT2, Intestinal Transporters for Vitamin C and Glucose, *J. Biol. Chem. 277*: 15252–15260.

39. Levine, M., Rumsey, S.C., Daruwala, R.C., Park, J.B., and Wang, Y. (1999) Criteria and Recommendations for Vitamin C Intake, *J. Am. Med. Assoc. 281*, 1415–1423.

40. Khaw, K.T., Bingham, S., Welch, A., Luben, R., Wareham, N., Oakes, S., and Day, N. (2001) Relation Between Plasma Ascorbic Acid and Mortality in Men and Women in EPIC-Norfolk Prospective Study: A Prospective Population Study, European Prospective Investigation into Cancer and Nutrition, *Lancet 357*, 657–663.

Chapter 3

Biochemical and Physiological Interactions of Vitamin C and Iron: Pro- or Antioxidant?

Ben-Zhan Zhu and Balz Frei

Linus Pauling Institute, Oregon State University, Corvallis, Oregon 97331

Vitamin C

Vitamin C is an essential micronutrient for humans, who have lost the ability to synthesize ascorbic acid due to a mutation in the gene coding for L-gulono-γ-lactone oxidase (1). A lack of vitamin C in the human diet causes the deficiency disease scurvy. The current Recommended Dietary Allowance (RDA) for vitamin C is 75 mg/d for women and 90 mg/d for men (2). The RDA for men is based on near-maximal neutrophil concentration and minimal urinary excretion of vitamin C at daily doses of between 60 and 100 mg (2,3); that for women is by extrapolation from the RDA for men on the basis of relative body mass (2).

The molecular mechanisms underlying the antiscorbutic effects of vitamin C are largely understood. For example, vitamin C is a co-substrate for enzymes involved in procollagen, carnitine, and catecholamine synthesis (1,4). Prolyl and lysyl hydroxylases, two enzymes involved in procollagen synthesis, require vitamin C for maximal activity (1). A deficiency of vitamin C results in a weakening of the collagenous triple helix structure with tooth loss, joint pains, and poor wound healing, all characteristic signs of scurvy (4). Two dioxygenases in the biosynthesis of carnitine also require vitamin C for maximal activity (1,4). Carnitine is essential for the transport of long-chain fatty acids into mitochondria for β-oxidation; not surprisingly, fatigue and lethargy are early symptoms of scurvy (3,4). Vitamin C also acts as a co-substrate for dopamine β-hydroxylase, which converts dopamine to norepinephrine (1,4).

The role of vitamin C as a co-substrate in the above enzyme reactions is to maintain the active center metal ions in the reduced, enzymatically active form (1). This same electron-donating activity of ascorbate also makes it a powerful antioxidant, i.e., vitamin C readily scavenges reactive oxygen species (ROS) (5–8) such as superoxide ($O_2^{\bullet-}$), hydroperoxyl radicals, aqueous peroxyl radicals, singlet oxygen, hypochlorous acid and ozone, reactive nitrogen species (RNS) (9–11) such as nitrogen dioxide and dinitrogen tetroxide, and antioxidant-derived radicals (5,12) such as thiyl and urate radicals (Table 3.1), thereby protecting biological macromolecules such as proteins, lipids, and DNA from oxidative damage (13–16). Vitamin C also acts as a co-antioxidant by regenerating α-tocopherol (vitamin E) from the α-tocopheroxyl radical in lipoproteins and membranes (15–17). This is a potentially important function because

TABLE 3.1
Reactive Oxygen Species and Reactive Nitrogen Species Scavenged by Ascorbate: Second-Order Rate Constants and One-Electron Reduction Potentials

Chemical species scavenged by ascorbate	Rate constant[a] $(mol/L)^{-1} s^{-1}$	Ref.	Redox couple	E°'[b] (mV)	ΔE°'[c] (mV)
Reactive oxygen species					
Hydroxyl radical ($^\bullet$OH)	1.1×10^{10}	(5)	$^\bullet$OH, H$^+$/H$_2$O	2310	2028
Aliphatic alkoxyl radical (RO$^\bullet$)	1.6×10^9	(5)	RO$^\bullet$, H$^+$/ROH	1600	1318
Alkylperoxyl radical (ROO$^\bullet$)	$1-2 \times 10^6$	(5)	ROO$^\bullet$, H$^+$/ROOH	1000	718
Superoxide anion/Hydroperoxyl radical (O$_2^{\bullet-}$/HO$_2^\bullet$)	1×10^5	(5)	O$_2^{\bullet-}$, 2H$^+$/H$_2$O$_2$	940	658
Hypochlorous acid (HOCl)	6×10^6	(6)			
Ozone (O$_3$)	6×10^7	(7)			
Singlet oxygen (^{1}O$_2$)	8.3×10^6	(8)			
Reactive nitrogen species					
Nitrogen dioxide ($^\bullet$NO$_2$)	1.2×10^9	(9)			
Dinitrogen trioxide/Dinitrogen tetroxide (N$_2$O$_3$/N$_2$O$_4$)	235	(10)			
Peroxynitrite/Peroxynitrous acid (ONOO$^-$/ONOOH)		(11)			
Antioxidant-derived radicals[d]					
Thiyl/Sulphenyl radical (RS$^\bullet$/RSO$^\bullet$)	6×10^8	(5)	RS$^\bullet$, H$^+$/RSH (cysteine)	920	638
Urate radical (UH$^{\bullet-}$)	1×10^6	(5)	UH$^{\bullet-}$, H$^+$/UH$_2^-$	590	308
α-Tocopheroxyl radical (α-TO$^\bullet$)	2×10^5	(5)	α-TO$^\bullet$, H$^+$/α-TOH	500	218
β-Carotene radical cation (β-C$^{\bullet+}$)		(12)			

[a] The approximate rate constant for the reaction with ascorbate at pH 7.4 is given, if known.

[b] The one-electron reduction potential at pH 7.0 is given, if known (from Ref. 15).

[c] The difference in one-electron reduction potentials for the reaction with ascorbate is shown (the one-electron reduction potential for the ascorbyl radical/ascorbate monoanion couple is 282 mV); if (E°' is positive, the reaction is "thermodynamically feasible," assuming equimolar concentrations of the reactants.

[d] A number of other small molecule antioxidants can be regenerated from their respective radical species by ascorbate.

in vitro experiments have shown that α-tocopherol in low-density lipoprotein (LDL) can act as a prooxidant in the absence of vitamin C (17–19), and evidence for a prooxidant effect of α-tocopherol *in vivo* is mounting (19).

There are several properties that make ascorbate such a strong physiologic antioxidant, including the low one-electron reduction (or redox) potentials of ascorbate (282 mV) and its one-electron oxidation product, the ascorbyl radical (−174 mV) (15). These low reduction potentials enable ascorbate and its radical to spontaneously react with and reduce most physiologically relevant radicals and oxidants, i.e., these reactions are energetically or thermodynamically feasible (Table 3.1). Indeed, it has been stated that vitamin C acts "as the terminal water-soluble small molecule antioxidant" in biological systems (15). Second, the second-order rate constants for the reactions of ascorbate and most physiologically relevant ROS and RNS are greater than 10^5 $(mol/L)^{-1} \cdot s^{-1}$ (Table 3.1), which makes these reactions highly competitive with those of the same ROS and RNS with biological macromolecules. Third, the ascorbyl radical formed from ascorbate upon one-electron oxidation is quite stable and unreactive, due to resonance stabilization of the unpaired electron (15). Importantly, the ascorbyl radical does not react with molecular oxygen (O_2) to form a more reactive peroxyl radical. Instead, the ascorbyl radical readily dismutates to ascorbate and dehydroascorbic acid (DHA). DHA is reduced back to ascorbate in biological systems, e.g., by glutathione, glutaredoxin, or thioredoxin reductase, or is rapidly and irreversibly hydrolyzed (1,20) (Scheme 3.1). Thus, a fourth reason that ascorbate is an effective physiologic antioxidant is that it can be regenerated from its oxidized forms, either by spontaneous chemical reactions or enzymatically.

Iron

Iron is the most abundant transition metal in biological systems. The total amount of iron in a normal adult human has been estimated to be ~4.5 g (21), most of it in hemoglobin. Iron is also one of the most abundant elements in the earth's crust; however, the large amount of iron in all living cells cannot be explained by this fact alone. The preference of iron as a biologically relevant metal is related to its unique physicochemical properties, allowing it to vary its oxidation state, redox potential, and elec-

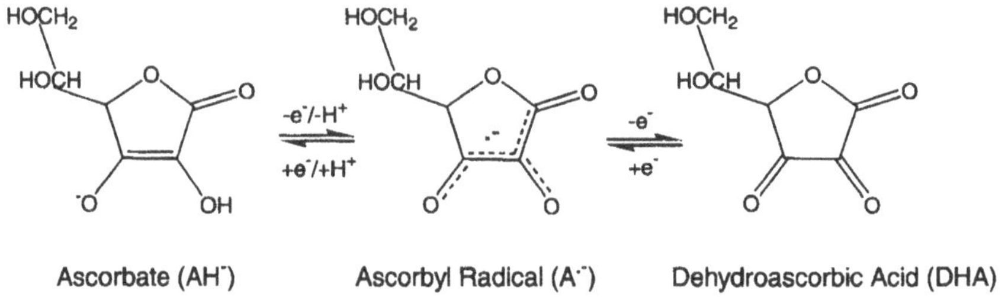

Scheme 3.1

tronic spin configuration in response to different coordinating environments. These properties enable iron, in contrast to many other metals, to play an essential role in a plethora of biological reactions. Iron exerts its functions either in the form of heme proteins or nonheme-containing proteins, such as iron-sulfur clusters. Iron-containing proteins catalyze many key reactions in energy metabolism, including respiration, O_2 delivery to tissues, DNA synthesis, and regulation of the citric acid cycle. Several other biological reactions, such as those with O_2 (monooxygenases and dioxygenases, see above) and peroxides (peroxidases, catalase and ferrioxidase), are also catalyzed by iron-containing enzymes (21).

In biological systems, iron is commonly found in three oxidation states, i.e., Fe(II), Fe(III), and in some cases Fe(IV). At physiologic pH, Fe(II) is water soluble, whereas Fe(III) precipitates as oxyhydroxide polymers. On the other hand, Fe(II) is unstable in aqueous solutions and tends to react with O_2 to form Fe(III). Thus, to use iron effectively as a cofactor in enzymatic systems, coordinating ligands such as O, N, S, or other metalloid atoms are used to shield iron from O_2 and the surrounding media (22,23). When O_2 is bound to iron in the active site of an enzyme, reactions of the bound O_2 and reduced oxygen intermediates become less energetically favorable. In addition, steric factors are well controlled in proteins, thus preventing the occurrence of undesirable side reactions between bound reactive intermediates and nonsubstrate molecules. The redox potential of iron is highly dependent on its coordination, and many enzymes "use" this property of iron as a fundamental means of controlling oxidation and reduction reactions of biomolecules.

Just as for O_2, which can be converted to ROS, the widespread use of iron in living organisms also gave rise to a paradox. On the one hand, by serving its multifunctional roles in biological systems, iron represents a great advantage for the complex chemical reactions of life; on the other hand, if not appropriately shielded, iron can readily participate in one-electron transfer reactions that can produce toxic free radicals. To overcome these problems, single-cell organisms first evolved molecules known as siderophores, which are secreted into the extracellular medium where they complex Fe(III) and are then assimilated by the cells *via* a receptor-mediated mechanism. In turn, multicellular organisms developed iron-binding proteins known as transferrins, which complex iron, transport it in the circulation (serum transferrin) or in other media (ovotransferrin, lactoferrin), and are taken up by peripheral cells, usually by a receptor-mediated mechanism.

Once transferrin is taken up and iron is released inside the cell, numerous mechanisms are in place whereby iron can be utilized in a form that is soluble under physiological conditions, bioavailable and nontoxic. "Nontoxic" means that iron is not able to elicit the well-known Fenton reaction, which gives rise to hydroxyl radicals ($^•OH$) from hydrogen peroxide (H_2O_2) and Fe(II) (see below, Reaction 3.1b). Thus, the complexation of iron within a cell necessitates that the iron storage protein, ferritin, must guard its iron in a form that will be assimilated into the many other proteins and molecules whose function depends on iron, but cannot be released indiscriminately to elicit oxidative damage and, thus, potentially pathological consequences.

Many studies have shown that iron can be released from iron-containing proteins under specific conditions. The release of iron from ferritin, the most concentrated source of iron [up to 4500 Fe(III) atoms per molecule], has been studied extensively (24). Any compound capable of reducing Fe(III) within the iron core of ferritin in the presence of a suitable chelator is capable of releasing iron from ferritin, particularly $O_2^{\bullet-}$, nitric oxide ($^{\bullet}NO$), and the "redox-cycling" class of xenobiotics. Ascorbate-mediated release of iron from ferritin seems to be due mainly to $O_2^{\bullet-}$ generated during oxidation of ascorbate catalyzed by iron bound to ferritin (25,26). In addition to this reductive release from ferritin, iron can also be released in an oxidative manner from heme-proteins or heme (27). It is thus conceivable that "free" iron is immediately conveyed into an intermediate, labile iron pool, and that this pool represents a steady-state exchangeable and readily chelatable iron compartment.

Studies of Biochemical Interactions Between Vitamin C and Iron

Autoxidation of Ascorbate and Other Biomolecules

It is widely believed that many biomolecules undergo autoxidation. Even though the reaction of O_2 with a biomolecule may be thermodynamically favorable, it does not occur at an appreciable rate because of a kinetic spin restriction of O_2 (23). For O_2 to react with a biomolecule, it must be activated either enzymatically or photochemically, or by sequential one-electron reductions to partially reduced oxygen species (i.e., ROS), including $O_2^{\bullet-}$, H_2O_2, and $^{\bullet}OH$. Transition metals, such as the reduced form of iron, are able to activate O_2. The sequential reduction of O_2 by iron is best exemplified by the iron-mediated Haber-Weiss reaction (28).

Coordination of iron to biomolecules almost always involves the d orbitals of the metal. Because O_2 can also bind to iron through the d orbitals of the metal (23), iron may simultaneously bind to biomolecules and O_2, effectively serving as a "bridge" between the two. It has been postulated that ascorbate, iron, and O_2 can form such a ternary complex (23). Ascorbate is both an iron chelator and an iron reductant; therefore, it can bind Fe(III) and subsequently reduce it to Fe(II). The Fe(II) can then reduce O_2 to $O_2^{\bullet-}$, with regeneration of Fe(III). Thus, it appears as if ascorbate autoxidizes, when in fact the reaction is catalyzed by iron. Accordingly, the "autoxidation" of ascorbate and numerous biogenic amines, such as epinephrine, is completely inhibited by strong metal chelating agents (29). Therefore, it has been suggested that biomolecules do not "autoxidize" but that the oxidation of biomolecules is mediated by trace amounts of transition metals, such as iron. In fact, it has been demonstrated (30) that the rate of ascorbate "autoxidation" in a given solution is proportional to the concentration of (contaminating) metal ions. To distinguish between Fe and Cu contamination, EDTA can be added to the assay solution, because Fe-EDTA is an excellent catalyst of ascorbate oxidation, whereas Cu-EDTA is a very poor catalyst. Measuring ascorbate oxidation spectrophotometrically at 265 nm, the iron levels in phosphate buffer can be estimated to a lower limit of ~100 nmol/L (30). Using electron spin resonance (ESR) spectroscopy, the detection limit can be further lowered to ~5 nmol/L (31). For this method, EDTA is added to the solution, which converts iron into a catalytic form, followed by the addition of ascorbate and measurement of the steady-state concentration of the ascorbyl radical by ESR. This method is useful not only because of its high sensitivity, but also if the solution to be assayed is colored or turbid and thus not suitable for standard colorimetric analysis of iron, or if only "loosely bound" iron is to be estimated (31).

The Ascorbate-Driven, Iron-Catalyzed Oxidation of Biological Macromolecules

In light of its important role as an antioxidant, it seems paradoxical that under certain conditions ascorbate can promote the generation of the very same ROS it is known to scavenge. This represents just another "paradox" of redox chemistry, which applies to any good reductant or antioxidant. Recognition of this property of ascorbate was an outgrowth of the pioneering studies of Udenfriend *et al.* (32) on the conversion of tyramine to hydroxytyramine by adrenal medullary homogenates. This work eventually led to the discovery that a nonenzymatic system comprised of ascorbate, O_2, ferrous iron, and EDTA catalyzes the hydroxylation of a number of aromatic and heterocyclic compounds with the formation of products that are similar to those generated in animals *in vivo*.

It is now generally agreed that the prooxidant activity of ascorbate is due to its capacity to reduce transition metal ions, such as iron or copper (Reaction 3.1a) (1,14,15,33). The reduced metal ions may react with H_2O_2 to form highly damaging $^\bullet OH$, a process known as Fenton chemistry (3.1b). Lipid hydroperoxides may also react with reduced metal ions to form lipid alkoxyl radicals (3.1c), which can initiate and propagate the chain reactions of lipid peroxidation (14,15).

$$AH^- + Fe(III) \rightarrow A^{\bullet-} + Fe(II) + H^+ \tag{3.1a}$$

$$H_2O_2 + Fe(II) \rightarrow {}^\bullet OH + {}^-OH + Fe(III) \tag{3.1b}$$

$$LOOH + Fe(II) \rightarrow LO^\bullet + {}^-OH + Fe(III) \tag{3.1c}$$

The "classical" prooxidant mixture (also called "Udenfriend system") of ascorbate, redox-active iron, and H_2O_2 (or lipid hydroperoxides, e.g., preexisting in membrane preparations) has been, and continues to be used to induce oxidative stress *in vitro* (1,14,15,33). The same or similar systems are also employed to induce oxidation in cell culture systems, whereby the redox-active metal ions often are supplied by the culture media. For example, Ham's F-10 medium contains trace amounts of transition metal ions, which are required for cell-mediated LDL oxidation by vascular cells (34). The ascorbate-driven, metal-catalyzed oxidation (MCO) system, because of its simplicity and because it mimics in every important aspect the properties of enzyme mixed function oxidation systems (35), has been used extensively to study oxidation of biological macromolecules, such as lipids, nucleic acids, and proteins.

Lipid peroxidation. The study of lipid oxidation, generally referred to as lipid peroxidation (because lipid peroxides are intermediates in the process), has been a topic of much research. Many investigators have proposed that iron is involved in the initiation of lipid peroxidation, although considerable controversy exists concerning its role in biological systems. One commonly proposed mechanism is that iron is responsible for catalyzing the generation of 'OH *via* Fenton chemistry and the Haber-Weiss reaction.

The key step in the initiation of lipid peroxidation is the abstraction of a hydrogen atom from the bis-allylic site of a polyunsaturated fatty acid. The 'OH is possibly the most powerful oxidant that can be formed in biological systems. As such, it can easily abstract a hydrogen atom from polyunsaturated fatty acids. It is, therefore, not surprising that 'OH has received so much consideration as the initiating species, although it is known that many other ROS and RNS can also initiate lipid peroxidation, e.g., hydroperoxyl radicals, peroxynitrite (ONOO⁻), and nitrogen dioxide. From a theoretical point of view, the ability of 'OH to initiate lipid peroxidation is unquestionable. However, it is imperative to realize that the indiscriminate, diffusion-limited reactivity of 'OH toward sugars, nucleotides, proteins, and any other biomolecules is a limitation to the idea that lipid peroxidation is initiated by 'OH. Some investigators have addressed this problem by proposing that 'OH is formed directly at the site of attack (36).

Another possible mechanism by which iron could be involved in initiating lipid peroxidation involves the formation of an Fe(III):Fe(II) complex. This mechanism has been proposed by Aust's group (37), and recently other researchers have arrived at similar conclusions (38,39). Interest in the initiation of lipid peroxidation by an iron complex started with the observation (37) that ADP:Fe(II) promoted the peroxidation of phospholipid liposomes only after a lag phase. Catalase, superoxide dismutase, and "'OH scavengers" did not extend the lag phase or inhibit the subsequent rate of lipid peroxidation, indicating that the reaction was not initiated by ROS. Interestingly, the lag phase was eliminated by the addition of ADP:Fe(III), which led to the proposal that the necessary species being generated during the lag phase was Fe(III). Furthermore, another study (40) showed that maximal rates of lipid peroxidation occurred when ~50% of the Fe(II) was oxidized, i.e., the Fe(II):Fe(III) ratio was ~1:1.

Consistent with the requirement for both Fe(II) and Fe(III), ascorbate, by reducing Fe(III) to Fe(II), stimulated iron-catalyzed lipid peroxidation; however, when the ascorbate concentration was high enough to reduce all of the Fe(III) to Fe(II), ascorbate inhibited lipid peroxidation (33). Exogenously added H_2O_2 also either stimulated or inhibited ascorbate-dependent, iron-catalyzed lipid peroxidation, apparently by altering the ratio of Fe(II):Fe(III). Thus, it appears that the prooxidant effect of ascorbate is related to its ability to promote the formation of the proposed Fe(II):Fe(III) complex and not due to ROS production, whereas the antioxidant effect of ascorbate may be due to complete reduction of Fe(III) to Fe(II) (33).

DNA oxidation. It has been suggested that the genotoxicity of many chemicals is enhanced by their ability to decompartmentalize cellular iron (41). Iron has been implicated as a causative agent in numerous cancers (28). One mechanism by which iron could be involved in the initiation or promotion of cancer is through the oxidation of DNA, causing mutations. DNA can be modified by free radicals resulting in single- and double-strand breaks, depurination, and depyrimidation, or chemical modification of the bases or phosphate-sugar backbone (42). Several ROS have been shown to oxidatively modify DNA, including $^{\bullet}OH$, singlet oxygen, and $ONOO^-$ (43). In contrast, $O_2^{\bullet-}$ and H_2O_2 are not capable of oxidizing DNA in the absence of adventitious metals (42), suggesting that the role of $O_2^{\bullet-}$ in DNA oxidation is simply as a constituent of the metal-driven Haber-Weiss reaction to produce $^{\bullet}OH$. In addition, it has been demonstrated that the addition of any chemical that will act as an alternate reactant for $^{\bullet}OH$, such as organic-based buffers or "$^{\bullet}OH$ traps," inhibits the oxidation of DNA (38). Conversely, the presence of chemicals that increase the iron-mediated production of $^{\bullet}OH$ will promote DNA oxidation (38).

Some researchers have postulated that the iron-mediated oxidation of DNA is a site-specific process (41,42). They propose that iron or an iron chelate binds to the DNA, either at phosphate on the backbone or to the purine or pyrimidine bases, where the iron can serve as a center for recurring formation of $^{\bullet}OH$, resulting in modification of the DNA (42,44,45). Experiments using purified DNA or isolated nuclei (46–48) confirm that in the presence of added metal ions, ascorbate acts as a prooxidant *in vitro*. In the absence of added metal ions, however, vitamin C inhibits oxidative DNA damage in purified DNA and cells (47,49–53), although there are a few exceptions (54–56). The latter are likely explained by "contaminating" metal ions in the cell culture media used.

The bleomycin-iron complex was the first well-studied system to damage DNA site specifically in a metal-dependent manner (44,45). Bleomycin-iron cleaves DNA to release *N*-propenal–substituted derivatives of thymine, cytosine, adenine, and guanine, which are believed to be responsible for some of the cytotoxic effects of bleomycin (45). The bleomycin-mediated cleavage of DNA is proposed to occur *via* a ternary bleomycin-DNA-iron complex. It is not known whether bleomycin binds to DNA first and then Fe(II) binds to the bleomycin-DNA complex, or whether a bleomycin-Fe(II) complex forms and then binds to DNA. Either way, oxidation of the complexed Fe(II) results in site-specific oxidation of DNA, presumably *via* $^{\bullet}OH$ production (57). In the presence of reducing agents such as ascorbate, Fe(III) is reduced back to Fe(II), thus continuing the oxidation of the DNA.

Protein oxidation. Studies of the metal-mediated denaturation of proteins were an outgrowth of investigations into the regulation of protein turnover in bacteria (35). These studies led to the discovery that degradation occurs when the protein has been oxidized (35). Similar to DNA, iron-mediated oxidation of protein may be a site-specific process. This notion is supported by the findings that the oxidation of proteins by MCO systems involves modification of only a few amino acid residues, in particular proline, histidine, arginine, lysine, and cysteine, whereas reactions of proteins with ROS generated by ionizing radiation are more or less random events, leading to the modification of many or all amino acid residues. Furthermore, metal ion-catalyzed oxidation reactions in proteins are not sensitive to inhibition by ROS scavengers (35). The site-specific nature of metal ion-catalyzed reactions is consistent with the fact that enzymes that require divalent metals for activity, and therefore must possess metal-binding sites, are particularly sensitive to inactivation by MCO systems. The concept of site specificity is also supported by the studies of Stadtman (35) showing that the ascorbate-mediated modification of glutamine synthetase involves the conversion of a single histidyl residue to an asparaginyl residue, and an arginyl residue to a glu-tamylsemialdehyde residue; furthermore, both of these amino acid residues are situated at the metal-binding site of the enzyme.

Oxidation of the amino acid side chain can lead to the conversion of some amino acids to carbonyl derivatives, loss of catalytic activity, and increased susceptibility of the protein to proteolytic degradation (35). Although it is very probable, it remains uncertain whether $^{\bullet}OH$ is the species responsible for the oxidation of proteins by MCO. Because the metal ion-catalyzed reactions lead to the conversion of some amino acid side chains (viz., prolyl, arginyl, and lysyl) to carbonyl derivatives, the concentration of protein carbonyl groups can be used as a measure of the extent of oxidative damage. Interestingly, protein carbonyl content increases with age and is associated with a number of pathological states (58).

The Crossover Effect for Ascorbate as an Anti- or Prooxidant

In general, low concentrations of ascorbate are required for prooxidant activity, whereas high concentrations are required for antioxidant activity. Thus, there is a "crossover effect," i.e., at a certain concentration, ascorbate "switches" from pro- to antioxidant activity. In the literature, a wide range of concentrations has been reported at which this switch occurs (5). The crossover effect can be rationalized as follows: In the presence of ascorbate, catalytic metals will initiate radical chain reactions. However, due to the hydrogen-donating activity of ascorbate, the chain length of these radical processes will be short as long as the ascorbate concentration is relatively high, resulting in little oxidative damage. As the ascorbate concentration decreases, the initiation processes are slowed somewhat, but more importantly, the antioxidant reactions of ascorbate are slowed as well. Thus, the chain length of these radical processes becomes longer and more oxidative damage will occur. It has been proposed (5) that the variability observed in the literature for the crossover effect is a result of the variability in the concentration and form of the catalytic metals present. Thus, at very low levels of catalytic metals, ascorbate will nearly always serve as antioxidant. However, if the levels of available metals should increase, then ascorbate may exert deleterious effects.

Studies on the Interaction of Vitamin C and Iron Under Physiologically Relevant Conditions

Vitamin C is known to increase the gastrointestinal absorption of nonheme iron by reducing it to a form that is more readily absorbed (59). It appears, however, that even at high intakes of vitamin C, iron uptake is tightly regulated in healthy people (59). Nevertheless, low dietary levels of vitamin C may be advantageous in cases of iron overload, such as homozygous hemochromatosis and β-thalassemia, because the excess iron can cause tissue damage (60,61). Individuals with iron overload generally have low plasma levels of vitamin C, possibly due to interaction with the elevated levels of "catalytic" iron in these individuals (60,61). Therefore, vitamin C administration has been claimed to be harmful in these patients (14,62). Iron overload has also been implicated in the sequelae of atherosclerosis, although the data are conflicting and inconsistent, and individuals with iron overload generally do not suffer from premature atherosclerosis (63,64). In addition, several vitamin C and iron cosupplementation studies, in both animals and humans, have shown that vitamin C inhibits, rather than promotes iron-dependent oxidative damage (summarized in Table 3.2).

Studies Using Plasma and Cultured Cells

In vitro experiments have shown that human serum and interstitial fluid strongly inhibit metal ion–dependent lipoprotein oxidation (65). These findings were attributed

TABLE 3.2
Role of Vitamin C in Iron-Mediated Oxidative Damage[a]

Study system	Challenge	Effects of vitamin C	Ref.
In vitro			
Human plasma	Iron	↓LOOH	(66)
Human plasma, lymph,	None	↔Hydroxyl radicals	(68)
synovial fluid	Iron-EDTA/H_2O_2	↑Hydroxyl radicals	
Human plasma	Iron/H_2O_2	↔Hydroxyl radicals	(69)
	Iron-EDTA	↑Hydroxyl radicals	
	Iron/H_2O_2	↓LOOH	[b]
3T3 fibroblasts	Iron	↔MDA	(84)
In vivo (animals)			
Guinea pig liver	Iron +	↓MDA	(73)
Ex vivo autoxidation			
Guinea pig plasma, liver	Iron	↓$F_{2\alpha}$-isoprostanes	(74)
Guinea pig serum	Iron	↑TBARS	(78)
In vivo (humans)			
Leukocytes	Iron (12 wk)	↓8-Oxoguanine, 8-oxoadenine, 5-hydroxyuracil, 5-chlorouracil ↑Thymine glycol, 5-hydroxycytosine	(79)
White blood cells	Iron (6 wk)	↔Total DNA damage ↓5-Hydroxymethyl uracil ↑5-Hydroxymethyl hydantoin, 5-hydroxy cytosine	(80)
Preterm infant plasma	(BDI), none	↔$F_{2\alpha}$-isoprostanes, protein carbonyls	(66)

[a]Abbreviations: ↑, increased damage; ↓, decreased damage; ↔, no change; BDI, bleomycin-detectable iron; LOOH, lipid hydroperoxides; MDA, malondialdehyde; TBARS, thiobarbituric acid-reactive substances.
[b]J. Suh, B.-Z. Zhu, and B. Frei, unpublished observations.

to the presence of metal binding proteins in these fluids, rather than vitamin C, because enzymatic removal of endogenous vitamin C did not alter the results. However, when sufficient exogenous iron (as ferrous ammonium sulfate) is added to plasma to saturate transferrin, resulting in the appearance of non-protein bound, bleomycin-detectable iron, endogenous and exogenous vitamin C strongly inhibits, rather than promotes lipid peroxidation (66) (Table 3.2). This finding is supported by an earlier study in which vitamin C acted as an antioxidant in serum to which excess copper had been added (67). Two other studies carried out with plasma, lymph, and synovial fluid showed that vitamin C can catalyze the formation of $^{\bullet}OH$, but only when the catalytically active form of iron, iron-EDTA, was added (68,69), not ferrous ammonium sulfate (69). Recently, we found that even when H_2O_2 is added to plasma, in addition to ascorbate and excess metal ions to constitute the complete Udenfriend system, ascorbate protects against iron- or copper-induced lipid peroxidation (J. Suh, B.-Z. Zhu, and B. Frei, unpublished observations). Ascorbate also did not enhance metal-dependent oxidation of plasma proteins under these conditions, as measured by protein carbonyl formation. These results demonstrate that even in the presence of high concentrations of transition metal ions and H_2O_2, ascorbate acts as an antioxidant that inhibits, and does not promote lipid and protein oxidation in biological fluids such as plasma.

Reports about the effects of ascorbate in cultured cells are conflicting, with some showing inhibition of cell death or apoptosis by ascorbate, and others suggesting that ascorbate is cytotoxic and induces apoptosis (70). Using three different cell types and two different culture media (Dulbecco's modified Eagle's medium and RPMI 1640), Clement et al. (71) found that the toxicity of ascorbate is due to ascorbate-mediated production of H_2O_2, to an extent that varies with the cell culture medium used. For example, in Dulbecco's modified Eagle's medium, 1 mmol/L ascorbate (a highly unphysiological, extracellular concentration) generated 161 ± 39 μmol/L H_2O_2 and induced apoptosis in 50% of HL60 cells, whereas in RPMI 1640 medium, only 83 ± 17 μmol/L H_2O_2 was produced and no apoptosis was observed. Apoptosis was prevented by catalase, and direct addition of H_2O_2 at the above concentrations to the cells mimicked the effects of ascorbate. These studies show that ascorbate itself is not toxic to cultured cells, and caution that the effects of ascorbate observed in cultured cells *in vitro* are of little or no relevance *in vivo*. The ability of ascorbate to interact with different cell culture media components, most probably contaminating metal ions, to produce H_2O_2 at different rates could account for most, if not all of the conflicting results reported (71,72).

Animal Supplementation Studies

Two animal studies have reported an antioxidant role for vitamin C in guinea pigs cosupplemented with vitamin C and iron (Table 3.2): (i) *ex vivo* autoxidation of liver microsomes obtained from iron-supplemented guinea pigs resulted in increased accumulation of malondialdehyde compared with microsomes obtained from control animals or animals cosupplemented with iron and vitamin C (73); and (ii) plasma and liver $F_{2\alpha}$-isoprostanes, markers of *in vivo* lipid peroxidation, were increased in vitamin C–deficient guinea pigs loaded with iron, but reduced by vitamin C supplementation (74). In the latter study, hepatic vitamin C levels, in contrast to iron levels, were inversely associated with hepatic $F_{2\alpha}$-isoprostane levels (74). Another recent study using rats showed an antioxidant effect of vitamin C when given before a paraquat challenge, but a prooxidant effect when given after the challenge, as determined by expiratory ethane (75). The prooxidant effect was attributed to the paraquat-mediated release of metal ions from damaged cells. That study (75), therefore, suggests that vitamin C may have different effects depending on when it is added to the system under study, as has been observed previously with copper-dependent lipid peroxidation in LDL (76,77). Another recent study (78) found that large doses of intravenous ascorbate increased the levels of loosely bound iron and *in vitro* oxidation of serum obtained from iron-loaded guinea pigs, but not control animals (Table 3.2). Susceptibility of LDL to *ex vivo* oxidation increased after vitamin C injection in the control group, but there was no further increase in the iron-loaded group.

Human Supplementation Studies

A study carried out in humans to assess the effects of simultaneous iron and vitamin C supplementation yielded mixed results with respect to formation of various types of oxidized DNA bases in leukocytes (Table 3.2). Reanalysis of the data from that study (79) shows an inverse association between the plasma concentration of vitamin C and total DNA base damage. In addition, there was a positive correlation between the concentration of plasma vitamin C and the percentage of transferrin saturation, possibly due to a vitamin C–dependent increase in iron bioavailability (59), but no correlation was observed between the percentage of transferrin saturation and total base damage. These correlations are analogous to those observed in the above study using guinea pigs and suggest that vitamin C acts as an antioxidant, rather than a prooxidant, *in vivo* in the presence of iron.

Decreased levels of serum vitamin C and increased levels of lipid and protein oxidation products have been detected in hemochromatosis and β-thalassemia patients (60,61), which was attributed to the iron overload condition. However, these conclusions are not supported by a study in preterm infants, who often have excess iron in their plasma (66). In that study, plasma levels of $F_{2\alpha}$-isoprostanes and protein carbonyls were not correlated with levels of bleomycin-detectable iron, even in the presence of high concentrations of vitamin C (Table 3.2). It was found recently (80) that supplementation of either vitamin C or vitamin C plus iron did not cause a rise in total oxidative DNA damage measured by gas chromatography-mass spectrometry. However, a significant decrease was observed in the levels of the purine base oxidation product, 8-oxoguanine, after ascorbate supplementation. 5-Hydroxymethyl uracil levels were also decreased by either ascorbate or ascorbate plus iron supplementation, relative to the presupplemental levels, but not relative to the placebo group. In addition, levels of 5-hydroxymethyl hydantoin and 5-hydroxy cytosine increased significantly by ascorbate plus iron supplementation relative to the presupplementation period (Table 3.2). However, no consistent or compelling evidence for a prooxidant effect of vitamin C supplementation, with or without iron cosupplementation, on DNA base damage was observed (80). In another study by the same group (81), iron supplementation failed to affect any of the iron status variables measured, including serum ferritin, transferrin-bound iron, and the percentage of saturation of transferrin, and there were no detrimental effects on oxidative damage to DNA in healthy individuals with high plasma ascorbate levels. Finally, a recent study investigated the effect of a daily combined iron (100 mg/d as fumarate) and vitamin C (500 mg/d as ascorbate) supplement on plasma lipid peroxidation in pregnant women during the third trimester (82). In the supplemented group, plasma iron levels were higher than in the control group and plasma levels of thiobarbituric-acid reactive substances (TBARS) were significantly enhanced, suggesting that pharmacologic doses of iron, associated with high vitamin C intakes, may result in oxidative damage *in vivo*. However, TBARS are a poor measure of lipid peroxidation in biological fluids because there are many interfering substances and the basal levels reported in human plasma vary widely (83).

Conclusions

It is evident that ascorbate can exhibit both antioxidant and prooxidant activities. Although its role as an antioxidant is well documented, there is little, if any evidence that it serves as a prooxidant under physiological conditions. A majority of the studies that specifically addressed the interaction of vitamin C with iron in physiological fluids or *in vivo* (Table 3.2) found either no effect of vitamin C or decreased oxidative damage. Vitamin C played a prooxidant role in biological fluids only if iron-EDTA was added *in vitro* (68,69).

It should be noted that the levels of ascorbate vary substantially from tissue to tissue, ranging from 30 to 120 μmol/L in the plasma of normal individuals to millimolar intracellular concentrations in eye lens, brain, lung, and adrenals. These concentrations are certainly sufficient to provoke radical generation. The limiting factor is most likely the availability of metal ions, which are absolutely required for the prooxidant activity of ascorbate to occur. Although the total iron concentration of most tissues is quite high, iron exists almost entirely tightly sequestered in protein complexes (*viz.*, transferrin, lactoferrin, hemoglobin, and ferritin) and is, therefore, not readily available for ROS generation (85,86). Conditions that facilitate the release of iron from these complexes most likely also promote ascorbate-mediated radical damage. Thus, the release of iron, which is often associated with tissue damage, would be expected to (secondarily) provoke ROS generation *via* the ascorbate-iron system.

On the basis of the studies on the biochemical interactions between ascorbate and iron, it is likely that the ratio of the concentration of iron and ascorbate will determine the ability of ascorbate to express prooxidant or antioxidant activity. At very low levels of catalytic iron under normal physiological conditions, ascorbate will act mainly as an antioxidant; however, if the levels of available iron increase significantly under certain pathologic conditions, ascorbate may exert prooxidant activity.

Acknowledgments

The work in the authors' laboratory is supported by National Institutes of Health grants HL60886 and AT00066 to B.F., ES11497 and ES00210 to B.-Z.Z.

References

1. Levine, M. (1986) New Concepts in the Biology and Biochemistry of Ascorbic Acid, *N. Engl. J. Med. 314*, 892–902.
2. Institute of Medicine (2000) *Dietary Reference Intakes for Vitamin C, Vitamin E, Selenium and Carotenoids,* pp. 95–185, National Academy Press, Washington.
3. Levine, M., Conry-Cantilena, C., Wang, Y., Welch, R.W., Washko, P.W., Dhariwal, K.R., Park, J.B., Lazarev, A., Graumlich, J.F., King, J., and Cantilena, L.R. (1996) Vitamin C Pharmacokinetics in Healthy Volunteers: Evidence for a Recommended Dietary Allowance, *Proc. Natl. Acad. Sci. USA 93*, 3704–3709.
4. Burri, B.J., and Jacob, R.A. (1997) Human Metabolism and the Requirement for Vitamin C, in *Vitamin C in Health and Disease* (Packer, L., and Fuchs, J., eds.) pp. 341–366, Marcel Dekker, New York.
5. Buettner, G.R., and Jurkiewicz, B.A. (1996) Catalytic Metals, Ascorbate and Free Radicals: Combinations to Avoid, *Radiat. Res. 145*, 532–541.
6. Folkes, L.K., Candeias, L.P., and Wardman, P. (1995) Kinetics and Mechanisms of Hypochlorous Acid Reactions, *Arch. Biochem. Biophys. 323*, 120–126.
7. Giamalva, D., Church, D.F., and Pryor, W.A. (1985) A Comparison of the Rates of Ozonation of Biological Antioxidants and Oleate and Linoleate Esters, *Biochem. Biophys. Res. Commun. 133*, 773–779.
8. Chou, P.T., and Khan, A.U. (1983) L-Ascorbic Acid Quenching of Singlet Delta Molecular Oxygen in Aqueous Media: Generalized Antioxidant Property of Vitamin C, *Biochem. Biophys. Res. Commun. 115*, 932–937.
9. Cooney, R.V., Ross, P.D., and Bartolini, G.L. (1986) *N*-Nitrosation and *N*-Nitration of Morpholine by Nitrogen Dioxide: Inhibition by Ascorbate, Glutathione and α-Tocopherol, *Cancer Lett. 32*, 83–90.
10. Licht, W.R., Tannenbaum, S.R., and Deen, W.M. (1988) Use of Ascorbic Acid to Inhibit Nitrosation: Kinetic and Mass Transfer Considerations for an *In Vitro* System, *Carcinogenesis 9*, 365–372.
11. Bartlett, D., Church, D.F., Bounds, P.L., and Koppenol, W.H. (1995) The Kinetics of the Oxidation of L-Ascorbic Acid by Peroxynitrite, *Free Radic. Biol. Med. 18*, 85–92.
12. Edge, R., and Truscott, T.G. (1997) Prooxidant and Antioxidant Reaction Mechanisms of Carotene and Radical Interactions with Vitamins E and C, *Nutrition 13*, 992–994.
13. Frei, B., England, L., and Ames, B.N. (1989) Ascorbate Is an Outstanding Antioxidant in Human Blood Plasma, *Proc. Natl. Acad. Sci. USA 86*, 6377–6381.
14. Halliwell, B. (1996) Vitamin C: Antioxidant or Pro-Oxidant in Vivo, *Free Radic. Res. 25*, 439–454.
15. Buettner, G.R. (1993) The Pecking Order of Free Radicals and Antioxidants: Lipid Peroxidation, α-Tocopherol, and Ascorbate, *Arch. Biochem. Biophys. 300*, 535–543.
16. Packer, L. (1997) Vitamin C and Redox Cycling Antioxidants, in *Vitamin C in Health and Disease* (Packer, L., and Fuchs, J., eds.) pp. 95–121, Marcel Dekker, New York.
17. Ingold, K.U., Bowry, V.W., Stocker, R., and Walling, C. (1993) Autoxidation of Lipids and Antioxidation by α-Tocopherol and Ubiquinol in Homogeneous Solution and in

Aqueous Dispersions of Lipids: Unrecognized Consequences of Lipid Particle Size as Exemplified by Oxidation of Human Low Density Lipoprotein, *Proc. Natl. Acad. Sci. USA 90*, 45–49.

18. Neuzil, J., Thomas, S.R., and Stocker, R. (1997) Requirement for, Promotion, or Inhibition by α-Tocopherol of Radical-Induced Initiation of Plasma Lipoprotein Lipid Peroxidation, *Free Radic. Biol. Med. 22*, 57–71.

19. Thomas, S.R., and Stocker, R. (2000) Molecular Action of Vitamin E in Lipoprotein Oxidation: Implications for Atherosclerosis, *Free Radic. Biol. Med. 28*, 1795–1805.

20. Carr, A. and Frei, B. (1999) Toward a New Recommended Dietary Allowance for Vitamin C Based on Antioxidant and Health Effects in Humans, *Am. J. Clin. Nutr. 69*, 1086–1107.

21. Emery, T. (1991) *Iron and Your Health: Facts and Fallacies*, CRC Press, Boca Raton.

22. Ryan, T.P., and Aust, S.D. (1992) The Role of Iron in Oxygen-Mediated Toxicities, *Crit. Rev. Toxicol. 22*, 119–141.

23. Miller, D.M., Buettner, G.R., and Aust, S.D. (1990) Transition Metals as Catalysts of "Autoxidation" Reactions, *Free Radic. Biol. Med. 8*, 95–108.

24. Minotti, G. (1993) Sources and Role of Iron in Lipid Peroxidation, *Chem. Res. Toxicol. 6*, 134–146.

25. Boyer, R.F., and McCleary, C.J. (1987) Superoxide Ion as a Primary Reductant in Ascorbate-Mediated Ferritin Iron Release, *Free Radic. Biol. Med. 3*, 389–395.

26. Roginsky, V.A., Barsukova, T.K., Bruchelt, G., and Stegmann, H.B. (1997) Iron Bound to Ferritin Catalyzes Ascorbate Oxidation: Effects of Chelating Agents, *Biochim. Biophys. Acta 1335*, 33–39.

27. Comporti, M., Signorini, C., Buonocore, G., Ciccoli, L. (2002) Iron Release, Oxidative Stress and Erythrocyte Ageing, *Free Radic. Biol. Med. 32*, 568–576.

28. Reilly, C.A., and Aust, S.D. (2000) Biological Oxidations Catalyzed by Iron Released from Ferritin, in *Toxicology of the Human Environment. The Critical Role of Free Radicals* (Rhodes, C.J., ed.) pp.155–190, Taylor and Francis, London.

29. Ryan, T.P., Miller, D.M., and Aust, S.D. (1993) The Role of Metals in the Enzymatic and Nonenzymatic Oxidation of Epinephrine, *J. Biochem. Toxicol 8*, 33–39.

30. Buettner, G.R. (1988) In the Absence of Catalytic Metals Ascorbate Does Not Autoxidize at pH 7: Ascorbate as a Test for Catalytic Metals, *J. Biochem. Biophys. Methods 16*, 27–40.

31. Buettner, G.R. (1990) Ascorbate Oxidation: UV Absorbance of Ascorbate and ESR Spectroscopy of the Ascorbyl Radical as Assays for Iron, *Free Radic. Res. Commun. 10*, 5–9.

32. Udenfriend, S., Clark, C.T., Axelrod, J., and Brodie, B.B. (1954) Ascorbic Acid in Aromatic Hydroxylation. I. A Model System for Aromatic Hydroxylation, *J. Biol. Chem. 208*, 731–739.

33. Miller, D.M., and Aust, S.D. (1989) Studies of Ascorbate-Dependent, Iron Catalyzed Lipid Peroxidation, *Arch. Biochem. Biophys. 271*, 113–119.

34. Martin, A., and Frei, B. (1997) Both Intracellular and Extracellular Vitamin C Inhibit Atherogenic Modification of LDL by Human Vascular Endothelial Cells, *Arterioscler. Thromb. Vasc. Biol. 17*, 1583–1590.

35. Stadtman, E.R. (1990) Metal Ion-Catalyzed Oxidation of Proteins: Biochemical Mechanism and Biological Consequences, *Free Radic. Biol. Med. 9*, 315–325.

36. Samumi, A., Chevion, M., and Czapski, G. (1981) Unusual Copper-Induced Sensitization of the Biological Damage Due to Superoxide Radicals, *J. Biol. Chem. 256*, 12632–12635.

37. Bucher, J.R., Tien, M., and Aust, S.D. (1983) The Requirement for Ferric in the Initiation of Lipid Peroxidation by Chelated Ferrous Iron, *Biochem. Biophys. Res. Commun. 111*, 777–784.
38. Djuric, Z., Potter, D.W., Taffe, B.G., and Strasburg, G.M. (2001) Comparison of Iron-Catalyzed DNA and Lipid Oxidation, *J. Biochem. Mol. Toxicol. 15*, 114–119.
39. Tang, L.X. and Shen, X. (1997) Effects of Additional Iron-Chelators on Fe(II)-Initiated Lipid Peroxidation: Evidence to Support the Fe(II)...Fe(III) Complex as the Initiator, *J. Inorg. Biochem. 68*, 265–272.
40. Minotti, G., and Aust, S.D. (1987) The Requirement for Iron(III) in the Initiation of Lipid Peroxidation by Iron(II) and Hydrogen Peroxide, *J. Biol. Chem. 262*, 1098–1104.
41. Bandy, A.S.B., Tsang, S., and Davison, A.J. (2001) DNA Breakage Induced by 1,2,4-Benzenetriol: Relative Contributions of Oxygen-Derived Active Species and Transition Metal Ions, *Free Radic. Biol. Med. 30*, 943–946.
42. Chevion, M. (1988) Site-Specific Mechanism for Free Radical Induced Biological Damage: The Essential Role of Redox-Active Transition Metals, *Free Radic. Biol. Med. 5*, 27–37.
43. Halliwell, B., and Gutteridge, J.M.C. (1999) *Free Radicals in Biology and Medicine*, Oxford University Press, Oxford.
44. Giloni, L., Takeshita, M., Johnson, F., Iden, C., and Grollman, A.P. (1981) Bleomycin-Induced Strand Scission of DNA, *J. Biol. Chem. 256*, 8608–8611.
45. Grollman, A.P., Takeshita, M., Pillai, K.M.R., and Johnson, F. (1985) Origin and Cytotoxic Properties of Base Propenols Derived from DNA, *Cancer Res. 45*, 1127–1131.
46. Drouin, R., Rodriguez, H., Gao, S.W., Gebreyes, Z., O'Connor, T.R., Holmquist, G.P., and Akman, S.A. (1996) Cupric Ion/Ascorbate/Hydrogen Peroxide-Induced DNA Damage: DNA-Bound Copper Ion Primarily Induces Base Modifications, *Free Radic. Biol. Med. 21*, 261–273.
47. Fischer-Nielsen, A., Poulsen, H.E., and Loft, S. (1992) 8-Hydroxydeoxyguanosine in Vitro: Effects of Glutathione, Ascorbate, and 5-Aminosalicylic Acid, *Free Radic. Biol. Med. 13*, 121–126.
48. Hu, M.L., and Shih, M.K. (1997) Ascorbic Acid Inhibits Lipid Peroxidation but Enhances DNA Damage in Rat Liver Nuclei Incubated with Iron Ions, *Free Radic. Res. 26*, 585–592.
49. Fiala, E.S., Sodum, R.S., Bhattacharya, M., and Li, H. (1996) (-)-Epigallocatechin Gallate, a Polyphenolic Tea Antioxidant, Inhibits Peroxynitrite-Mediated Formation of 8-Oxodeoxyguanosine and 3-Nitrotyrosine, *Experientia 52*, 922–926.
50. Wei, H., Cai, Q., Tian, L., and Lebwohl, M. (1998) Tamoxifen Reduces Endogenous and UV Light-Induced Oxidative Damage to DNA, Lipid, and Protein In Vitro and In Vivo, *Carcinogenesis 19*, 1013–1018.
51. Noroozi, M., Angerson, W.J., and Lean, M.E.J. (1998) Effects of Flavonoids and Vitamin C on Oxidative DNA Damage to Human Lymphocytes, *Am. J. Clin. Nutr. 67*, 1210–1218.
52. Fischer-Nielsen, A., Loft, S., and Jensen, K.G. (1993) Effect of Ascorbate and 5-Aminosalicylic Acid on Light-Induced 8-Hydroxydeoxyguanosine Formation in V79 Chinese Hamster Cells, *Carcinogenesis 14*, 2431–2433.
53. Pflaum, M., Kielbassa, C., Garmyn, M., and Epe, B. (1998) Oxidative DNA Damage Induced by Visible Light in Mammalian Cells: Extent, Inhibition by Antioxidants and Genotoxic Effects, *Mutat. Res. 408*, 137–146.
54. Singh, N.P. (1997) Sodium Ascorbate Induces DNA Single-Strand Breaks in Human Cells In Vitro, *Mutat. Res. 375*, 195–203.

55. Anderson, D., Yu, T.W., Phillips, B.J., and Schmezer, P. (1994) The Effect of Various Antioxidants and Other Modifying Agents on Oxygen-Radical-Generated DNA Damage in Human Lymphocytes in the COMET Assay, *Mutat. Res. 307*, 261–271.

56. Green, M.H.L., Lowe, J.E., Waugh, A.P.W., Aldridge, K.E., Cole, J., and Arlett, C.F. (1994) Effect of Diet and Vitamin C on DNA Strand Breakage in Freshly Isolated Human White Blood Cells, *Mutat. Res. 316*, 91–102.

57. Sausville, E.A., Peisacti, J., and Horwitz, B. (1978) Effect of Chelating Agents and Metal Ions on the Degradation of DNA by Bleomycin, *Biochemistry 17*, 2740–2746.

58. Berlett, B.S., and Stadtman, E.R. (1997) Protein Oxidation in Aging, Disease, and Oxidative Stress, *J. Biol. Chem. 272*, 20313–20316.

59. Bendich, A., and Cohen, M. (1990) Ascorbic Acid Safety: Analysis of Factors Affecting Iron Absorption, *Toxicol. Lett. 51*, 189–201.

60. Young, I.S., Trouton, T.G., Torney, J.J., McMaster, D., Callender, M.E., and Trimble, E.R. (1994) Antioxidant Status and Lipid Peroxidation in Hereditary Haemochromatosis, *Free Radic. Biol. Med. 16*, 393–397.

61. Livrea, M.A., Tesoriere, L., Pintaudi, A.M., Calabrese, A., Maggio, A., Freisleben, H.J., D'Arpa, D., D'Anna, R., and Bongiorno, A. (1996) Oxidative Stress and Antioxidant Status in β-Thalassemia Major: Iron Overload and Deletion of Lipid-Soluble Antioxidants, *Blood 88*, 3608–3614.

62. Herbert, V. (1994) The Antioxidant Supplement Myth, *Am. J. Clin. Nutr. 60*, 157–158.

63. Kiechl, S., Willeit, J., Egger, G., Poewe, W., and Oberhollenzer, F. (1997) Body Iron Stores and the Risk of Carotid Atherosclerosis: Prospective Results from the Bruneck Study, *Circulation 96*, 3300–3307.

64. Franco, R.F., Zago, M.A., Trip, M.D., Cate, H., van den Ende, A., Prins, M.H., Kastelein, J.J., and Reitsma, P.H. (1998) Prevalence of Hereditary Haemochromatosis in Premature Atherosclerotic Vascular Disease, *Br. J. Haematol. 102*, 1172–1175.

65. Dabbagh, A.J., and Frei, B. (1995) Human Suction Blister Interstitial Fluid Prevents Metal Ion-Dependent Oxidation of Low Density Lipoprotein by Macrophages and in Cell-Free Systems, *J. Clin. Investig. 96*, 1958–1966.

66. Berger, T.M., Polidori, M.C., Dabbagh, A., Evans, P.J., Halliwell, B., Morrow, J.D., Roberts, L.J., and Frei, B. (1997) Antioxidant Activity of Vitamin C in Iron-Overloaded Human Plasma, *J. Biol. Chem. 272*, 15656–15660.

67. Dasgupta, A., and Zdunek, T. (1992) In Vitro Lipid Peroxidation of Human Serum Catalyzed by Cupric Ion: Antioxidant Rather than Prooxidant Role of Ascorbate, *Life Sci. 50*, 875–882.

68. Winterbourn, C.C. (1981) Hydroxyl Radical Production in Body Fluids: Roles of Metal Ions, Ascorbate and Superoxide, *Biochem. J. 198*, 125–131.

69. Minetti, M., Forte, T., Soriani, M., Quarisima, V., Menditto, A., and Ferrari, M. (1992) Iron-Induced Ascorbate Oxidation in Plasma as Monitored by Ascorbate Free Radical Formation: No Spin-Trapping Evidence for the Hydroxyl Radical in Iron-Overload Plasma, *Biochem. J. 282*, 459–465.

70. Halliwell, B. (2001) Vitamin C and Genomic Stability, *Mutat. Res. 475*, 29–35.

71. Clement, M.V., Ramalingam, J., Long, L.H., and Halliwell, B. (2001) The In Vitro Cytotoxicity of Ascorbate Depends on the Culture Medium Used to Perform the Assay and Involves Hydrogen Peroxide, *Antioxid. Redox Signal. 3*, 157–163.

72. Halliwell, B., Clement, M.V., Ramalingam, J., and Long, L.H. (2000) Hydrogen Peroxide. Ubiquitous in Cell Culture and in Vivo? *IUBMB Life 50*, 251–257.

73. Collis, C.S., Yang, M., Diplock, A.T., Hallinan, T., and Rice-Evans, C.A. (1997) Effects of Co-Supplementation of Iron with Ascorbic Acid on Antioxidant-Pro-Oxidant Balance in the Guinea Pig, *Free Radic. Res. 27*, 113–121.
74. Chen, K., Suh, J., Carr, A.C., Morrow, J.D., Zeind, J., and Frei, B. (2000) Vitamin C Suppresses Oxidative Lipid Damage In Vivo, Even in the Presence of Iron Overload, *Am. J. Physiol. 279*, E1406–E1412.
75. Kang, S.A., Jang, Y.J., and Park, H. (1998) In Vivo Dual Effect of Vitamin C on Paraquat-Induced Lung Damage: Dependence on Released Metals from the Damaged Tissue, *Free Radic. Res. 28*, 93–107.
76. Retsky, K.L., and Frei, B. (1995) Vitamin C Prevents Metal Ion-Dependent Initiation and Propagation of Lipid Peroxidation in Human Low-Density Lipoprotein, *Biochim. Biophys. Acta 1257*, 279–287.
77. Otero, P., Viana, M., Herrera, E., and Bonet, B. (1997) Antioxidant and Prooxidant Effects of Ascorbic Acid, Dehydroascorbic Acid and Flavonoids on LDL Submitted to Different Degrees of Oxidation, *Free Radic. Res. 27*, 619–626.
78. Kapsokefalou, M., and Miller, D.D. (2001) Iron Loading and Large Doses of Intravenous Ascorbic Acid Promote Lipid Peroxidation in Whole Serum in Guinea Pigs, *Br. J. Nutr. 85*, 681–687.
79. Rehman, A., Collis, C.S., Yang, M., Kelly, M., Diplock, A.T., Halliwell, B., and Rice-Evans, C. (1998) The Effects of Iron and Vitamin C Co-Supplementation on Oxidative Damage to DNA in Healthy Volunteers, *Biochem. Biophys. Res. Commun. 246*, 293–298.
80. Proteggente, A.R., Rehman, A., Halliwell, B., and Rice-Evans, C.A. (2000) Potential Problems of Ascorbate and Iron Supplementation: Pro-Oxidant Effect in Vivo? *Biochem. Biophys. Res. Commun. 277*, 535–540.
81. Proteggente, A.R., England, T.G., Rice-Evans, C.A., and Halliwell, B. (2001) Iron Supplementation and Oxidative Damage to DNA in Healthy Individuals with High Plasma Ascorbate, *Biochem. Biophys. Res. Commun. 288*, 245–251.
82. Lachili, B., Hininger, I., Faure, H., Arnaud, J., Richard, M.J., Favier, A., and Roussel, A.M. (2001) Increased Lipid Peroxidation in Pregnant Women After Iron and Vitamin C Supplementation, *Biol. Trace Elem. Res. 83*, 103–110.
83. Moore, K., and Roberts, L.J., II (1998) Measurement of Lipid Peroxidation, *Free Radic. Res. 28*, 659–671.
84. Collis, C.S., Yang, M., Peach, S.J., Diplock, A.T., and Rice-Evans, C. (1996) The Effects of Ascorbic Acid and Iron Co-Supplementation on the Proliferation of 3T3 Fibroblasts, *Free Radic. Res. 25*, 87–93.
85. Frei, B., Stocker, R., and Ames, B.N. (1988) Antioxidant Defenses and Lipid Peroxidation in Human Blood Plasma, *Proc. Natl. Acad. Sci. USA 85*, 9748–9752.
86. Halliwell, B., and Gutteridge, J.M.C. (1986) Oxygen Free Radicals and Iron in Relation to Biology and Medicine: Some Problems and Concepts, *Arch. Biochem. Biophys. 246*, 501–514.

Chapter 4

Vitamin C and Cancer

Seon Hwa Lee, Tomoyuki Oe, and Ian A. Blair

Center for Cancer Pharmacology, University of Pennsylvania School of Medicine, Philadelphia, PA 19104–6160

Introduction

The role of vitamin C in treating and preventing cancer remains very controversial in spite of a number of studies that have been conducted over the last 25 years. In 1974, Cameron and Pauling (1) suggested that vitamin C might play a role in the supportive care of cancer patients. A study conducted by Cameron and Campbell (2) on 50 terminal cancer patients appeared to support this concept. They showed that large doses of vitamin C given orally, intravenously, or by a combination of both routes could significantly improve survival in advanced cancer patients. In a subsequent report by these authors, 10 of the original patients with unusual cancers were replaced with 10 new ones randomly selected from the records of vitamin C–treated patients from the same area. Furthermore, almost 50% of the 1000 control subjects were replaced because data on some of the initial control population were considered to be unreliable (3). The second study appeared to show an even greater benefit of vitamin C treatment. Because of the potential for bias in these nonrandomized studies, a group of researchers at the Mayo Clinic performed a randomized study to evaluate the efficacy of vitamin C in patients with advanced cancer (4). A second community-based study was reported at the American Society of Clinical Oncologists in 1983 (5). Finally, a third placebo-controlled study was conducted in cancer patients at the Mayo Clinic and reported in 1986 (6). Similar results were obtained in all three studies, i.e., there was no significant effect of vitamin C therapy on patient comfort or survival.

Vitamin C is an efficient water-soluble one-electron reducing agent that would be predicted to have efficacy in preventing oxidative DNA damage. A number of studies have in fact shown a reduction in oxidative damage to DNA, although many of these studies used methodology that was flawed (7). In contrast, cancer chemoprevention studies have consistently shown that there is no benefit from vitamin C supplementation in terms of cancer outcome (8,9). This contrasts with studies that consistently show a reduction in disease risk with diets that are rich in antioxidants such as vitamin C (10).

It is thought that lipid peroxidation of polyunsaturated fatty acids (PUFA) plays an important role in the degenerative diseases of aging such as cancer (11,12). The formation of lipid hydroperoxides from PUFA by free radical processes is a complex process, which leads to a number of different regioisomers and stereoisomers (13). It would be predicted that vitamin C would inhibit this pathway. However, lipoxygenase (LOX)-mediated oxidation of PUFA results in the formation of lipid hydroperoxides (14), and vitamin C would not be predicted to inhibit this pathway. Lipid hydroperoxides undergo homolytic decomposition to bifunctional electrophiles (15) that can react with DNA (12). Therefore, we were interested in determining whether vitamin C could mediate lipid hydroperoxide–mediated DNA damage.

In previous studies, we examined the homolytic decomposition of 13-[S-(Z,E)]-9,11-hydroperoxyoctadecadienoic acid (13-HPODE; a prototypic n-6 PUFA lipid hydroperoxide) in the presence of the DNA bases 2'-deoxyadenosine (dAdo) and 2'-deoxyguanosine (dGuo). From structures of the resulting DNA-adducts, we proposed that the major covalent modifications arose through generation of 4-oxo-2-nonenal from 13-HPODE. The same adducts were formed when DNA bases were treated with synthetic 4-oxo-2-nonenal (16,17). Subsequently, 4-oxo-2-nonenal was confirmed as a major breakdown product of homolytic lipid hydroperoxide decomposition (18), a finding that was recently confirmed by Spiteller *et al.* (19). Surprisingly, 4-hydroperoxy-2-nonenal was also characterized as a product of 13-HPODE decomposition by our laboratory and by Schneider *et al.* (18,20). 4-Hydroperoxy-2-nonenal was subsequently shown to be a precursor in the formation of 4-oxo-2-nonenal and 4-hydroxy-2-nonenal (21). The other major bifunctional electrophile identified in homolytic 13-HPODE decomposition was 4,5-epoxy-2(E)-decenal (21), a recently identified product from the autoxidation of arachidonic acid (22).

An environmental contaminant in water and food [see Ref. (23) for discussion], *trans,trans*-2,4-decadienal is also a product of lipid peroxidation through α-cleavage of the alkoxy radicals derived from 9-hydroperoxy-(E,E)-10,12-octadecadienoic acid or 11-hydroperoxy-(Z,Z,E,E)-eicosa-5,8,12,14-tetraenoic acid (19,22). Recent studies have shown that the reaction of peroxide-treated *trans,trans*-2,4-decadienal with dAdo or dGuo results in the formation of 1,N^6-etheno-dAdo (23) and 1,N^2-etheno-dGuo (24), respectively. We reasoned that 4,5-epoxy-2(E)-decenal could have been formed when *trans,trans*-2,4-decadienal was treated with peroxides (22,25) and that this bifunctional electrophile was in fact the precursor to the formation of etheno-adducts from lipid hydroperoxides.

Formation of Lipid Hydroperoxides

Lipid hydroperoxides are formed nonenzymatically by reactive oxygen species (ROS) such as superoxide ($O_2^{\bullet -}$), peroxide (O_2^{2-}), and hydroxyl radical (HO$^{\bullet}$). The endogenous pathways for ROS generation include normal mitochondrial aerobic respiration, phagocytosis of bacteria or virus-containing cells, peroxisomal-mediated degradation of fatty acids, and cytochrome P_{450}-mediated metabolism of xenobiotics DNA (26). Mixtures of vitamin C and transition metal ions (27), and the conversion of catechols to quinones (28) can also cause ROS formation from molecular oxygen. Antioxidant defense systems *in vivo* that can detoxify ROS include the following: superoxide dismutase, catalase, and reduced glutathione (GSH)-dependent peroxidases (26). Also, endogenous processes such as the sequestration of hydrogen peroxide–generating enzymes or the chelation of free transition metal ions by transferrin, ferritin, and ceruloplasmin can protect against ROS-mediated damage. However, it is always possible that cellular macromolecules and lipids can be damaged by the ROS that escape from these defense systems. It has been suggested that ROS generation is a major contributor to the degenerative diseases of aging, including cardiovascular disease, cancer, immune-system decline, and brain dysfunction (26). ROS-mediated formation of lipid hydroperoxides is a complex process, which involves initiation by the abstraction of a *bis*-allylic methylene hydrogen atom of PUFA followed by addition of molecular oxygen (13). This results in the formation of 9- and 13-hydroperoxyoctadecadienoic acid (HPODE) isomers from linoleic acid, the major n-6 PUFA present in plasma lipids. Lipid hydroperoxides can also be formed enzymatically from LOX (14) and cyclooxygenases (COX) (29) with much greater stereoselectivity than is observed in the free radical mechanism (Fig. 4.1). Human 15-LOX convert linoleic acid mainly to 13-HPODE (30). COX-1 and COX-2 produce mainly 9-[R-(E,Z)]-10,12-hydroperoxyoctadecadienoic acid (9-HPODE) and 13-HPODE (31). The other C_{18} PUFA including linolenic acid (n-3) and dihomo-γ-linolenic acid (n-6) and all C_{20} PUFA undergo 15-LOX–mediated conversion to hydroperoxides. The products that arise from 5-LOX and COX-derived metabolism of C_{20} PUFA are prostaglandins, thromboxanes, and leukotrienes rather than lipid hydroperoxides.

Transition Metal Ion-Mediated Decomposition of Lipid Hydroperoxides

Lipid hydroperoxides undergo transition metal ion–induced decomposition to the α,β-unsaturated aldehyde genotoxins that can react with DNA (12). We determined previously that 4-oxo-2-nonenal was a major product from homolytic decomposition of 13-HPODE (18), which was confirmed by Spiteller *et al.* (19). 4-Hydroperoxy-2-nonenal was also characterized as a product of 13-HPODE decomposition (18,20). We recently developed a liquid chromatography (LC)/atmospheric pressure chemical ionization (APCI)/mass spectrometry (MS) methodology to identify the α,β-unsaturated aldehy-

13(S)-Z,E-HPODE   linoleic acid   9(R)-E,Z-HPODE

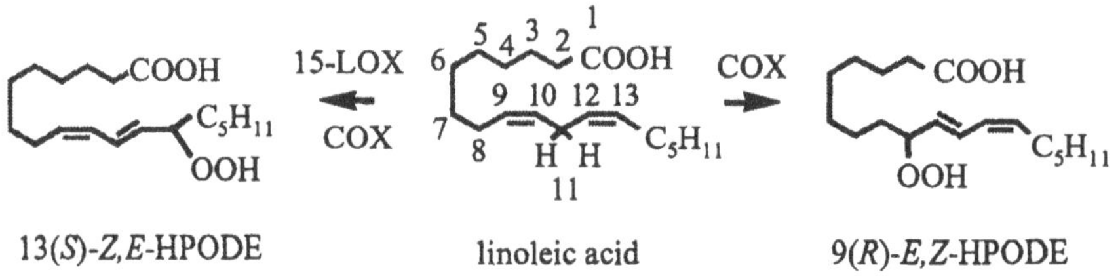

Fig. 4.1. Enzymatic formation of lipid hydroperoxide. Abbreviations: 13(S)-Z,E-HPODE, 13-[S-(Z,E)]-9,11-hydroperoxyoctadecadienoic acid; LOX, lipoxygenase; COX, cyclooxygenase.

dic bifunctional electrophiles that could potentially be formed during homolytic lipid hydroperoxide decomposition (21). Using this methodology, the aldehydes resulting from Fe^{II}-mediated (50 or 500 µmol/L) decomposition of 13-HPODE (400 µmol/L) were analyzed. Four major products were observed in the ultraviolet (UV; 226 nm) chromatogram at 10.6, 11.0, 12.3, and 17.1 min, respectively (Fig. 4.2A-e). The LC/MS characteristics of the earliest eluting aldehyde with a protonated molecular ion (MH$^+$) at *m/z* 169 were identical to authentic *trans*-4,5-epoxy-2(*E*)-decenal. *cis*-4,5-Epoxy-2(*E*)-decenal was also observed in the ion chromatogram for *m/z* 169 with a retention time of 11.7 min (Fig. 4.2A-a). The aldehyde with a retention time of 11.0 min and MH$^+$ at *m/z* 155 was identified as 4-oxo-2-nonenal (Fig. 4.2A-b). The LC/MS characteristics of the most abundant aldehyde with [MH–OH]$^+$ at *m/z* 156 were identical to authentic 4-hydroperoxy-2-nonenal (Fig. 4.2A-c). The last eluting product with MH$^+$ at *m/z* 157 was identified as 4-hydroxy-2-nonenal (Fig. 4.2A-d). At a higher concentration of Fe^{II} (500 µmol/L), the major products were *trans*-4,5-epoxy-2(*E*)-decenal, 4-oxo-2-nonenal, and 4-hydroxy-2-nonenal (Fig. 4.2B). Under these conditions, the 4-hydroperoxy-2-nonenal was undetectable by MS or UV (Fig. 4.2B-c,e).

The initial formation of 4-hydroperoxy-2-nonenal at lower Fe^{II} concentrations and subsequent decline at higher concentrations suggested that it was a precursor to the formation of 4-oxo-2-nonenal and 4-hydroxy-2-nonenal. To test this possibility, authentic 4-hydroperoxy-2-nonenal was treated with increasing concentrations of Fe^{II}. As the concentration of Fe^{II} increased, there was a gradual decline in the amount of

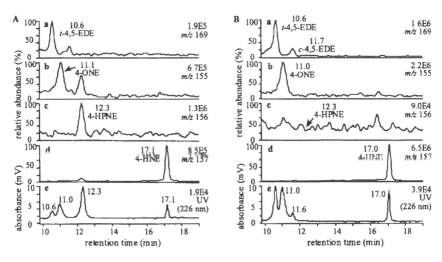

Fig. 4.2. Decomposition of 13-[*S*-(*Z,E*)]-9,11-hydroperoxyoctadecadienoic acid (13-HPODE) with transition metal ions. Abbreviations: 4,5-EDE, 4,5-epoxy-2(*E*)-decenal; 4-ONE, 4-oxo-2-nonenal; 4-HPNE, 4-hydroperoxy-2-nonenal; 4-HNE, 4-hydroxy-2-nonenal; UV, ultraviolet.

4-hydroperoxy-2-nonenal and a concomitant increase in the formation of 4-oxo-2-nonenal and 4-hydroxy-2-nonenal. Stoichiometric amounts of 4-oxo-2-nonenal and 4-hydroxy-2-nonenal were formed in a ratio that was identical to that from the reaction of 13-HPODE with Fe^{II}. These findings confirmed that one of major products in homolytic 13-HPODE decomposition was 4,5-epoxy-2(E)-decenal, which was recently identified as a product from the autoxidation of arachidonic acid (22). We have also shown that 4-hydroperoxy-2-nonenal was an important precursor to the formation of 4-oxo-2-nonenal and 4-hydroxy-2-nonenal.

Transition Metal Ions, Vitamin C, and ROS

Transition metal ion–mediated Haber Weiss reactions are known to produce ROS, which then cause oxidative damage to the DNA. Vitamin C is used as an antioxidant because it can prevent such damage (32). When vitamin C reacts with hydrogen peroxide, the vitamin C radical anion and a hydroxyl radical are produced. The reactive hydroxyl radical is then detoxified by reaction with vitamin C radical anion or vitamin C itself to give a water molecule. If the vitamin C radical anion reacts with the hydroxyl radical, dehydro-vitamin C is also formed. However, biological buffers contain substantial amounts of transition metal ions and it is paradoxical that ROS formation is enhanced in the presence of vitamin C (27). We have assessed the transition metal ion contamination in typical aqueous buffers by monitoring the decline in absorbance of vitamin C solutions at 265 nm as suggested by Buettner and Jurkiewicz (27). As can be seen in Figure 4.3A, normal 3-(N-morpholino)propanesulfonic acid (MOPS) buffer solutions cause a substantial decline in absorbance at 265 nm over 2 h resulting from oxidation of vitamin C to its dehydro form. Vitamin C proved to be quite stable in Chelex-treated MOPS buffer solutions (Fig. 4.3A). However, when vitamin C was dissolved in Chelex-treated MOPS buffer and 500 nmol/L Cu^{II} was added, the same decline in absorbance was observed as in non-Chelex–treated buffer. This indicated that normal buffers contained the equivalent of 500 nmol/L Cu^{II} in transition metal ion contamination. Chelex-treated MOPS buffer containing vitamin C showed no decline in absorbance at 265 nm unless it contained ≥50 nmol/L of added Cu^{II} (Fig. 4.3B). This suggested that the maximum amount of Cu^{II} in Chelex-treated MOPS buffer was <50 nmol/L. Concentrations of Cu and Fe in the buffer were determined subsequently by graphite furnace atomic absorption spectrophotometry and inductively coupled plasma/MS, respectively. Chelex treatment of the MOPS buffer reduced the Cu from 234 to 16 nmol/L and Fe from 250 to <36 nmol/L. No other transition metal ions were detected.

Vitamin C–Induced Decomposition of Lipid Hydroperoxides

When vitamin C reacts with a lipid hydroperoxide, an alkoxy radical is formed by a mechanism analogous to the formation of hydroxyl radicals from hydrogen peroxide (Fig. 4.4). The alkoxy radical could then be detoxified by the vitamin C radical anion

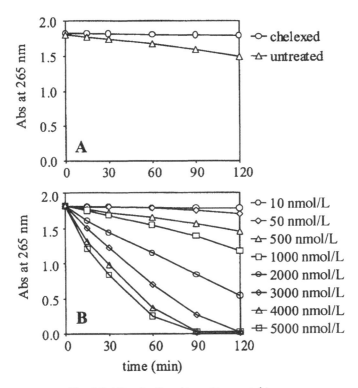

Fig. 4.3. Vitamin C and transition metal ions.

or vitamin C itself to give a nontoxic hydroxy acid. This is similar to the formation of water from hydroxyl radicals. Alternatively, it could rearrange in a manner analogous to alkoxy radicals formed from transition metal ions. Therefore, the same bifunctional electrophiles produced by transition metal ion–mediated decomposition of lipid hydroperoxides could potentially be formed by vitamin C. To test this possibility, 13-HPODE was treated with vitamin C in the transition metal ion–free MOPS buffer at 37°C. The same bifunctional electrophiles as those formed by transition metal ion–mediated decomposition of lipid hydroperoxides were produced by vitamin C (21). Although the MOPS buffer contained <50 nmol/L of transition metal ions, we determined whether Cu or Fe had an additive effect on the vitamin C–mediated decomposition of 13-HPODE. When 1 μmol/L of either Cu^{II} or Fe^{II} was added to the reaction mixture, there was no increased formation of bifunctional electrophiles from 13-HPODE. This confirmed that trace amounts of these transition metal ions did not substantially affect the reaction between vitamin C and 13-HPODE. In separate experiments, we established that synergistic effects between vitamin C and transition metal ions occurred only at concentrations that were two orders of magnitude greater than the two present in the Chelex-treated MOPS buffer (10–20 μmol/L).

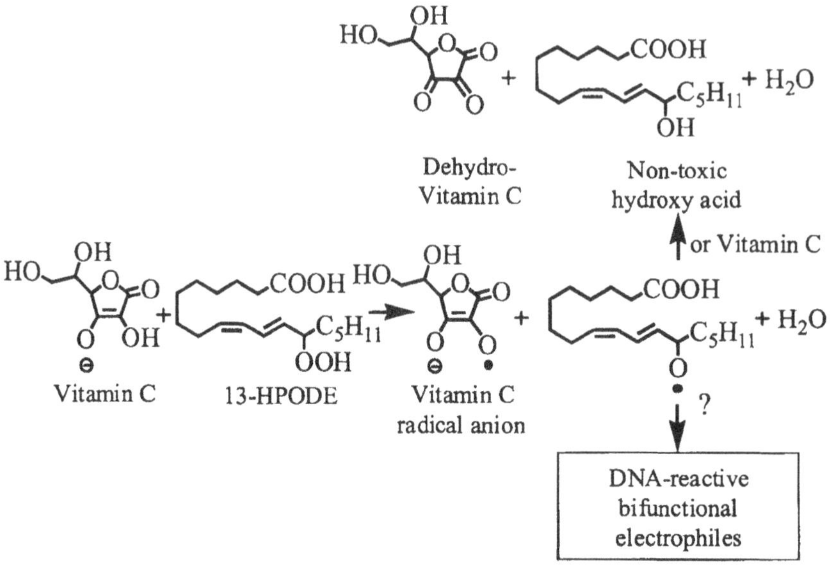

Fig. 4.4. Potential formation of DNA-reactive bifunctional electrophiles by vitamin C-mediated decomposition of 13-[S-(Z,E)]-9,11-hydroperoxyoctadecadienoic acid (13-HPODE).

The lipid hydroperoxide–derived α,β-unsaturated aldehydes were formed by two quite different pathways (Fig. 4.5). The first pathway is based on that described previously by Pryor and Porter (33) and is initiated by alkoxy radical formation. Complex rearrangements of the alkoxy radical, together with the addition of molecular oxygen results in the formation of 4-oxo-2-nonenal, 4-hydroxy-2-nonenal, and 4,5-epoxy-2(E)-decenal. The formation of 4-hydroperoxy-2-nonenal cannot be rationalized by any previously proposed mechanism. However, we have now established that 4-hydroperoxy-2-nonenal undergoes both transition metal ion– and vitamin C–mediated breakdown to 4-oxo-2-nonenal and 4-hydroxy-2-nonenal, which provides an additional route to these genotoxins.

DNA-Adducts from Lipid Hydroperoxides

The compound *trans,trans*-2,4-decadienal is an α,β-unsaturated aldehydic decomposition product from 9-hydroperoxy-(E,E)-10,12-octadecadienoic acid or 11-hydroperoxy-(Z,Z,E,E)-eicosa-5,8,12,14-tetraenoic acid (19,22). Recent studies have shown that the reaction of peroxide-treated *trans,trans*-2,4-decadienal with dAdo or dGuo results in the formation of 1,N^6-etheno-dAdo (23) and 1,N^2-etheno-dGuo (24), respectively. We reasoned that *trans*-4,5-epoxy-2(E)-decenal could have been formed when *trans,trans*-2,4-decadienal was treated with peroxides (22,25) and that this bifunctional electrophile was in fact the precursor to the formation of etheno-adducts from lipid

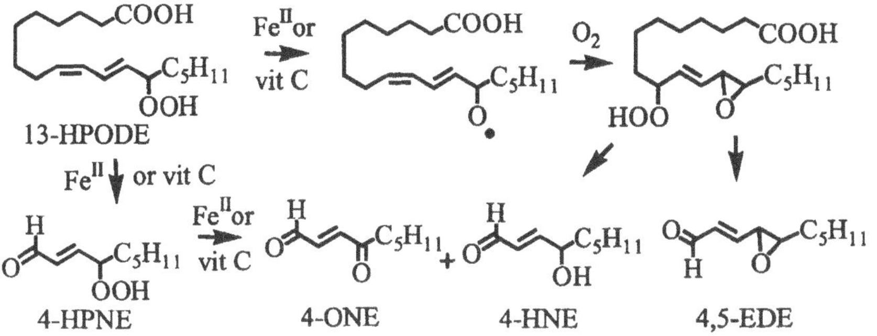

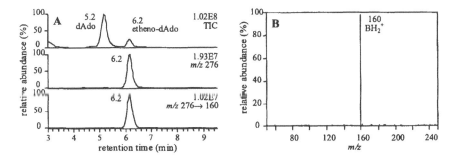

Fig. 4.5. Formation of 4-oxo-2-nonenal (4-ONE), 4-hydroperoxy-2-nonenal (4-HPNE), 4-hydroxy-2-nonenal (4-HNE), and 4,5-epoxy-2(E)-decenal (4,5-EDE) from 13-[S-(Z,E)]-9,11-hydroperoxyoctadecadienoic acid (13-HPODE).

hydroperoxides. The reaction of dAdo and dGuo with *trans*-4,5-epoxy-2(E)-decenal formed unsubstituted etheno-adducts (34). The structure of 1,N^6-etheno-dAdo was confirmed by LC/MS analysis (Fig. 4.6A) and by multiple tandem mass spectrometry (MS^n) analyses (Fig. 4.6B). The structure of 1,N^2-etheno-dGuo was established by hydrolysis to the corresponding guanine (Gua)-adduct and comparison with authentic etheno-Gua isomers. N^2,3-etheno-Gua and 1,N^2-etheno-Gua eluted with retention times of 8.0 min and 8.7 min, respectively (Fig. 4.7A). The APCI/MS spectra of 1,N^2-etheno-Gua and N^2,3-etheno-Gua were identical (data not shown). MS^2 analysis of 1,N^2-etheno-Gua (m/z 176→) gave rise to product ions at m/z 148 [MH$^+$–CO] and m/z 121 [MH$^+$–CO–CNH]. For N^2,3-etheno-Gua, no product ions were observed when

Fig. 4.6. Analysis of 1,N^6-etheno-dAdo from the reaction of *trans*-4,5-epoxy-2(E)-decenal with dAdo. (A) Liquid chromatography/mass spectrometry chromatograms showing the total ion current chromatogram (total ion current; upper), the ion chromatogram for MH$^+$ (m/z 276; middle), and selected reaction monitoring chromatogram for MH$^+$ (m/z 276) → BH$_2^+$ (m/z 160; lower). (B) Product ion spectrum of MH$^+$ (m/z 276). Abbreviation: TIC, total ion current.

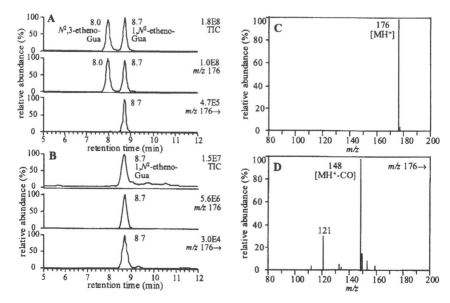

Fig. 4.7. Liquid chromatography/mass spectrometry (LC/MS) analysis of 1,N^2-etheno-Gua and N^2,3-etheno-Gua. (A) Authentic standards showing the total ion current (TIC; upper), ion chromatograms for MH$^+$ (m/z 176; middle) and TIC for MS/MS analysis of MH$^+$ (m/z 176 →; lower). (B) Etheno-Gua from the reaction of *trans*-4,5-epoxy-2(*E*)-decenal with dGuo showing the TIC (upper), the ion chromatogram for MH$^+$ (m/z 176; middle), and TIC for MS/MS analysis of MH$^+$ (m/z 176 →; lower). (C) MS spectrum for etheno-Gua from reaction of *trans*-4,5-epoxy-2(*E*)-decenal with dGuo. (D) Product ion spectrum of MH$^+$ (m/z 176) for etheno-Gua from reaction of *trans*-4,5-epoxy-2(*E*)-decenal with dGuo.

MS^2 analysis of m/z 176 was conducted. The dGuo adduct from the reaction with *trans*-4,5-epoxy-2(*E*)-decenal was isolated by preparative high-performance liquid chromatography and subjected to acid hydrolysis (1N HCl, 100°C, 1 h). After depurination, the retention time of the adduct was 8.7 min (Fig. 4.7B) and the MS spectrum exhibited an intense MH$^+$ ion at m/z 176 (Fig. 4.7C). The MS^2 spectrum showed two major product ions at m/z 148 [MH$^+$–CO] and m/z 121 [MH$^+$–CO–CHN] (Fig. 4.7D). This confirmed that the adduct was 1,N^2-etheno-dGuo rather than the isomeric N^2,3-etheno-dGuo.

Transition metal ion–free buffers were used in the reactions of dAdo and dGuo with *trans*-4,5-epoxy-2(*E*)-decenal and the pH was maintained at 7.4. Under these conditions, *trans*-4,5-epoxy-2(*E*)-decenal was quite stable; thus, the unsubstituted etheno-adducts could not have been formed from further breakdown products of the epoxide. 2,3-Epoxyoctanal, used in the synthesis of *trans*-4,5-epoxy-2(*E*)-decenal, is much more efficient at converting both dAdo and dGuo to unsubstituted etheno-adducts (data not shown). Therefore, we considered the possibility that the *trans*-4,5-

epoxy-2(*E*)-decenal was contaminated with 2,3-epoxyoctanal. A normal phase LC/MS assay was developed that would detect trace amounts of 2,3-epoxyoctanal. LC/APCI/selected ion monitoring (SIM)/MS analysis of the two epoxides employed MH$^+$ for 2,3-epoxyoctanal and *trans*-4,5-epoxy-2(*E*)-decenal at *m/z* 143 and *m/z* 169, respectively. Under these conditions, there was a clear separation of the two epoxides with no interfering signals at the retention times of either analyte. A calibration curve was obtained by analyzing standard solutions containing known amounts of 2,3-epoxyoctanal and *trans*-4,5-epoxy-2(*E*)-decenal (20 µg). A typical regression line was $y = 6.0 \times 10^6 + 194385$, $r^2 = 0.9995$ (Fig. 4.8). The amount of 2,3-epoxyoctanal in the authentic *trans*-4,5-epoxy-2(*E*)-decenal (20.0 µg, 2 µg on column) was then determined. The 2,3-epoxyoctanal was below the detection limit of the assay (<2 ng on column, <0.1%). At this level of contamination, there would have been no significant contribution from 2,3-epoxyoctanal to the formation of unsubstituted etheno-adducts. Furthermore, *trans*-4,5-epoxy-2(*E*)-decenal was incubated in Chelex-treated MOPS buffer (pH 7.4) for 24, 48, or 72 h at 37°C. We confirmed that the LC/MS response for *trans*-4,5-epoxy-2(*E*)-decenal at each time point was identical to that observed before the sample was placed in the incubator, and 2,3-epoxyoctanal was not formed during prolonged incubations of *trans*-4,5-epoxy-2(*E*)-decenal.

In a recent series of experiments, we explored the potential for unsubstituted etheno-dAdo and etheno-dGuo formation to occur in intact DNA. Calf thymus DNA (1.5 mg, 5.03 µmol) in 100 mmol/L MOPS containing 150 mmol/L NaCl (pH 7.4, 500 µL) was treated with *trans*-4,5-epoxy-2(*E*)-decenal (10 mg, 60.0 µmol). The reaction mixture was incubated at 37°C for 24 h after sonication for 15 min. Samples were placed in ice for 30 min, the DNA was precipitated by adding ice-cold ethanol, and the DNA pellet was removed. It was then hydrolyzed enzymatically under very mild conditions using methodology that we developed previously (17). The hydrolysate was applied directly to a solid-phase extraction cartridge and the etheno-adducts were

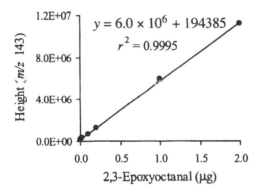

Fig. 4.8. Standard curve for 2,3-epoxyoctanal in ether obtained by normal phase liquid chromatography/atmospheric pressure chemical ionization/selected ion monitoring/mass spectrometry (*m/z* 143).

eluted with methanol/water. The eluates were evaporated to dryness under nitrogen and dissolved in water. Using both LC/MS and LC/selected reaction monitoring (SRM)/MS, it was possible to detect $1,N^2$-etheno-dGuo (Fig. 4.9A) and $1,N^6$-etheno-dAdo (Fig. 4.9B) in the DNA hydrolysate. The MS^2 spectrum of etheno-adducts from *trans*-4,5-epoxy-2(*E*)-decenal–modified calf thymus DNA showed exclusive product ion at m/z 176 (retention time of 15.2 min, BH_2^+ for $1,N^2$-etheno-dGuo, Fig. 4.9C) and at m/z 160 (retention time of 18.0 min, BH_2^+ for $1,N^6$-etheno-dAdo, Fig. 4.9D). These were identical to the LC/MS characteristics of authentic $1,N^2$-etheno-dGuo and $1,N^6$-etheno-dAdo. Based on the signal of authentic standards, the signal for $1,N^2$-etheno-dGuo corresponded to 1.3 adducts/10^5 normal bases and the signal for $1,N^6$-etheno-dAdo corresponded to 3.7 adducts/10^5 normal bases.

Summary

$1,N^6$-Etheno-dAdo has been detected in human tissues (35) as well as in the liver of vinyl chloride–treated rats (36). $1,N^6$-Etheno-dAdo is highly mutagenic in mammalian cells and much more mutagenic than lesions that arise from oxidative damage such as 7,8-dihydro-8-oxo-dGuo (37). This most likely stems from the ability of atypical

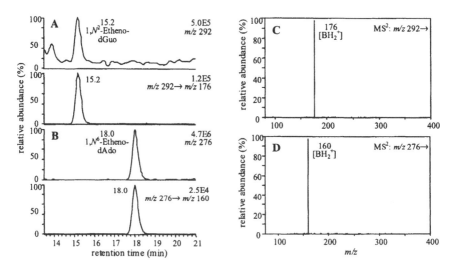

Fig. 4.9. Liquid chromatography/mass spectrometry (LC/MS) and LC/selected reaction monitoring (SRM)/MS analysis of the DNA hydrolysate after the treatment of calf thymus DNA with *trans*-4,5-epoxy-2(*E*)-decenal for 24 h at 37°C. (A) The ion chromatogram for MH+ of $1,N^2$-etheno-dGuo (m/z 292; upper), and SRM chromatogram for MH+ (m/z 292) $\rightarrow BH_2^+$ (m/z 176; lower). (B) The ion chromatogram for MH+ of $1,N^6$-etheno-dAdo (m/z 276; upper), and SRM chromatogram for MH+ (m/z 276) $\rightarrow BH_2^+$ (m/z 160; lower). (C) Product ion spectrum for MH+ (m/z 292) of $1,N^2$-etheno-dGuo. (D) Product ion spectrum for MH+ (m/z 276) of $1,N^6$-etheno-dAdo.

mammalian DNA-polymerases to perform translesional synthesis, which results in A to T transversions (38). N^2,3-Etheno-Gua was also found in the liver DNA of rats treated with vinyl chloride (39). 1,N^2-Etheno-Gua was identified in chloroacetalde-hyde-treated DNA (40) and both N^2,3-etheno-Gua and 1,N^2-etheno-Gua were isolated from 2-halooxirane-treated DNA (41). 1,N^2-Etheno-dGuo was shown to be mutagenic in mammalian cells (42), although the mutation profile was much more complex than for 1,N^6-etheno-dAdo (37). *In vitro* studies with DNA bases have demonstrated that peroxide-treatment of 4-hydroxy-2-nonenal and *trans,trans*-2,4-decadienal results in the formation of etheno DNA-adducts (23,24,43). However, it is not clear that such reactions could occur *in vivo* because of the competition between detoxication by glu-tathione-*S*-transferases and aldo-keto reductases (44) and activation by peroxidation. We have now shown that the reaction of dAdo and dGuo with *trans*-4,5-epoxy-2(*E*)-decenal results in the formation of unsubstituted etheno-adducts (Fig. 4.10), which provides an important link between a primary product of lipid peroxidation and a mutagenic DNA-lesion (37,42) that was detected in human tissue DNA (35).

We have shown that lipid hydroperoxide-derived 4-oxo-2-nonenal reacts with dGuo and dAdo to give substituted etheno-adducts (45,46). The mechanism for the formation of the etheno-dGuo adducts involves highly regioselective nucleophilic addition of N^2 of the dGuo to the C-1 aldehyde of 4-oxo-2-nonenal followed by reac-tion of N1 at C2 of the resulting α,β-unsaturated ketone. The intermediate ethano-

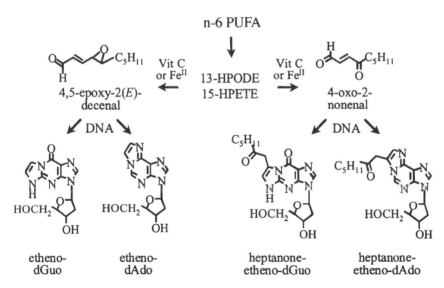

Fig. 4.10. Formation of etheno-adducts from lipid hydroperoxide–derived 4-oxo-2-none-nal and *trans*-4,5-epoxy-2(*E*)-decenal. Abbreviations: PUFA, polyunsaturated fatty acids; 13-HPODE, 13-[*S*-(*Z*,*E*)]-9,11-hydroperoxyoctadecadienoic acid; 15-HPETE, 15*S*-hydro-peroxy-5,8,11,13(*Z*,*Z*,*Z*,*E*)-eicosatetraenoic acid.

adducts then dehydrate to a single substituted etheno-adduct (Fig. 4.10) (16). Similarly, initial nucleophilic addition of N^6 of dAdo to the C-1 aldehyde of 4-oxo-2-nonenal is followed by reaction of N1 at C2 of the resulting α,β-unsaturated ketone to generate a mixture of two ethano-adducts that interconvert. The ethano-adducts subsequently dehydrate to give a single heptanone etheno-dAdo adduct (Fig. 4.10) (17,47). The reaction of 4-oxo-2-nonenal with calf thymus DNA was also shown to result in the formation of heptanone-etheno-dGuo and dAdo adducts (17). There was almost three times as much of the dGuo adduct as the dAdo adduct at all time points up to 12 h after the addition of the 4-oxo-2-nonenal.

Reactions between dGuo and the α,β-unsaturated aldehydes, 4-hydroxy-2-nonenal and malondialdehyde (MDA, another breakdown product of lipid peroxidation) result in the formation of exocyclic propano adducts (48). In contrast to 4-oxo-2-nonenal, Michael addition occurs initially at the β-carbon, which is followed by nucleophilic addition of dGuo at the aldehyde carbon. When the α,β-unsaturated aldehyde has a substituent at the β-carbon such as in 4-hydroxy-2-nonenal, the steric hindrance inhibits nucleophilic attack from N1. Kinetic control of the reaction favors the regioisomer in which N^2 is attached to the β-carbon atom and N1 is attached to the aldehyde carbon. This results in the formation of two pairs of diastereomeric hexanol-1,N^2-propano-dGuo adducts from 4-hydroxy-2-nonenal and pyrimido[1,2-α]purin-10(3H)-one from MDA. When MDA reacts with dAdo, an acyclic N^6-oxopropenyl adduct is generated (49,50). This adduct is thought to be a consequence of initial 1,4-addition to the β-hydroxyacrolein form of MDA followed by dehydration. 4-Hydroxy-2-nonenal does not appear to react very efficiently with dAdo. However, the epoxide derivative, 2,3-epoxy-4-hydroxynonanal, has been shown to form substituted and unsubstituted etheno-adducts with dAdo through the same mechanism by which heptanone-etheno-dAdo adducts are formed from 4-oxo-2-nonenal (43,51,52). 1,N^2-Etheno-dGuo, a substituted 1,N^2-etheno-adduct, and a tetracyclic adduct containing two 5-membered fused rings at N1 and N^2 atoms of guanine are also formed from the reaction between 2,3-epoxy-4-hydroxynonanal and dGuo (51,53). DNA-adducts have not yet been identified from the reaction of DNA bases with the potential genotoxic bifunctional electrophile 4-hydroperoxy-2-nonenal.

This study provides a mechanism for vitamin C–mediated decomposition of lipid hydroperoxides to genotoxic bifunctional electrophiles without the need for free transition metal ions. Future studies will focus on determining whether vitamin C can induce such genotoxin formation *in vivo*.

Acknowledgments

We acknowledge support of National Institutes of Health grant RO1-CA91016.

References

1. Cameron, E., and Pauling, L. (1974) The Orthomolecular Treatment of Cancer. I. The Role of Ascorbic Acid in Host Resistance, *Chem.-Biol. Interact. 9*, 273–283.
2. Cameron, E., and Campbell, A. (1974) The Orthomolecular Treatment of Cancer. II.

Clinical Trial of High-Dose Ascorbic Acid Supplements in Advanced Human Cancer, *Chem.-Biol. Interact. 9*, 285–315.

3. Cameron, E., and Pauling, L. (1976) Supplemental Ascorbate in the Supportive Treatment of Cancer: Prolongation of Survival Times in Terminal Human Cancer, *Proc. Natl. Acad. Sci. USA 73*, 3685–3689.

4. Creagan, E.T., Moertel, C.G., O'Fallon, J.R., Schutt, A.J., O'Connell, M.J., Rubin, J., and Frytak, S. (1979) Failure of High-Dose Vitamin C (Ascorbic Acid) Therapy to Benefit Patients with Advanced Cancer, *N. Engl. J. Med. 301*, 687–690.

5. Tschetter, L., Creagan, E. T., O'Fallon, J. R., Schutt, A. J., Krook, J. E., Windschitl, H. E., Reuter, N. P., and Pfeifle, D.M. (1983) A Community-Based Study of Vitamin C (Ascorbic Acid) Therapy in Patients with Advanced Cancer, *Proc. Am. Soc. Clin. Oncol. 2*, 92 (abstr.).

6. Moertel, C.G., Fleming, T.R., Creagan, E.T., Rubin, J., O'Connell, M.J., and Ames, M.M. (1985) High-Dose Vitamin C Versus Placebo in the Treatment of Patients with Advanced Cancer Who Have Had No Prior Chemotherapy, *N. Engl. J. Med. 312*, 137–141.

7. Helbock, H.J., Beckman, K.B., Shigenaga, M.K., Walter, P.B., Woodall, A.A., Yeo, H.C., and Ames, B.N. (1998) DNA Oxidation Matters: The HPLC-Electrochemical Detection Assay of 8-Oxo-deoxyguanosine and 8-Oxo-guanine, *Proc. Natl. Acad. Sci. USA 95*, 288–293.

8. Lippman, S.M., Lee, J.J., and Sabichi, A.L. (1998) Cancer Chemoprevention: Progress and Promise, *J. Natl. Cancer Inst. 90*, 1514–1528.

9. Ruffin, M.T. IV, and Rock, C.L. (2001) Do Antioxidants Still Have a Role in the Prevention of Human Cancer? *Curr. Oncol. Rep. 3*, 306–313.

10. Dribble, D. (1999) Antioxidant Consumption and Risk of Coronary Heart Disease: Emphasis on Vitamin C, Vitamin E, and β-Carotene, *Circulation 99*, 591–595.

11. Ames, B.N., Gold, L.S., and Willett, W.C. (1995) The Causes and Prevention of Cancer, *Proc. Natl. Acad. Sci. USA 92*, 5258–5265.

12. Marnett, L.J. (2000) Oxyradicals and DNA damage, *Carcinogenesis 21*, 361–370.

13. Porter, N.A., Caldwell, S.E., and Mills, K.A. (1995) Mechanisms of Free Radical Oxidation of Unsaturated Lipids, *Lipids 30*, 277–290.

14. Brash, A.R. (1999) Lipoxygenases: Occurrence, Functions, Catalysis, and Acquisition of Substrate, *J. Biol. Chem. 274*, 23679–23682.

15. Esterbauer, H., Schaur, R.J., and Zollner, H. (1991) Chemistry and Biochemistry of 4-Hydroxynonenal, Malonaldehyde and Related Aldehydes, *Free Radic. Biol. Med. 11*, 81 128.

16. Rindgen, D., Nakajima, M., Wehrli, S., Xu, K., and Blair, I.A. (1999) Covalent Modifications to 2'-Deoxyguanosine by 4-Oxo-2-nonenal, a Novel Product of Lipid Peroxidation, *Chem. Res. Toxicol. 12*, 1195–1204.

17. Lee, S.H., Rindgen, D., Bible, R.A., Hajdu, E., and Blair, I.A. (2000) Characterization of 2'-Deoxyadenosine Adducts Derived from 4-Oxo-2-nonenal, a Novel Product of Lipid Peroxidation, *Chem. Res. Toxicol. 13*, 565–574.

18. Lee, S.H., and Blair, I.A. (2000) Characterization of 4-Oxo-2-nonenal as a Novel Product of Lipid Peroxidation, *Chem. Res. Toxicol. 13*, 698–702.

19. Spiteller, P., Kern, W., Reiner, J., and Spiteller, G. (2001) Aldehydic Lipid Peroxidation Products Derived from Linoleic Acid, *Biochim. Biophys. Acta*, 1531, 188–208.

20. Schneider, C., Tallman, K.A., Porter, N.A., and Brash, A.R. (2001) Two Distinct

Pathways of Formation of 4-Hydroxynonenal. Mechanisms of Nonenzymatic Transformation of the 9- and 13-Hydroperoxides of Linoleic Acid to 4-Hydroxyalkenals, *J. Biol. Chem. 276*, 20831–20838.

21. Lee, S.H., Oe, T., and Blair, I.A. (2001) Vitamin C-Induced Decomposition of Lipid Hydroperoxides to Endogenous Genotoxins, *Science 292*, 2083–2086.

22. Lin, J., Fay, L.B., Welti, D.H., and Blank, I. (2001) Quantification of Key Odorants Formed by Autoxidation of Arachidonic Acid Using Isotope Dilution Assay, *Lipids 36*, 749–756.

23. Carvalho, V.M., Asahara, F., Di Mascio, P., de Arruda Campos, I.P., Cadet, J., and Medeiros, M.H.G. (2000) Novel 1,N^6-Etheno-2′-deoxyadenosine Adducts from Lipid Peroxidation Products, *Chem. Res. Toxicol. 13*, 397–405.

24. Loureiro, A.P.M., Di Mascio, P., Gomes, O.F., and Medeiros, M.H.G. (2000) *trans,trans*-2,4-Decadienal-Induced 1,N^2-Etheno-2′-deoxyguanosine Adduct Formation, *Chem. Res. Toxicol. 13*, 601–609.

25. Zamora, R., and Hidalgo, F.J. (1995) Linoleic Acid Oxidation in the Presence of Amino Compounds Produces Pyrroles by Carbonyl Amine Reactions, *Biochim. Biophys. Acta 1258*, 319–327.

26. Ames, B.N., Shigenaga, M.K., and Hagen, T.M. (1993) Oxidants, Antioxidants, and the Degenerative Diseases of Aging, *Proc. Natl. Acad. Sci. USA 90*, 7915–7922.

27. Buettner, G.R., and Jurkiewicz, B.A. (1996) Catalytic Metals, Ascorbate and Free Radicals: Combinations to Avoid, *Radiat. Res. 145*, 532–541.

28. Bolton, J.L., Trush, M.A., Penning, T.M., Dryhurst, G., and Monks, T.J. (2000) Role of Quinones in Toxicology, *Chem. Res. Toxicol. 13*, 136–160.

29. Laneuville, O., Breuer, D.K., Xu, N., Huang, Z.H., Gage, D.A., Watson, J.T., Lagarde, M., DeWitt, D.L., and Smith, W.L. (1995) Fatty Acid Substrate Specificities of Human Prostaglandin-Endoperoxide H Synthase-1 and -2, *J. Biol. Chem. 270*, 19330–19336.

30. Kamitani, H., Geller, M., and Eling, T. (1998) Expression of 15-Lipoxygenase by Human Colorectal Carcinoma Caco-2 Cells During Apoptosis and Cell Differentiation, *J. Biol. Chem. 273*, 21569–21577.

31. Hamberg, M. (1998) Stereochemistry of Oxygenation of Linoleic Acid Catalyzed by Prostaglandin-Endoperoxide H Synthase-2, *Arch. Biochem. Biophys. 349*, 376–380.

32. Proteggente, A.R., Rehman, A., Halliwell, B., and Rice-Evans, C.A. (2000) Potential Problems of Ascorbate and Iron Supplementation: Pro-Oxidant Effect In Vivo? *Biochem. Biophys. Res. Commun. 277*, 535–540.

33. Pryor,W.A., and Porter, N.A . (1990) Suggested Mechanisms for the Production of 4-Hydroxy-2-nonenal from the Autoxidation of Polyunsaturated Fatty Acids, *Free Radic. Biol. Med. 8*, 541–543.

34. Lee, S.H., Oe, T., and Blair, I.A. (2002) 4,5-Epoxy-2(*E*)-decenal-Induced Formation of 1,N^6-Etheno-2′-deoxyadenosine and 1,N^2-Etheno-2′-deoxyguanosine Adducts, *Chem. Res. Toxicol. 15*, 300–304.

35. Doerge, D.R., Churchwell, M.I., Fang, J.L., and Beland, F.A. (2000) Quantification of Etheno DNA Adducts Using Liquid Chromatography, On-Line Sample Processing, and Electrospray Tandem Mass Spectrometry, *Chem. Res. Toxicol. 13*, 1259–1264.

36. Swenberg, J.A., Fedtke, N., Ciroussel, F., Barbin, A., and Bartsch, H. (1992) Etheno Adducts Formed in DNA of Vinyl Chloride-Exposed Rats Are Highly Persistent in Liver, *Carcinogenesis 13*, 727–729.

37. Levine, R.L., Yang, I.-Y., Hossain, M., Pandya, G., Grollman, A.P., and Moriya, M.

(2000) Mutagenesis Induced by a Single 1,N^6-Ethenodeoxyadenosine Adduct in Human Cells, *Cancer Res. 60*, 4098–4104.

38. Levine, R.L., Miller, H., Grollman, A., Ohashi, E., Ohmori, H., Masutani, C., Hanaoka, F., and Moriya, M. (2001) Translesion DNA Synthesis Catalyzed by Human Pol η and Pol κ Across 1,N^6-Ethenodeoxyadenosine, *J. Biol. Chem. 276*, 18717–18721.

39. Fedtke, N., Boucheron, J.A., Walker, V.E., and Swenberg, J.A. (1990) Vinyl Chloride-Induced DNA Adducts. II: Formation and Persistence of 7-(2'-Oxoethyl)guanine and N^2,3-Ethanoguanine in Rat Tissue DNA, *Carcinogenesis 11*, 1287–1292.

40. Kusmierek, J.T., and Singer, B. (1992) 1,N^2-Ethenodeoxyguanosine: Properties and Formation in Chloroacetaldehyde-Treated Polynucleotides and DNA, *Chem. Res. Toxicol. 5*, 634–638.

41. Guengerich, F.P., Persmark, M., and Humphreys, W.G. (1993) Formation of 1,N^2- and N^2,3-Ethenoguanine from Halooxiranes: Isotopic Labeling Studies and Isolation of a Hemiaminal Derivative of N^2-(2-Oxoethyl)guanine, *Chem. Res. Toxicol. 6*, 635–648.

42. Akasaka, S. and Guengerich, F.P. (1999) Mutagenicity of Site-Specifically Located 1,N^2-Ethenoguanine in Chinese Hamster Ovary Cell Chromosomal DNA, *Chem. Res. Toxicol. 12*, 501–507.

43. Chen, H.-J.C., and Chung, F.-L. (1994) Formation of Etheno Adducts in Reactions of Enals Via Autoxidation, *Chem. Res. Toxicol. 7*, 857–860.

44. Burczynski, M.E., Sridhar, G.R., Palackal, N.T., and Penning, T.M. (2001) The Reactive Oxygen Species- and Michael Acceptor-Inducible Human Aldo-Keto Reductase AKR1C1 Reduces the α,β-Unsaturated Aldehyde 4-Hydroxy-2-nonenal to 1,4-Dihydroxy-2-nonene, *J. Biol. Chem. 276*, 2890–2897.

45. Blair, I.A. (2001) Lipid Hydroperoxide-Mediated DNA Damage, *Exp. Gerontol. 36*, 1473–1481.

46. Lee, S.H., and Blair, I.A. (2001) Oxidative DNA Damage and Cardiovascular Disease, *Trends Cardiovasc. Med. 9*, 148–155.

47. Rindgen, D., Lee, S.H., Nakajima, M., and Blair, I.A. (2000) Formation of a Substituted 1,N^6-Etheno-2'-deoxyadenosine Adduct by Lipid Hydroperoxide-Mediated Generation of 4-Oxo-2-nonenal, *Chem. Res. Toxicol. 13*, 846–852.

48. Burcham, P.C. (1998) Genotoxic Lipid Peroxidation Products: Their DNA Damaging Properties and Role in Formation of Endogenous DNA Adducts, *Mutagenesis 13*, 287–305.

49. Stone, K., Ksebati, M.B., and Marnett, L.J. (1990) Investigation of the Adducts Formed by Reaction of Malondialdehyde with Adenosine, *Chem. Res. Toxicol. 3*, 33–38.

50. Chaudhary, A.K., Reddy, P.G., Blair, I.A., and Marnett, L.J. (1996) Characterization of an N^6-Oxopropenyl-2'-deoxyadenosine Adduct in Malondialdehyde-Modified DNA Using Liquid Chromatography/Electrospray Ionization Tandem Mass Spectrometry, *Carcinogenesis 17*, 1167–1170.

51. Sodum, R.S., and Chung, F.-L. (1991) Stereoselective Formation of In Vitro Nucleic Acid Adducts by 2,3-Epoxy-4-hydroxynonanal, *Cancer Res. 51*, 137–143.

52. Chung, F.-L., Chen, H.-J.C., and Nath, R.G. (1996) Lipid Peroxidation as a Potential Endogenous Source for the Formation of Exocyclic DNA Adducts, *Carcinogenesis 17*, 2105–2111.

53. Sodum, R.S., and Chung, F.-L. (1989) Structural Characterization of Adducts Formed in the Reaction of 2,3-Epoxy-4-hydroxynonanal with Deoxyguanosine, *Chem. Res. Toxicol. 2*, 23–28.

Chapter 5

Ascorbic Acid and Endothelial NO Synthesis

Regine Heller[a] and Ernst R. Werner[b]

[a]Institute of Molecular Cell Biology, Friedrich-Schiller-University of Jena, D-99089 Erfurt, Germany and [b]Institute for Medical Chemistry and Biochemistry, University of Innsbruck, A-6020 Innsbruck, Austria

Introduction

Endothelium-derived nitric oxide (NO) was originally discovered as a vasodilator product and is now known as a central regulator of vascular homeostasis and a principal factor involved in the antiatherosclerotic properties of endothelial cells (1–4). Experimental studies have shown that NO interferes with key events of atherosclerosis such as monocyte and leukocyte adhesion to the endothelium, platelet-endothelium interactions, smooth muscle cell proliferation, and increased endothelial permeability. In agreement with these findings, a dysfunction of the endothelium with a decreased generation of NO caused accelerated atherosclerosis in experimental models (5). Moreover, a reduced NO-dependent vasodilation was detectable in patients with atherosclerosis or with a cardiovascular risk profile even when the coronary vessels were angiographically still normal (6,7). All major risk factors for atherosclerosis including hypercholesterolemia, hypertension, hyperhomocysteinemia, and cigarette smoking have been associated with impaired vascular NO synthesis (8–11). Because these conditions are also correlated with increased oxidative stress (12), particularly increased production of superoxide radicals and elevated levels of oxidized lipoproteins, which can directly inactivate NO (13,14), antioxidants have been thought to improve endothelial dysfunction. Accordingly, epidemiologic studies demonstrated that a diet high in antioxidant vitamins is associated with lower cardiovascular morbidity and mortality (15).

With regard to ascorbic acid, it has been shown that plasma or leukocyte ascorbate levels are reduced in patients with an unstable coronary syndrome or angiographically documented coronary artery disease, respectively (16,17). Moreover, a number of clinical studies have demonstrated that ascorbic acid can reverse NO-dependent endothelial dysfunction present in coronary or peripheral arteries of patients with atherosclerosis and several conditions that predispose patients to atherosclerosis (18–37). Protective ascorbic acid effects have been seen with different stimuli of NO-dependent vasodilation such as acetylcholine or metacholine (18, 20,22–27,33,35–37), flow (19,21,22,28–30,32) or arginine infusion (31), and with both ascorbate infusion (18,20–28,31–37) and oral ascorbate supplements (19,26,29,30). Concerning the underlying mechanisms, several possibilities have been discussed (38–40). Ascorbic acid could interfere with the oxidation of low density lipoprotein (LDL) (41,42) or protect endothelial NO synthesis from the effects of oxidized LDL (14,43–47).

Ascorbate could also enhance endothelial-dependent vasodilation by sparing intracellular thiols (48), which in turn may stabilize NO through the formation of biologically active *S*-nitrosothiols (49). The latter likely serve as a reservoir of NO in plasma (50,51) and, interestingly, ascorbate has been shown to release NO from *S*-nitrosothiols and to improve the delivery of NO to the vasculature (52–54). Given the importance of the superoxide anion as a mechanism of endothelial dysfunction (40), several investigators have also assumed that ascorbic acid exerts its beneficial effect by scavenging superoxide anion and protecting NO from inactivation. A recent report, however, showed that ascorbate concentrations ≥10 mmol/L would be required to compete efficiently with the reaction of NO and superoxide (55). These concentrations are potentially achievable in plasma by ascorbate infusion and may account for the beneficial effects of ascorbic acid seen in the respective studies (18,20–28,31–37). Plasma ascorbate levels after oral supplementation are in the range of 100 μmol/L (26,30), however, and are unlikely to prevent NO inactivation by superoxide. Thus, if superoxide scavenging is involved in the beneficial effect of physiologic doses of ascorbic acid, it should possibly occur in the intracellular milieu where ascorbate concentrations are likely in the low millimolar range (42,56).

Little is known about the effects of antioxidants and especially of ascorbic acid on the synthesis of NO in endothelial cells. NO is generated from the conversion of L-arginine to L-citrulline by the enzymatic action of an NADPH-dependent NO synthase (NOS) that requires Ca^{2+}/calmodulin, FAD, FMN, and tetrahydrobiopterin as cofactors (57–60). The endothelial NOS isoform (eNOS) is constitutively expressed and activated upon cell stimulation with calcium-mobilizing agonists and fluid shear stress (60). Optimal NO formation has been shown to be dependent on the availability of intracellular cofactors (61–63) and the membrane localization of the enzyme (64). eNOS activity is regulated at the transcriptional level and by a variety of modifications such as acylation, which enables membrane targeting, and phosphorylation, which is involved in shear stress–dependent enzyme activation. Moreover, protein-protein interactions support either activation or inactivation of eNOS (65). Within plasmalemmal caveolae, eNOS is quantitatively associated with caveolin, the structural coat component of these microdomains (64). This complex formation has been shown to inhibit enzyme activity, and the inhibitory effect was reversed upon binding of Ca^{2+}/calmodulin. The activation of eNOS is also facilitated by interactions of the enzyme with heat shock protein 90 and with dynamin. On the other hand, eNOS binding to the bradykinin B2 receptor participates in its inactivation (65).

Generally, changes in the intracellular redox state could affect NO generation at different levels. Oxidized LDL, for example, which is likely to induce oxidative stress in the cells, has been shown to inhibit the NO-dependent vasorelaxation (66), and the underlying mechanisms are thought to involve a decrease in eNOS expression (43), an uncoupling of Gi protein-dependent signal transduction (44), a limited availability of L-arginine (45), and changes in the subcellular eNOS localization (46,47). An alteration of the intracellular redox state might also affect the availability of reduced cofactors for eNOS. In particular, tetrahydrobiopterin seems to be a cofactor that can limit

NO synthesis (67–70). Tetrahydrobiopterin acts as a redox-active cofactor (71–74) and additionally has profound effects on the structure of eNOS, including the ability to shift the heme iron to its high spin state, the promotion of arginine binding and the stabilization of the active dimeric form of the enzyme (58,59). A number of experimental (75–78) and clinical studies (79–83) have shown that low tetrahydrobiopterin levels were associated with decreased NO formation and impaired endothelium-dependent relaxation and, conversely, that tetrahydrobiopterin supplementation was capable of restoring NO production and endothelium-dependent vasodilation.

The present study was designed to examine whether ascorbic acid affects NO synthesis in human endothelial cells and to investigate possible underlying mechanisms. We show that saturated ascorbic acid levels in endothelial cells are necessary to protect the eNOS cofactor tetrahydrobiopterin from inactivation and to provide optimal conditions for cellular NO synthesis. A detailed description of the methods and results presented here was published earlier (84,85).

Materials and Methods

Materials. Medium 199 (M199), human serum (HS), fetal calf serum (FCS), and collagenase were from BioWhittaker Europe (Verviers, Belgium). L-[2,3,4,5-^{3}H]Arginine monohydrochloride (61 Ci/mmol), L-[U-^{14}C]arginine monohydrochloride (303 mCi/mmol), L-[*carboxyl*-^{14}C]ascorbic acid (16 mCi/mmol), and [^{3}H]cGMP Biotrak radioimmunoassay systems were purchased from Amersham Pharmacia Biotech (Freiburg, Germany). Tumor necrosis factor-α (TNF-α) and interferon-γ (IFN-γ) were from Pharma Biotechnology (Hannover, Germany). NADPH, tetrahydrobiopterin, sepiapterin, and L-nitroarginine methylester (L-NAME) were obtained from Alexis Corporation (Läufelfingen, Switzerland). All other biochemical reagents were purchased from Sigma Chemical (Deisenhofen, Germany). Endotoxin contamination of ascorbic acid solutions was measured with the coagulation *Limulus* amebocyte lysate assay and was proved to be below the detection limit of the kit (0.05 U/mL).

Cell cultures. Human umbilical cord vein endothelial cells (HUVEC) were prepared with 0.05% collagenase and cultured in M199 containing 15% FCS, 5% HS, and 7.5 μg/mL endothelial cell growth supplement. Experiments were carried out with monolayers of the first to second passage. Preincubation of cells with L-ascorbic acid, L-gulonolactone, dehydroascorbic acid, sepiapterin, 2,4-diamino-6-hydroxypyrimidine (DAHP), or the combination of TNF-α, IFN-γ, and lipopolysaccharide (LPS) was performed in culture medium. Cell stimulation with ionomycin (2 μmol/L, 15 min) or thrombin (1 U/mL, 15 min) was carried out in Hepes buffer (10 mmol/L Hepes, 145 mmol/L NaCl, 5 mmol/L KCl, 1 mmol/L MgSO$_4$, 10 mmol/L glucose, 1.5 mmol/L CaCl$_2$, pH 7.4) in the absence of ascorbic acid, sepiapterin, DAHP, or cytokines.

Measurement of citrulline and cGMP formation. Endothelial cells were stimulated with ionomycin or thrombin in Hepes buffer (pH 7.4) containing 10 μmol/L

L-[³H]arginine (0.33 Ci/mmol) for the measurement of citrulline formation or 0.5 mmol/L isobutylmethylxanthine for cGMP determinations. The [³H]citrulline generated was separated from [³H]arginine by cation exchange chromatography [Dowex AG50WX-8 (Na⁺ form)] and quantified by liquid scintillation counting (84,85). The cGMP accumulated was measured in cellular extracts by radioimmunoassay following the instructions of the manufacturer.

[¹⁴C]Ascorbic acid uptake in endothelial cells. Endothelial cells were incubated in culture medium containing 100 μmol/L [¹⁴C]ascorbic acid (16 mCi/mmol). After various times, cells were washed with cold Hepes buffer (pH 7.4) containing 100 μmol/L phloretin and lysed with 100 mmol/L NaOH, 2% Na_2CO_3, and 1% sodium dodecylsulfate. The radioactivity of cell lysates was measured by liquid scintillation counting.

Determination of eNOS activity. Experiments were performed with tetrahydrobiopterin-free eNOS that was expressed in yeast *Pichia pastoris* and purified as described (86). The assay solution (100 μL) contained 50 mmol/L Tris-HCl buffer (pH 7.4), 0.3 μg eNOS, 100 μmol/L L-[³H]arginine (100,000 cpm), 0.5 mmol/L $CaCl_2$, 0.2 mmol/L NADPH, 5 μmol/L FAD, 5 μmol/L FMN, 10 μg/mL calmodulin, 10 nmol/L–100 μmol/L tetrahydrobiopterin, and 0.2 mmol/L CHAPS. The [³H]citrulline generated was separated from [³H]arginine by ion exchange columns and quantified as described above.

Measurement of biopterin derivatives. Culture medium was collected and endothelial monolayers were detached with trypsin/EDTA (0.05%/0.02%, vol/vol). Aliquots of 5 × 10⁶ cells and 1-mL aliquots of medium were oxidized with 0.02 mol/L KI/I_2 in 0.1 mol/L HCl or in 0.1 mol/L NaOH for 1 h in the dark. The precipitates were removed by centrifugation and excess iodine was destroyed by the addition of 0.02 mol/L ascorbic acid. Quantification of biopterin in supernatants was performed by high-performance liquid chromatography (HPLC) as described (87). The amount of 5,6,7,8-tetrahydrobiopterin was calculated from the difference in biopterin concentrations measured after oxidation in acid (total biopterins) and base (7,8-dihydrobiopterin + biopterin). Additionally, nonoxidized supernatants were used to determine biopterin.

Determination of GTP cyclohydrolase I (GTP-CH I) expression and activity. Extraction of total RNA from endothelial cells, electrophoresis on 1% agarose/6% formaldehyde gels, Northern blotting, and hybridization of the blots with 10⁶ cpm/mL [³²P]dCTP-labeled probe for human GTP cyclohydrolase I, obtained by polymerase chain reaction using consensus primers to GTP cyclohydrolase I from *Escherichia coli*, mouse and human, were performed according to standard protocols. GTP-CH I activity in cytosolic fractions from endothelial cells was measured as described (87).

Statistical analysis. All data are given as means ± SEM, n = 3–5 independent experiments. To determine the statistical significance of the described results, analysis of variance with Bonferroni's correction for multiple comparisons or Student's t-test for paired data was performed. A *P* value of < 0.05 was accepted as significant.

Results and Discussion

Ascorbic Acid Potentiates NO Synthesis in Endothelial Cells

NO synthesis in our study was measured as the formation of citrulline, which is produced stoichiometrically with NO, and as an accumulation of intracellular cGMP, which is generated when NO activates the soluble guanylate cyclase of the cells. Both parameters are increased after cell stimulation with calcium-mobilizing agonists or shear stress, and this increase can be prevented by eNOS inhibition. To investigate whether ascorbic acid in concentrations corresponding to physiologically achievable plasma levels (26,30,88) affects agonist-induced NO synthesis, we preincubated endothelial cells for 24 h with 0.1–100 μmol/L ascorbate before cell stimulation with ionomycin or thrombin. Figure 5.1 shows that both agonist-induced citrulline and cGMP formation were increased in a dose-dependent fashion by pretreatment of cells with the compound, thus indicating a potentiation of endothelial NO synthesis by ascorbic acid. These results were obtained not only in HUVEC but also in coronary artery endothelial cells (data not shown, see Ref. 84). Our data are also in good agreement with findings reported by Huang *et al.* (89) who measured a potentiation of A23187-induced cGMP accumulation in ascorbate-pretreated porcine aortic endothelial cells, whereas ascorbic acid did not affect the eNOS-independent cGMP formation.

Because cell stimulation was performed in the absence of extracellular ascorbate, the effect on endothelial NO synthesis was most likely due to an increase of intracellular ascorbic acid concentrations. Indeed, under normal culture conditions, cells are unlikely to be saturated with ascorbic acid because its concentration in culture media is generally low. Using 100 μmol/L [^{14}C]-labeled ascorbate, we found an uptake of the compound into endothelial cells that was time dependent and saturated between 12 and 24 h (Fig. 5.2A). Assuming that ascorbate in nonsupplemented cells was negligible, the maximal intracellular ascorbate concentration was 21.5 ± 3.7 nmol/mg protein as calculated from the specific radioactivity of the added compound. Thus, these data confirm that endothelial cells can accumulate ascorbic acid in the low millimolar range at a medium concentration related to the normal plasma concentration of the antioxidant (42,56). A saturation of intracellular ascorbate levels at an extracellular concentration of 100 μmol/L might also explain the lack of further NO synthesis potentiation with higher ascorbic acid supplements in our study. No differences in ionomycin-induced citrulline production were seen between a 24-h pretreatment of endothelial cells with 100 μmol/L or 1 mmol/L ascorbate (561 ± 40 or 574 ± 40 fmol [^{3}H]citrulline/mg cell protein, respectively, n = 3). The ascorbic acid uptake is most probably mediated by sodi-

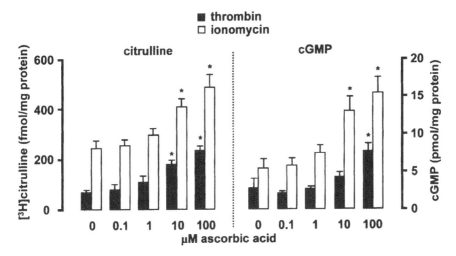

Fig. 5.1. Influence of ascorbic acid on ionomycin- or thrombin-induced citrulline (left panel) and cGMP (right panel) formation. Endothelial cells from human umbilical veins were preincubated for 24 h with 0.1–100 μmol/L ascorbic acid in culture medium. Subsequently, cells were stimulated with ionomycin (2 μmol/L, 15 min) or thrombin (1 U/mL, 15 min) in Hepes buffer (pH 7.4) containing 10 μmol/L L-[^{3}H]arginine (0.33 Ci/mmol) for the measurement of citrulline formation or 0.5 mmol/L isobutyl-methylxanthine for cGMP determinations. The [^{3}H]citrulline generated was separated from [^{3}H]arginine by cation exchange chromatography and quantified by liquid scintillation counting. The accumulated cGMP was measured in cellular extracts by radioimmunoassay. Data are shown as agonist-induced increases in [^{3}H]citrulline formation or cGMP production calculated from the differences between stimulated and unstimulated cells (means ± SEM, n = 4); cells with and without ascorbate pretreatment were compared, *$P \leq 0.05$.

um-dependent transporters, which have recently been cloned from rat and human cDNA libraries (90,91). Although these transporters have not yet been characterized in endothelial cells, the involvement of an active transport mechanism for ascorbic acid in these cells has already been demonstrated (56). Interestingly, the time dependence of the ascorbate effect on endothelial NO synthesis followed a kinetics similar to the ascorbate uptake (Fig. 5.2B), thus emphasizing the importance of intracellular ascorbate accumulation for the observed effects of ascorbic acid on NO synthesis. Moreover, our data suggest that the reductive capacity of the compound may be essential for the potentiation of NO formation by ascorbate. The molecular structure of L-ascorbic acid consists of an unsaturated γ-lactone ring with an enediol configuration conjugated with a carbonyl group (Fig. 5.3). L-Gulonolactone, an ascorbic acid precursor molecule, is lacking the redox-active enediol configuration and cannot be transformed into ascorbic acid in human cells

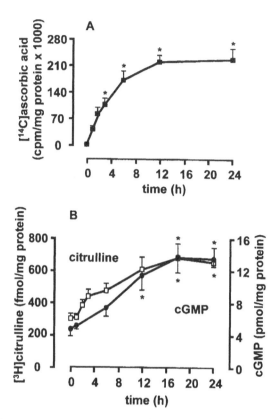

Fig. 5.2. Time-dependence of the uptake of [^{14}C]ascorbic acid and the effect of ascorbate on ionomycin-induced citrulline and cGMP formation. Endothelial cells were preincubated with 100 µmol/L [^{14}C]ascorbic acid (16 mCi/mmol) (A) or 100 µmol/L unlabeled ascorbate (B) for the indicated times. Then, cells were washed, solubilized, and analyzed for the cell-associated radioactivity (A). Alternatively, cells were stimulated with ionomycin (2 µmol/L, 15 min) and either citrulline or cGMP formation was measured (B). Data are shown as cpm incorporated [^{14}C]ascorbic acid/mg cell protein (means ± SEM, n = 3) (A) or as ionomycin-induced increases in [^{3}H]citrulline or cGMP production calculated from the differences between stimulated and unstimulated cells (means ± SEM, n = 4) (B). *$P < 0.05$ vs. untreated control cells. Reprinted with permission from Ref. 84.

due to the absence of the enzyme gulonolactone oxidase. Accordingly, it did not affect ionomycin-induced citrulline or cGMP synthesis when incubated with endothelial cells. On the other hand, dehydroascorbic acid, which is partially converted back to ascorbate by glutathione-dependent reactions (92), exerted a partial stimulatory effect (Fig. 5.3).

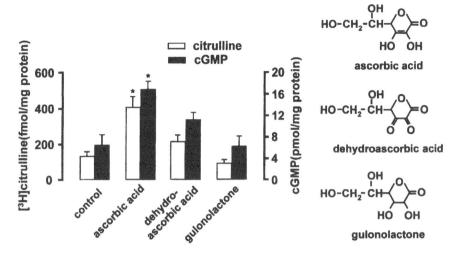

Fig. 5.3. Comparison of the effects of ascorbic acid, dehydroascorbic acid, and gulono-lactone on ionomycin-induced citrulline and cGMP formation. Endothelial cells from umbilical veins were preincubated for 24 h with 100 µmol/L of the respective compounds, stimulated for 15 min with 2 µmol/L ionomycin, and processed for either citrulline or cGMP measurement. Data are shown as agonist-induced increases in [³H]citrulline formation or cGMP production calculated from the differences between stimulated and unstimulated cells (means ± SEM, n = 4), *P < 0.05 vs. untreated control cells. The right panel shows the molecular structures of the compounds tested.

Ascorbic Acid Enhances the Availability of the NOS Cofactor Tetrahydrobiopterin

The data reported above demonstrated that ascorbic acid potentiates agonist-induced NO formation in cultured endothelial cells in a dose- and time-dependent fashion. The effect was saturated within physiologically relevant concentrations, related to an intracellular ascorbate accumulation and dependent on the redox-active enediol group of ascorbic acid. We next performed experiments to investigate mechanisms responsible for the observed effects. In agreement with other studies (89,93), preincubation of endothelial cells with ascorbate neither induced the expression of eNOS nor affected its subcellular distribution between membrane and cytosolic fractions (data not shown, details in Ref. 84). Similarly, an increased availability of the eNOS substrate L-arginine did not account for the potentiation of NO synthesis because ascorbic acid did not improve the cellular uptake of this amino acid (data not shown, details in Ref. 84). However, we found that the effect of ascorbate on agonist-stimulated citrulline and cGMP production was mimicked by pretreatment of the cells with increasing concentrations of sepiapterin (0.001–10 µmol/L, 24 h) (Fig. 5.4). This compound is readily taken up by cells and converted into tetrahydrobiopterin *via* a salvage pathway (94). Its potentiating

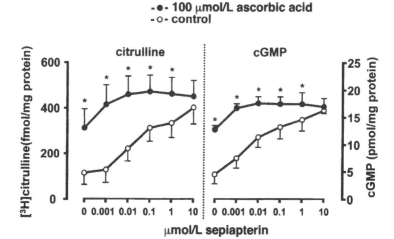

Fig. 5.4. Influence of sepiapterin on ascorbic acid–induced potentiation of citrulline and cGMP formation. Endothelial cells were preincubated for 24 h with 0.001–10 μmol/L sepiapterin in culture medium in the absence or presence of 100 μmol/L ascorbic acid. Subsequently, cells were stimulated in Hepes buffer (pH 7.4) for 15 min with 2 μmol/L ionomycin and either citrulline or cGMP formation was measured. Data are shown as agonist-induced [^{3}H]citrulline formation or cGMP production calculated from the differences between stimulated and unstimulated cells (means ± SEM, n = 4); cells with and without ascorbate pretreatment were compared, *$P < 0.05$. Reprinted with permission from Ref. 85.

effect on endothelial NO formation indicates that eNOS is not saturated with its cofactor tetrahydrobiopterin thus confirming previous studies from cultured endothelial cells (68,70). Interestingly, sepiapterin abolished the potentiating effect of ascorbic acid on NO production in a concentration-dependent manner (Fig. 5.4), suggesting that ascorbate exerts its effect on NO synthesis only under suboptimal intracellular tetrahydrobiopterin concentrations. Accordingly, we hypothesized that ascorbic acid may either enhance the availability of tetrahydrobiopterin in endothelial cells or increase its affinity for eNOS.

To test the latter possibility, we performed experiments with tetrahydrobiopterin-free eNOS expressed in and purified from *Pichia pastoris* as described recently (86). The enzyme was inactive in the absence of exogenous tetrahydrobiopterin. The addition of the pteridine (1 nmol/L–100 μmol/L) stimulated the formation of citrulline in a concentration-dependent manner with a 50% effective concentration (EC_{50}) of 0.31 ± 0.036 μmol/L and a maximal effect at ~100 μmol/L (Fig. 5.5). The presence of 100 μmol/L ascorbic acid in the assay solution resulted only in a slight decrease of the EC_{50} to 0.16 ± 0.014 μmol/L without significant increase in maximal enzyme activity. From these data, we concluded that ascorbate

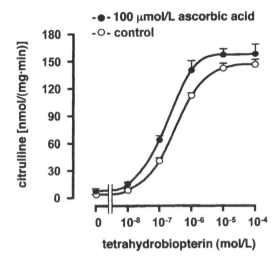

Fig. 5.5. Influence of ascorbic acid on tetrahydrobiopterin dependence of the activity of the endothelial NOS isoform (eNOS). Enzyme assays were performed in the absence or presence of 100 μmol/L ascorbate with tetrahydrobiopterin-free recombinant eNOS purified from the yeast *Pichia pastoris.* The assay solution contained 50 mmol/L Tris-HCl buffer (pH 7.4), 0.3 μg/100 μL eNOS, 100 μmol/L L-[^{3}H]arginine (100,000 cpm), 0.5 mmol/L $CaCl_2$, 0.2 mmol/L NADPH, 5 μmol/L FAD, 5 μmol/L FMN, 10 μg/mL calmodulin, 10 nmol/L–100 μmol/L tetrahydrobiopterin and 0.2 mmol/L CHAPS. After 10 min at 37°C, the mixture was subjected to cation exchange chromatography and [^{3}H]citrulline was quantified by liquid scintillation counting (means ± SEM, n = 3). Reprinted with permission from Ref. 85.

does not substantially modify the pterin affinity of the enzyme. Interestingly, ascorbate did not activate eNOS in the absence of exogenous tetrahydrobiopterin, indicating that it does not act as a cofactor for eNOS itself.

We next investigated the influence of ascorbic acid on intracellular tetrahydrobiopterin concentrations. We found that the preincubation of endothelial cells with ascorbic acid (100 μmol/L, 24 h) increased intracellular tetrahydrobiopterin levels from 0.38 ± 0.04 to 1.14 ± 0.09 pmol/mg protein (n = 20). The effect of ascorbic acid on endothelial tetrahydrobiopterin levels was concentration dependent and saturable at 100 μmol/L (Fig. 5.6). Thus, there is a close relationship between intracellular ascorbic acid accumulation, the potentiation of agonist-induced citrulline and cGMP synthesis, and the increase of tetrahydrobiopterin levels induced by ascorbate, suggesting that intracellular tetrahydrobiopterin concentration and, consequently, NO formation are critically dependent on the tissue levels of ascorbate. From the data presented in our study, we can speculate that intracellular ascorbate levels of ~2 mmol/L and tetrahydrobiopterin levels in the range of 200 nmol/L provide optimal reaction conditions for NO formation in endothelial cells.

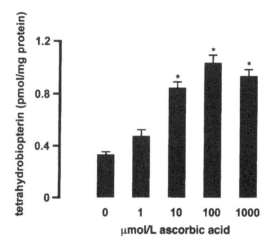

Fig. 5.6. Effect of ascorbic acid on intracellular tetrahydrobiopterin levels. Endothelial cells from human umbilical veins were preincubated for 24 h with 1 μmol/L–1 mmol/L ascorbic acid in culture medium. Aliquots of 5×10^6 cells were oxidized with 0.02 mol/L KI/I_2 in 0.1 mol/L HCl or 0.1 mol/L NaOH and the resulting biopterin was quantified by reversed-phase high-performance liquid chromatography. Tetrahydrobiopterin was calculated from the difference in biopterin concentration after oxidation in acid and base. Data are shown as means ± SEM from 5 experiments. *$P < 0.05$ vs. untreated control cells. Reprinted with permission from Ref. 85.

Our results are in agreement with two recent reports showing that an enhanced eNOS activity after ascorbic acid pretreatment was associated with an increase of intracellular tetrahydrobiopterin levels in porcine aortic endothelial cells and HUVEC (89,93). Moreover, conditions that are thought to be associated with tetrahydrobiopterin deficiency (i.e., coronary artery disease or smoking) have been characterized by low ascorbic acid levels in plasma or leukocytes (16,17,95,96) and are known to be associated with an impaired NO-dependent vasodilation. Taken together, these data led us to the suggestion that increased availability of tetrahydrobiopterin could be a common mechanism underlying the improvement of endothelial dysfunction in patients with chronic oral ascorbic acid administration.

Ascorbic Acid Protects Tetrahydrobiopterin from Oxidation

The findings presented so far have related the potentiating effect of ascorbic acid on endothelial NO synthesis to an increase in intracellular tetrahydrobiopterin levels. We next attempted to understand whether the improved availability of tetrahydrobiopterin was due to an enhanced synthesis or to a decreased degradation of the compound. Tetrahydrobiopterin is synthesized *de novo* from GTP by the sequential action of three enzymes, GTP cyclohydrolase I, 6-pyruvoyl-tetrahydropterin synthase, and sepiapterin reductase (Fig. 5.7). GTP cyclohydrolase I has been shown to be the rate-limiting enzyme of the *de novo* pathway and to be regulated by cytokines such as TNF-α, IFN-γ, and interleukin-1β in a number of cell types including endothelial cells (68,70,97,98). We first investigated whether ascorbic acid affects the expression and activity of GTP cyclohydrolase I. Because the expression of this enzyme in endothelial cells is generally low, experiments were carried out with cells pretreated without and with cytokines (250 u/mL TNF-α, 250 u/ml IFN-γ and 1 μg/mL LPS, 24 h) to induce enzyme expression. Figure 5.8A shows that the mRNA expression of GTP cyclohydrolase I was upregulated by cytokines but no differences were seen between ascorbate-treated cells and their respective controls. Similarly, ascorbic acid did not affect GTP cyclohydrolase I activity when added to cytosolic fractions of cytokine-treated cells, suggesting that the compound does not act as a direct cofactor of the enzyme (Fig. 5.8B). These results suggest that the effect of ascorbic acid on intracellular tetrahydrobiopterin level is not due to an increased synthesis of the compound. Accordingly, inhibition of tetrahydrobiopterin formation by DAHP, an inhibitor of GTP cyclohydrolase I (99), did not prevent the ascorbate-mediated increase of the pteridine although it substantially decreased tetrahydrobiopterin levels in both control and ascorbic acid-treated endothelial cells (Fig. 5.9). In parallel, ionomycin-stimulated formation of citrulline and cGMP was decreased upon pretreatment of the cells with

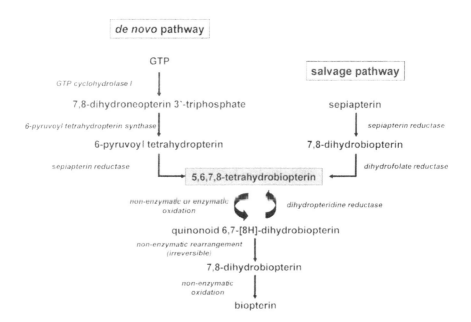

Fig. 5.7. Biosynthetic pathways and oxidative degradation of 5,6,7,8-tetrahydrobiopterin.

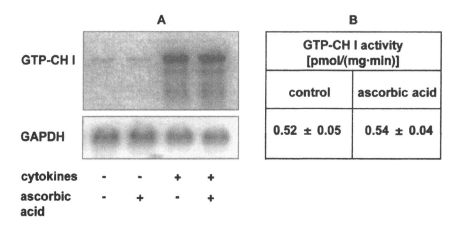

Fig. 5.8. Effect of ascorbic acid on GTP cyclohydrolase mRNA expression and activity. (A) Total RNA was extracted from endothelial cells preincubated for 24 h with 100 μmol/L ascorbic acid in the absence or presence of a mixture of cytokines (250 U/mL tumor necrosis factor-α, 250 U/mL interferon-γ) and lipopolysaccharide (1 μg/mL). After electrophoresis on 1% agarose/6% formaldehyde gels (20 μg/lane) the RNA was blotted on nylon membranes and hybridized overnight with [^{32}P]dCTP-labeled probes for human GTP cyclohydrolase I (GTP-CH I) and human glyceraldehyde-3-phosphate dehydrogenase (GAPDH). One typical experiment out of three is shown. (B) GTP-CH I activity was measured in cytosolic fractions from cytokine-treated endothelial cells in the absence or presence of 100 μmol/L ascorbate. Data are expressed as pmol neopterin/(mg cytosolic protein·min) (means ± SEM, n = 3). Panel (A) was reprinted with permission from Ref. 85.

DAHP in controls and in the presence of ascorbic acid, but the potentiating effect of ascorbate was maintained (Fig. 5.9).

Because ascorbic acid did not affect tetrahydrobiopterin synthesis, we speculated that it might act by preventing the degradation of the compound. 5,6,7,8-Tetrahydrobiopterin is oxidized intracellularly to the quinonoid 6,7[8H]-dihydrobiopterin, which spontaneously rearranges to 7,8-dihydrobiopterin (100). The latter is further degraded to biopterin (Fig. 5.7). To investigate whether ascorbic acid prevents degradation of tetrahydrobiopterin, the levels of tetrahydrobiopterin, 7,8-dihydrobiopterin, and biopterin in cells and cell supernatants were measured and balanced on the basis of pmol pteridines/dish. The experiments were performed in cells coincubated with ascorbate (1–100 μmol/L ascorbic acid, 24 h) and cytokines (250 U/mL TNF-α, 250 U/mL IFN-γ and 1 μg/mL LPS, 24 h) to increase pteridine production. Fig. 5.10 shows that the sum of all biopterin derivatives in cells and media was not influenced by ascorbate, confirming the lack of an effect on pterin biosynthesis. However, the ascorbic acid–induced increase of intracellular tetrahydrobiopterin was paralleled by a decrease of 7,8-dihydrobiopterin + biopterin in cells and cell supernatants. Interestingly, ~90% of the dihydrobiopterin + biopterin formed in endothelial cells

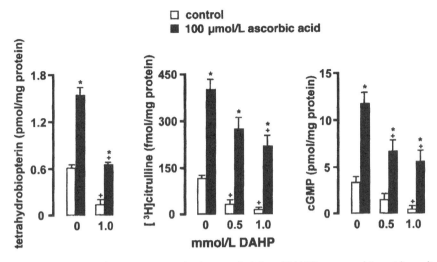

Fig. 5.9. Influence of 2,4-diamino-6-hydroxypyrimidine (DAHP) on ascorbic acid–mediated increase of tetrahydrobiopterin levels and potentiation of citrulline and cGMP formation. Endothelial cells were preincubated for 24 h with 0.5–1.0 mmol/L DAHP in culture medium in the absence or presence of 100 μmol/L ascorbic acid. Subsequently, cells were stimulated for 15 min in Hepes buffer (pH 7.4) with 2 μmol/L ionomycin and processed for either citrulline or cGMP measurement. Additionally, biopterin levels were measured by reversed-phase high-performance liquid chromatography after oxidation of cells with 0.02 mol/L KI/I_2 in 0.1 mol/L HCl or 0.1 mol/L NaOH; tetrahydrobiopterin was calculated from the difference. Data are shown as means ± SEM (n = 4). Controls and DAHP-treated cells ([+]) and cells with and without ascorbate pretreatment ([*]) were compared, [+][*]$P < 0.05$. Reprinted with permission from Ref. 85.

was released into the medium in control cells, whereas tetrahydrobiopterin was not detectable in supernatants of both control and ascorbate-treated cells. Our data suggest that a chemical stabilization of the fully reduced pterin by ascorbate is the underlying mechanism for the increased intracellular tetrahydrobiopterin concentration. The stabilizing function of ascorbate is most likely due to a chemical reduction of the quinonoid 6,7[8H]-dihydrobiopterin to tetrahydrobiopterin, which had already been demonstrated for other reducing compounds such as dithioerythritol and NADPH. Moreover, Toth and co-workers (101) confirmed in a recent study that ascorbate mediated the reductive reversal of the autoxidation process of tetrahydrobiopterin. Another possible mechanism is the regeneration of tetrahydrobiopterin from its trihydrobiopterin radical by ascorbate, which has been shown in *in vitro* experiments (102). This action could possibly also support eNOS activation because the trihydrobiopterin radical is formed in the catalytic mechanism of NOS. However, the protective effect of ascorbate in our study was prominent in intact cells and only minimal with the purified eNOS, although this was possibly due to the presence of reducing

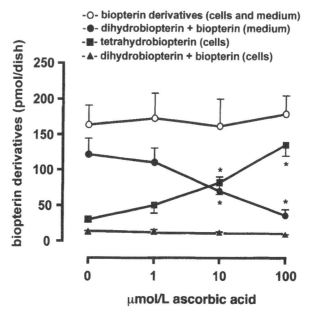

Fig. 5.10. Effect of ascorbic acid on tetrahydrobiopterin stability in intact cells. Endothelial cells were preincubated for 24 h with 250 U/mL tumor necrosis factor-α, 250 U/mL interferon-γ and 1 μg/mL lipopolysaccharide in the absence or presence of 1–100 μmol/L ascorbic acid. Aliquots of 2×10^6 cells and 1-mL aliquots of cell supernatants were oxidized with 0.02 mol/L KI/I_2 in 0.1 mol/L HCl or 0.1 mol/L NaOH and the resulting biopterin was quantified by reversed-phase high-performance liquid chromatography. Biopterin levels after oxidation in base indicate the amount of dihydrobiopterin + biopterin, whereas tetrahydrobiopterin was calculated from the difference in biopterin concentration after oxidation in acid and base. To balance biopterin derivatives in cells and medium, pteridine levels were calculated in pmol/dish. Data are shown as means ± SEM from 3 experiments. *$P < 0.05$ vs. levels of biopterin derivatives in untreated control cells. Reprinted with permission from Ref. 85.

agents in the enzyme assay. On the other hand, a stimulating ascorbate effect has been seen in lysates from HUVEC and in bovine eNOS preparations with higher ascorbate concentrations and under different assay conditions (89,93). It is not yet known which agents cause an increased tetrahydrobiopterin oxidation *in vivo*. *In vitro* studies have shown that tetrahydrobiopterin is a primary target for peroxynitrite-catalyzed oxidation but does not significantly react with hydrogen peroxide (103). Interestingly, a cell membrane-permeable superoxide dismutase mimetic was not able to increase tetrahydrobiopterin levels in endothelial cells, although a significant effect of ascorbic acid was observed in that study (93). Thus, superoxide anion might not be involved in tetrahydrobiopterin oxidation in endothelial cells and moreover, the stabilizing effect of ascorbate on tetrahydrobiopterin may not involve superoxide scavenging. Finally, in this context, it is worth noting that the increased availability of tetrahydrobiopterin in endothelial cells saturated with ascorbic acid will not only potentiate NO formation but also decrease superoxide formation by eNOS by preventing uncoupling of oxygen reduction and arginine oxidation (86,104,105), thereby further decreasing oxidative stress in endothelial cells.

Summary

There is strong indication that a decreased bioavailability of endothelium-derived NO predisposes to atherosclerosis and related disease states. Among other factors, oxidative stress might be responsible for the development of endothelial dysfunction, implying that atherogenesis and progression of atherosclerosis can be inhibited by antioxidants. This concept is supported by the consistent finding that ascorbic acid can facilitate endothelium-dependent vasodilation. The data presented and discussed in this paper provide evidence that one of the possible mechanisms by which ascorbate might prevent or ameliorate endothelial dysfunction is its potentiating effect on endothelial NO synthesis. This is due to the ability of ascorbic acid to protect tetrahydrobiopterin, an essential cofactor of eNOS, from oxidation and requires saturated intracellular ascorbate levels as well as the reductive capacity of the compound. The role of ascorbic acid in increasing eNOS activity and possibly preventing endothelial dysfunction could provide a rationale for optimizing its dietary intake or for oral ascorbate supplementation and must be confirmed in clinical long-term studies. In this context, it seems to be important that not only plasma levels but also tissue concentrations (in leukocytes for example) of ascorbic acid will be measured to evaluate the body status of ascorbate and its relationship to NO synthesis and bioavailability.

Acknowledgments

This work was supported by a grant from the Interdisziplinäres Zentrum für Klinische Forschung, Klinikum der Friedrich-Schiller-Universität Jena (to R.H.) and by Austrian research funds "Zur Förderung der wissenschaftlichen Forschung" Project 16188 (to E.R.W.).

References

1. Harrison, D.G. (1997) Cellular and Molecular Mechanisms of Endothelial Dysfunction, *J. Clin. Invest. 100*, 2153–2157.
2. Ignarro, L.J., Cirino, G., Casini, A., and Napoli, C. (1999) Nitric Oxide as a Signaling Molecule in the Cardiovascular System: An Overview, *J. Cardiovasc. Pharmacol. 34*, 879–886.
3. Napoli, C., and Ignarro, L.J. (2001) Nitric Oxide and Atherosclerosis, *Nitric Oxide 5*, 88–97.
4. Maxwell, A.J. (2002) Mechanisms of Dysfunction of the Nitric Oxide Pathway in Vascular Disease, *Nitric Oxide 6*, 101–124.

5. Cayatte, A.J., Placino, J.J., Horten, K., and Cohen, R.A. (1994) Chronic Inhibition of Nitric Oxide Production Accelerates Neointima Formation and Impairs Endothelial Function in Hypercholesterolemic Rabbits, *Arterioscler. Thromb. 14*, 753–759.

6. Zeiher, A.M., Drexler, H., Wollschläger, H., and Just, H.J. (1991) Modulation of Coronary Vasomotor Tone in Humans: Progressive Endothelial Dysfunction with Different Early Stages of Coronary Atherosclerosis, *Circulation 83*, 391–401.

7. McLenachan, J.M., Williams, J.K., Fish, R.D., Ganz, P., and Selwyn, A.P. (1991) Loss of Flow-Mediated Endothelium-Dependent Dilation Occurs Early in the Development of Atherosclerosis, *Circulation 84*, 1273–1278.

8. Stroes, E.S., Koomans, H.A., de Bruin, T.W., and Rabelink, T.J. (1995) Vascular Function in the Forearm of Hypercholesterolaemic Patients off and on Lipid-Lowering Medication, *Lancet 346*, 467–471.

9. Panza, J.A., Garcia, G.E., Kilcoyne, C.M., Quyyumi, A.A. and Cannon, R.O. (1995) Impaired Endothelium-Dependent Vasodilation in Patients with Essential Hypertension. Evidence That Nitric Oxide Abnormality Is Not Localized to a Single Signal Transduction Pathway, *Circulation 91*, 1732–1738.

10. Celermajer, D.S., Sorensen, K., Ryalls, M., Robinson, J., Thomas, O., Leonard, J.V., and Deanfield, J.E. (1993) Impaired Endothelial Function Occurs in the Systemic Arteries of Children with Homozygous Homocystinuria but Not in Their Heterozygous Parents, *J. Am. Coll. Cardiol. 22*, 854–858.

11. Celermajer, D.S., Adams, M.R., Clarkson, P., Robinson, J., McCredie, R., Donald, A., and Deanfield, J.E. (1996) Passive Smoking and Impaired Endothelium-Dependent Arterial Dilation in Healthy Young Adults, *N. Engl. J. Med. 334*, 150–154.

12. Keaney, J.F., and Vita, J.A. (1995) Atherosclerosis, Oxidative Stress, and Antioxidant Protection in Endothelium-Derived Relaxing Factor Action, *Prog. Cardiovasc. Dis. 38*, 129–154.

13. Gryglewski, R.J., Palmer, R.M.J., and Moncada, S. (1986) Superoxide Anion Is Involved in the Breakdown of Endothelium-Derived Relaxing Factor, *Nature 320*, 454–456.

14. Chin, J.H., Azhar, S., and Hoffman, B.B. (1992) Inactivation of Endothelial Derived Relaxing Factor by Oxidized Lipoproteins, *J. Clin. Investig. 82*, 10–18.

15. Diaz, M.N., Frei, B., Vita, J.A., and Keaney, J.F. (1997) Antioxidants and Atherosclerotic Heart Disease, *N. Engl. J. Med. 337*, 408–416.

16. Vita, J.A., Keaney, J.F., Raby, K.E., Morrow, J.D., Freedman, J.E., Lynch, S., Koulouris, S.N., Hankin, B.R., and Frei, B. (1998) Low Plasma Ascorbic Acid Independently Predicts the Presence of an Unstable Coronary Syndrome, *J. Am. Coll. Cardiol. 31*, 980–986.

17. Ramirez, J., and Flowers, N.C. (1980) Leukocyte Ascorbic Acid and Its Relationship to Coronary Artery Disease in Man, *Am. J. Clin. Nutr. 33*, 2079–2087.

18. Ting, H.H., Timimi, F.K., Boles, K.S., Creager, S.J., Ganz, P., and Creager, M.A. (1996) Vitamin C Improves Endothelium-Dependent Vasodilation in Patients with Non-Insulin-Dependent Diabetes Mellitus, *J. Clin. Invest. 97*, 22–28.

19. Levine, G.N., Frei, B., Koulouris, S.N., Gerhardt, M.D., Keaney, J.F., and Vita, J.A. (1996) Ascorbic Acid Reverses Endothelial Vasomotor Dysfunction in Patients with Coronary Artery Disease, *Circulation 93*, 1107–1113.

20. Heitzer, T., Just, H., and Münzel, T. (1996) Antioxidant Vitamin C Improves Endothelial Dysfunction in Chronic Smokers, *Circulation 94*, 6–9.

21. Motoyama, T., Kawano, H., Kugiyama, K., Hirashima, O., Ohgushi, M., Yoshimura, M., Ogawa, H., and Yasue, H. (1997) Endothelium-Dependent Vasodilation in the Brachial Artery Is Impaired in Smokers: Effect of Vitamin C, *Am. J. Physiol. 273*, H1644-H1650.
22. Solzbach, U., Hornig, B., Jeserich, M., and Just, H. (1997) Vitamin C Improves Endothelial Dysfunction of Epicardial Coronary Arteries in Hypertensive Patients, *Circulation 96*, 1513-1519.
23. Ting, H.H., Timimi, F.K., Haley, E.A., Roddy, M.-A., Ganz, P., and Creager M.A. (1997) Vitamin C Improves Endothelium-Dependent Vasodilation in Forearm Resistance Vessels of Humans with Hypercholesterolemia, *Circulation 95*, 2617-2622.
24. Timimi, F.K., Ting, H.H., Haley, E.A., Roddy, M.-A., Ganz, P., and Creager, M.A. (1998) Vitamin C Improves Endothelium-Dependent Vasodilation in Patients with Insulin-Dependent Diabetes Mellitus, *J. Am. Coll. Cardiol. 31*, 552-557.
25. Kugiyama, K., Motoyama, T., Hirashima, O., Ohgushi, M., Soejima, H., Misumi, K., Kawano, H., Miyao, Y., Yoshimura, M., Ogawa, H., Matsumura, T., and Sugiyama, S. (1998) Vitamin C Attenuates Abnormal Vasomotor Reactivity in Spasm Coronary Arteries in Patients with Coronary Spastic Angina, *J. Am. Coll. Cardiol. 32*, 103-109.
26. Hornig, B., Arakawa, N., Kohler, C., and Drexler, H. (1998) Vitamin C Improves Endothelial Function of Conduit Arteries in Patients with Chronic Heart Failure, *Circulation 97*, 363-368.
27. Taddei, S., Virdis, A., Ghiadoni, L., Magagna, A., and Salvetti, A. (1998) Vitamin C Improves Endothelium-Dependent Vasodilation by Restoring Nitric Oxide Activity in Essential Hypertension, *Circulation 97*, 2222-2229.
28. Ito, K., Akita, H., Kanazawa, K., Yamada, S., Terashima, M., Matsuda, Y., and Yokoyama, M. (1998) Comparison of Effects of Ascorbic Acid on Endothelium-Dependent Vasodilation in Patients with Chronic Congestive Heart Failure Secondary to Idiopathic Dilated Cardiomyopathy Versus Patients with Effort Angina Pectoris Secondary to Coronary Artery Disease, *Am. J. Cardiol. 82*, 762-767.
29. Chambers, J.C., McGregor, A., Jean-Marie, J., Obeid, O.A., and Kooner, J.S. (1999) Demonstration of Rapid Onset Vascular Endothelial Dysfunction After Hyperhomocysteinemia. An Effect Reversible with Vitamin C Therapy, *Circulation 99*, 1156-1160.
30. Gokce, N., Keaney, J.F., Frei, B., Holbrook, M., Olesiak, M., Zachariah, B.J., Leeuwenburgh, C., Heinecke, J.W., and Vita, J.A. (1999) Long-Term Ascorbic Acid Administration Reverses Endothelial Vasomotor Dysfunction in Patients with Coronary Artery Disease, *Circulation 99*, 3234-3240.
31. Tousoulis, D., Davies, G., and Toutouzas, P. (1999) Vitamin C Increases Nitric Oxide Availability in Coronary Atherosclerosis, *Ann. Intern. Med. 131*, 156-157.
32. Kaufmann, P.A., Gnecchi-Ruscone, T., di Terlizzi, M., Schäfers, K.P., Lüscher, T.F., and Camici, P.G. (2000) Coronary Heart Disease in Smokers. Vitamin C Restores Coronary Microcirculatory Function, *Circulation 102*, 1233-1238.
33. Sherman, D.L., Keaney, J.F., Biegelsen, E.S., Duffy, S.J., Coffman, J.D., and Vita, J.A. (2000) Pharmacological Concentrations of Ascorbic Acid Are Required for the Beneficial Effect on Endothelial Vasomotor Function in Hypertension, *Hypertension 35*, 936-941.
34. Schindler, T.H., Magosaki, N., Jeserich, M., Olschewski, M., Nitzsche, E., Holubarsch, C., Solzbach, U., and Just, H. (2000) Effect of Ascorbic Acid on Endothelial

Dysfunction of Epicardial Coronary Arteries in Chronic Smokers Assessed by Cold Pressor Testing, *Cardiology 94*, 239–246.

35. Natali, A., Sironi, A.M., Toschi, E., Camastra, S., Sanna, G., Perissinotto, A., Taddei, S., and Ferrannini, E. (2000) Effect of Vitamin C on Forearm Blood Flow and Glucose Metabolism in Essential Hypertension, *Arterioscler. Thromb. Vasc. Biol. 20*, 2401–2406.

36. Beckman, J.A., Goldfine, A.B., Gordon, M.B., and Creager, M.A. (2001) Ascorbate Restores Endothelium-Dependent Vasodilation Impaired by Acute Hyperglycemia in Humans, *Circulation 103*, 1618–1623.

37. Richartz, B., Werner, G.S., Ferrari, M., and Figulla, H.R. (2001) Reversibility of Coronary Endothelial Vasomotor Dysfunction in Idiopathic Dilated Cardiomyopathy: Acute Effects of Vitamin C, *Am. J. Cardiol. 88*, 1001–1005.

38. May, J.M. (2000) How Does Ascorbic Acid Prevent Endothelial Dysfunction? *Free Radic. Biol. Med. 28*, 1421–1429.

39. Carr, A., and Frei, B. (2000) The Role of Natural Antioxidants in Preserving the Biological Activity of Endothelium-Derived Nitric Oxide, *Free Radic. Biol. Med. 28*, 1806–1814.

40. Tomasian D., Keaney, J.F., and Vita, J.A. (2000) Antioxidants and the Bioactivity of Endothelium-Derived Nitric Oxide, *Cardiovasc. Res. 47*, 426–435.

41. Negre-Salvayre, A., Mabile, L., Delchambre, J., and Salvayre R. (1995) α-Tocopherol, Ascorbic Acid and Rutin Inhibit Synergistically the Copper-Promoted LDL-Oxidation and the Cytotoxicity of Oxidized LDL to Cultured Endothelial Cells, *Biol. Trace Elem. Res. 47*, 81–91.

42. Martin, A., and Frei, B. (1997) Both Intracellular and Extracellular Vitamin C Inhibit Atherogenic Modification of LDL by Human Endothelial Cells, *Arterioscler. Thromb. Vasc. Biol. 17*, 1583–1590.

43. Liao, J.K., Shin, W.S., Lee, W.Y., and Clark, S.L. (1995) Oxidized Low-Density Lipoproteins Decrease the Expression of Endothelial Nitric Oxide Synthase, *J. Biol. Chem. 270*, 319–324.

44. Liao, J.K., and Clark, S.L. (1995) Regulation of G Protein Alpha i2 Subunit Expression by Oxidized Low-Density Lipoprotein, *J. Clin. Investig. 95*, 1457–1463.

45. Vergnani, L., Hatrik, S., Ricci, F., Passaro, A., Manzoli, N., Zulianin, G., Brovkovych, V., Fellin, R., and Malinski, T. (2000) Effect of Native and Oxidized Low-Density Lipoproteins on Endothelial Nitric Oxide and Superoxide Production. Key Role of L-Arginine Availability, *Circulation 101*, 1261–1266.

46. Blair, A., Shaul, P.W., Yuhanna, I.S., Conrad, P.A., and Smart, E.J. (1999) Oxidized Low Density Lipoprotein Displaces Endothelial Nitric-Oxide Synthase (eNOS) from Plasmalemmal Caveolae and Impairs eNOS Activation, *J. Biol. Chem. 274*, 32512–32519.

47. Nuzskowski, A., Gräbner, R., Marsche, G., Unbehaun, A., Malle, E., and Heller, R. (2001) Hypochlorite-Modified Low Density Lipoprotein Inhibits Nitric Oxide Synthesis in Endothelial Cells via an Intracellular Dislocalization of Endothelial Nitric-Oxide Synthase, *J. Biol. Chem. 276*, 14212–14221.

48. Meister A. (1994) Glutathione-Ascorbic Acid Antioxidant System in Animals, *J. Biol. Chem. 269*, 9397–9400.

49. Stamler, J.S., and Singel, D.J. (1992) Biochemistry of Nitric Oxide and Its Redox-Activated Forms, *Science 258*, 1898–1902.

50. Stamler, J.S., Jaraki, O.A., Osborne, J.A., Simon, D.I., Keaney, J.F., Singel, D.J., Valeri, C.R., and Loscalzo, J. (1992) Nitric Oxide Circulates in Mammalian Plasma Primarily as an *S*-Nitroso Adduct of Serum Albumin, *Proc. Natl. Acad. Sci. USA 89*, 7674–7677.

51. Scharfstein, J.S., Keaney, J.F., Slivka, A., Welch, G.N., Vita, J.A., Stamler, J.S., and Loscalzo, J. (1994) In Vivo Transfer of Nitric Oxide Between a Plasma Protein-Bound Reservoir and Low Molecular Weight Thiols, *J. Clin. Investig. 94*, 1432–1439.

52. Kashiba-Iwatsuki, M., Yamaguchi, M., and Inoue, M. (1996) Role of Ascorbic Acid in the Metabolism of *S*-Nitroso-glutathione, *FEBS Lett. 389*, 149–152.

53. Scorza, G., Pietraforte, D., and Minetti, M. (1997) Role of Ascorbate and Protein Thiols in the Release of Nitric Oxide from *S*-Nitroso-albumin and *S*-Nitroso-glutathione in Human Plasma, *Free Radic. Biol. Med. 22*, 633–642.

54. Xu, A., Vita, J.A., and Keaney, J.F. (2000) Ascorbic Acid and Glutathione Modulate the Biological Activity of *S*-Nitrosoglutathione, *Hypertension 36*, 291–295.

55. Jackson, T.S., Xu, A., Vita, J.A., and Keaney, J.F. (1998) Ascorbate Prevents the Interaction of Superoxide and Nitric Oxide Only at Very High Physiological Concentrations, *Circ. Res. 83*, 916–922.

56. Ek, A., Ström, K., and Cotgreave, I.A. (1995) The Uptake of Ascorbic Acid into Human Umbilical Vein Endothelial Cells and Its Effect on Oxidant Insult, *Biochem. Pharmacol. 50*, 1339–1346.

57. Mayer, B., and Hemmens, B. (1997) Biosynthesis and Action of Nitric Oxide in Mammalian Cells, *Trends Biochem. Sci. 22*, 477–481.

58. Gorren, A.C.F., and Mayer, B. (1998) The Versatile and Complex Enzymology of Nitric Oxide Synthase, *Biochemistry 63*, 734–743.

59. Alderton, W.K., Cooper, C.E., and Knowles, R.G. (2001) Nitric Oxide Synthases: Structure, Function and Inhibition, *Biochem. J. 357*, 593–615.

60. Govers, R., and Rabelink, T.J. (2001) Cellular Regulation of Endothelial Nitric Oxide Synthase, *Am. J. Physiol. 280*, F193–F206.

61. Cosentino, F., and Lüscher, T.F. (1999) Tetrahydrobiopterin and Endothelial Nitric Oxide Synthase Activity, *Cardiovasc. Res. 43*, 274–278.

62. Katusic, Z.S. (2001) Vascular Endothelial Dysfunction: Does Tetrahydrobiopterin Play a Role? *Am. J. Physiol. 281*, H981–H986.

63. Tiefenbacher, C.P. (2001) Tetrahydrobiopterin: A Critical Cofactor for eNOS and a Strategy in the Treatment of Endothelial Dysfunction? *Am. J. Physiol. 280*, H2448–H2488.

64. Feron, O. (1999) Intracellular Localization and Activation of Endothelial Nitric Oxide Synthase, *Curr. Opin. Nephrol. Hypertens. 8*, 55–59.

65. Kone, B.C. (2000) Protein-Protein Interactions Controlling Nitric Oxide Synthases, *Acta Physiol. Scand. 168*, 27–31.

66. Kugiyama, K., Kerns, S.A., Morrisett, J.D., Roberts, R., and Henry, P.D. (1990) Impairment of Endothelial-Dependent Arterial Relaxation by Lysolecithin in Modified Low-Density Lipoproteins, *Nature 344*, 160–162.

67. Schmidt, K., Werner, E.R., Mayer, B., Wachter, H., and Kukovetz, E.R. (1992) Tetrahydrobiopterin-Dependent Formation of Endothelium-Derived Relaxing Factor (Nitric Oxide) in Aortic Endothelial Cells, *Biochem. J. 281*, 297–300.

68. Werner-Felmayer, G., Werner, E.R., Fuchs, D., Hausen, A., Reibnegger, G., Schmidt, K., Weiss, G., and Wachter, H. (1993) Pteridine Synthesis in Human Endothelial Cells.

Impact on Nitric Oxide-Mediated Formation of Cyclic GMP, *J. Biol. Chem. 268*, 1842–1846.

69. Schoedon, G., Schneemann, M., Blau, N., Edgell, C.-J.S., and Schaffner, A. (1993) Modulation of Human Endothelial Cell Tetrahydrobiopterin Synthesis by Activating and Deactivating Cytokines, *Biochem. Biophys. Res. Commun. 196*, 1343–1348.

70. Rosenkranz-Weiss, P., Sessa, W.C., Milstien, S., Kaufman, S., Watson, C.A., and Pober, J.S. (1994) Regulation of Nitric Oxide Synthesis by Proinflammatory Cytokines in Human Umbilical Vein Endothelial Cells. Elevations in Tetrahydrobiopterin Levels Enhance Endothelial Nitric Oxide Synthase Specific Activity, *J. Clin. Investig. 93*, 2236–2243.

71. Bec, N., Gorren, A.C.F., Voelker, C., Mayer, B., and Lange, R. (1998) Reaction of Neuronal Nitric-Oxide Synthase with Oxygen at Low Temperature, *J. Biol. Chem. 273*, 13502–13508.

72. Hurshman, A.R., Krebs, C., Edmondson, D.E., Huynh, B.H., and Marletta, M.A. (1999) Formation of a Pterin Radical in the Reaction of the Heme Domain of Inducible Nitric Oxide Synthase with Oxygen, *Biochemistry 38*, 15689–15696.

73. Witteveen, C.F.B., Giovanelli, J., and Kaufman, S. (1999) Reactivity of Tetrahydrobiopterin Bound to Nitric-Oxide Synthase, *J. Biol. Chem. 274*, 29755–29762.

74. Gorren, A.C., Bec, N., Schrammel, A., Werner, E.R., Lange, R., and Mayer, B. (2000) Low-Temperature Optical Absorption Spectra Suggest a Redox Role for Tetrahydrobiopterin in Both Steps of Nitric Oxide Synthase Catalysis, *Biochemistry 39*, 11763–11770.

75. Tiefenbacher, C.P., Chilian, W.M., Mitchell, M., and DeFily, D.V. (1996) Restoration of Endothelium-Dependent Vasodilation After Reperfusion Injury by Tetrahydrobiopterin, *Circulation 94*, 1423–1429.

76. Kinoshita, H., Milstien, S., Wambi, C., and Katusic, Z.S. (1997) Inhibition of Tetrahydrobiopterin Biosynthesis Impairs Endothelium-Dependent Relaxations in Canine Basilar Artery, *Am. J. Physiol. 273*, H718–H724.

77. Pieper, G.M. (1997) Acute Amelioration of Diabetic Dysfunction with a Derivative of the Nitric Oxide Synthase Cofactor, Tetrahydrobiopterin, *J. Cardiovasc. Pharmacol. 29*, 8–15.

78. Meininger, C., Marinos, R.S., Hatakeyama, K., Martinez-Zaguilan, R., Rojas, J.D., Kelly, K.A., and Wu, G. (2000) Impaired Nitric Oxide Production in Coronary Endothelial Cells of the Spontaneously Diabetic BB Rat Is Due to Tetrahydrobiopterin Deficiency, *Biochem. J. 349*, 353–356.

79. Stroes, E., Kastelein, J., Cosentino, F., Erkelens, W., Wever, R., Koomans, H., Lüscher, T., and Rabelink, T. (1997) Tetrahydrobiopterin Restores Endothelial Function in Hypercholesterolemia, *J. Clin. Investig. 99*, 41–46.

80. Heitzer, T., Brockhoff, C., Mayer, B., Warnholtz, A., Mollnau, H., Henne, S., Meinertz, T., and Münzel, T. (2000) Tetrahydrobiopterin Improves Endothelium-Dependent Vasodilation in Chronic Smokers, *Circ. Res. 86*, E36–E41.

81. Maier, W., Cosentino, F., Lutolf, R.B., Fleisch, M., Seiler, C., Hess, O.M., Meier, B., and Lüscher, T.F. (2000) Tetrahydrobiopterin Improves Endothelial Function in Patients with Coronary Artery Disease, *J. Cardiovasc. Pharmacol. 35*, 173–178.

82. Ueda, S., Matsuoka, H., Miyazaki, H., Usui, M., Okuda, S., and Imaizumi, T. (2000) Tetrahydrobiopterin Restores Endothelial Function in Long-Term Smokers, *J. Am. Coll. Cardiol. 35*, 71–75.

83. Tiefenbacher, C.P., Bleeke, T., Vahl, C., Amann, K., Vogt, A., and Kübler, W. (2000) Endothelial Dysfunction of Coronary Resistance Arteries Is Improved by Tetrahydrobiopterin in Atherosclerosis, *Circulation 102*, 2172–2179.

84. Heller, R., Münscher-Paulig, F., Gräbner, R., and Till, U. (1999) L-Ascorbic Acid Potentiates Nitric Oxide Synthesis in Endothelial Cells, *J. Biol. Chem. 274*, 8254–8260.

85. Heller, R., Unbehaun, A., Schellenberg, B., Mayer, B., Werner-Felmayer, G., and Werner, E.R. (2001) L-Ascorbic Acid Potentiates Endothelial Nitric Oxide Synthesis Via a Chemical Stabilization of Tetrahydrobiopterin, *J. Biol. Chem. 276*, 40–47.

86. Leber, A., Hemmens, B., Klösch, B., Goessler, W., Raber, G., Mayer, B., and Schmidt, K. (1999) Characterization of Recombinant Human Endothelial Nitric-Oxide Synthase Purified from the Yeast *Pichia pastoris*, *J. Biol. Chem. 274*, 37658–37664.

87. Werner, E.R., Wachter, H., and Werner-Felmayer, G. (1997) Determination of Tetrahydrobiopterin Biosynthetic Activities by High-Performance Liquid Chromatography with Fluorescence Detection, *Methods Enzymol. 281*, 53–61.

88. Levine, M., Conry-Cantilena, C., Wang, Y., Welch, R.W., Washko, P.W., Dhariwal, K.R., Park, J.B., Lazarev, A., Graumlich, J.F., King, J., and Cantilena, L. (1996) Vitamin C Pharmacokinetics in Healthy Volunteers: Evidence for a Recommended Dietary Allowance, *Proc. Natl. Acad. Sci. USA 93*, 3704–3709.

89. Huang, A., Vita, J.A., Venema, R.C., and Keaney, J.F. (2000) Ascorbic Acid Enhances Endothelial Nitric-Oxide Synthase Activity by Increasing Intracellular Tetrahydrobiopterin, *J. Biol. Chem. 275*, 17399–17406.

90. Tsukaguchi, H., Tokui, T., Mackenzie, B., Berger, U.V., Chen, X.-Z., Wang, Y., Brubaker, R.F., and Hediger, M.A. (1999) A Family of Mammalian Na^+-Dependent L-Ascorbic Acid Transporters, *Nature 399*, 70–75.

91. Daruwala, R., Song, J., Koh, W.S., Rumsey, S.C., and Levine, M. (1999) Cloning and Functional Characterization of the Human Sodium-Dependent Vitamin C Transporters hSVCT1 and hSVCT2, *FEBS Lett. 460*, 480–484.

92. Winkler, B.S., Orseili, S.M., and Rex, T.S. (1994) The Redox Couple Between Glutathione and Ascorbic Acid: A Chemical and Physiological Perspective, *Free Radic. Biol. Med. 17*, 333–349.

93. Baker, T.A., Milstien, S., and Katusic, Z.S. (2001) Effect of Vitamin C on the Availability of Tetrahydrobiopterin in Human Endothelial Cells, *J. Cardiovasc. Pharmacol. 37*, 333–338.

94. Nichol, C.A., Smith, G.K., and Duch, D.S. (1985) Biosynthesis and Metabolism of Tetrahydrobiopterin and Molybdopterin, *Annu. Rev. Biochem. 54*, 729–764.

95. Schectman, G., Byrd, J.C., and Gruchow, H.W. (1989) The Influence of Smoking on Vitamin C Status in Adults, *Am. J. Public Health 79*, 158–162.

96. Langlois, M., Duprez, D., Delanghe, J., De Buyzere, M., and Clement, D.L. (2001) Serum Vitamin C Concentration Is Low in Peripheral Arterial Disease and Is Associated with Inflammation and Severity of Atherosclerosis, *Circulation 103*, 1863–1868.

97. Hattori, Y., Nakanishi, N., Kasai, K., and Shimoda, S.-I. (1997) GTP Cyclohydrolase I mRNA Induction and Tetrahydrobiopterin Synthesis in Human Endothelial Cells, *Biochim. Biophys. Acta 1358*, 61–66.

98. Katusic, Z.S., Stelter, A., and Milstien, S. (1998) Cytokines Stimulate Cyclohydrolase I Gene Expression in Cultured Human Umbilical Vein Endothelial Cells, *Arterioscler. Thromb. Vasc. Biol. 18*, 27–32.

99. Gal, E.M., Nelson, J.M., and Sherman, A.D. (1978) Biopterin: III. Purification and Characterization of Enzymes Involved in the Cerebral Synthesis of 7,8-Dihydrobiopterin, *Neurochem. Res. 3*, 69–88.

100. Kaufman S. (1993) New Tetrahydrobiopterin-Dependent Systems, *Annu. Rev. Nutr. 13*, 261–286.

101. Toth, M., Kukor, Z., and Valent, S. (2002) Chemical Stabilization of Tetrahydrobiopterin by L-Ascorbic Acid: Contribution to Placental Endothelial Nitric Oxide Synthase Activity, *Mol. Hum. Reprod. 8*, 271–280.

102. Patel, K.B., Stratford, M.R., Wardman, P., and Everett, S.A. (2002) Oxidation of Tetrahydrobiopterin by Biological Radicals and Scavenging of the Trihydrobiopterin Radical by Ascorbate, *Free Radic. Biol. Med. 32*, 203–211.

103. Milstien, S., and Katusic, Z. (1999) Oxidation of Tetrahydrobiopterin by Peroxynitrite: Implications for Vascular Endothelial Function, *Biochem. Biophys. Res. Commun. 263*, 681–684.

104. Vasquez-Vivar, J., Kalyanaraman, B., Martasek, P., Hogg, N., Masters, B.S., Karoui, H., Tordo, P., and Pritchard, K.A. (1998) Superoxide Generation by Endothelial Nitric Oxide Synthase: The Influence of Cofactors, *Proc. Natl. Acad. Sci. USA 95*, 9220–9225.

105. Xia, Y., Tsai, A.-L., Berka, V., and Zweier, J.L. (1998) Superoxide Generation from Endothelial Nitric-Oxide Synthase. A Ca^{2+}/Calmodulin-Dependent and Tetrahydrobiopterin Regulatory Process, *J. Biol. Chem. 273*, 25804–25808.

Chapter 6

Serum Ascorbic Acid and Disease Prevalence in U.S. Adults: The Third National Health and Nutrition Examination Survey (NHANES III)

Joel A. Simon

University of California, San Francisco and the San Francisco VA Medical Center, San Francisco, CA

Introduction

Ascorbic acid is an essential nutrient required for multiple biologic functions. During the course of evolution, humans along with other primates lost the ability for the hepatic biosynthesis of ascorbic acid due to a mutation in L-gulonolactone oxidase, the enzyme that controls the conversion of glucose to ascorbic acid (1,2). This inborn error of carbohydrate metabolism for our species has resulted in the dependence of humans on dietary consumption to achieve blood and tissue levels sufficient for the maintenance of health (1). The clinical features of scurvy, resulting from severe ascorbic acid deficiency, are well described (3). In addition to preventing scurvy, ascorbic acid functions as a water-soluble antioxidant (4) and has other important biologic actions that may be unrelated to its antioxidant properties.

To examine the relation of ascorbic acid to a number of health conditions, we undertook an examination of data collected in the Third National Health and Nutrition Examination Survey (NHANES III), a cross-sectional survey based on a probability sample of the U.S. population conducted between 1988 and 1994 by the National Center for Health Statistics (5). Although NHANES III collected dietary intake information using a 24-h recall and a food-frequency questionnaire, it importantly also measured serum ascorbic acid levels using modern high-performance liquid chromatography (HPLC) (6), an improvement over the previous colorimetric assay employed in NHANES II (7). Having serum levels of ascorbic acid, which reflect dietary and supplement intake, permitted a more precise and accurate estimation of the relation between ascorbic acid and health conditions than might typically be available using dietary intake estimations only. This chapter will review and summarize the findings of our research efforts using the NHANES III data set and will present some new findings relating to the association between ascorbic acid and nontraditional cardiovascular disease (CVD) risk factors.

Serum Ascorbic Acid and Cardiovascular Disease

Ascorbic acid may reduce the risk of CVD through several mechanisms. First, ascorbic acid is a highly effective water-soluble antioxidant (4) capable of inhibiting lipid peroxidation (8–12), which has been hypothesized to be an important factor in atherogenesis (13,14). Some studies have reported that ascorbic acid may increase concentrations of high density lipoprotein cholesterol and decrease concentrations of total cholesterol, at least under certain conditions (15–18). Ascorbic acid also promotes endothelial prostacyclin (19–21) [a prostaglandin that decreases vascular tone and inhibits platelet aggregation (22)] and nitric oxide production, thereby resulting in vasodilatation (23, 24).

An association between ascorbic acid status and CVD, however, has been reported inconsistently (15,25). James Lind's (3) descriptions of sudden cardiac death among sailors with scurvy in 1757 provide the earliest intriguing evidence that ascorbic acid status may influence coronary heart disease (CHD) events. Recently published longitudinal data from the NHANES II Mortality Study, in fact, suggest that low to marginally low serum levels of ascorbic acid may indeed be a risk factor for CHD mortality (25).

In our examination of data from the earlier NHANES II, conducted between 1976 and 1980, we reported that serum ascorbic acid levels were independently associated with a decreased prevalence of self-reported CHD (defined as angina and myocardial infarction) and stroke; a 0.5 mg/dL increase in serum ascorbic acid level was associated with an 11% reduction in CHD and stroke prevalence (26). In NHANES III, we detected an interaction between serum ascorbic acid concentration and alcohol intake; thus, we performed analyses stratified by drinking status (27). Among participants who reported no alcohol consumption, serum ascorbic acid concentrations were not associated with CVD prevalence. However, among participants who consumed alcohol, serum ascorbic acid concentrations consistent with tissue saturation (1.0–3.0 mg/dL) were associated with a decreased prevalence of angina [multivariate odds ratio (OR) = 0.48; 95% confidence interval (CI) 0.23–1.03; P for trend = 0.06], but were not significantly associated with myocardial infarction or stroke prevalence. Because the metabolism of alcohol may be linked to ascorbic acid status, a biologic interaction is indeed plausible (28). Blood ascorbic acid concentrations have been reported to be strongly correlated with the activity of hepatic alcohol dehydrogenase, the principal enzyme in alcohol metabolism (29), and both alcohol and ascorbic acid are known to have vasodilatory properties (19,30). It is conceivable, therefore, that alcohol and ascorbic acid may interact synergistically to affect angina, as we found in NHANES III.

Serum Ascorbic Acid and Gallbladder Disease

Gallbladder disease is highly prevalent among adult Americans, and as many as 20 million Americans are estimated to have gallstones (31). Most gallstones are composed either partially or entirely of cholesterol (31–33) and form when bile that is supersaturated with cholesterol becomes destabilized (34). In the guinea pig, an animal that like humans lacks the ability to biosynthesize ascorbic acid (35), ascorbic acid affects the activity of cholesterol 7α-hydroxylase, the enzyme regulating the rate-limiting step in the catabolism of cholesterol to bile acids (36–39). Ascorbic acid supplementation increases cholesterol 7α-hydroxylase activity by as much as 15-fold compared with ascorbic acid–deficient guinea pigs (39), which typically develop cholesterol gallstones (40–43). Because of the animal evidence and the observation that gallbladder disease risk factors in humans are frequently associated with ascorbic acid status, I hypothesized in 1993 that ascorbic acid status might also be a risk factor for human gallbladder disease (44).

To ascertain whether ascorbic acid status is associated with gallbladder disease in humans, and particularly with the presence of asymptomatic gallstones, we analyzed data collected in NHANES III that included serum ascorbic acid levels and information on gallbladder disease among >13,000 American adults. A total of 11% of women and 4% of men had a history of clinical gallbladder disease (i.e., either symptomatic gallstones or a cholecystectomy). Of the NHANES III participants without a history of clinical gallbladder disease (or abdominal pain consistent with gallbladder disease), 8% of women and 6% of men had asymptomatic gallstones. Among women, serum ascorbic acid level was inversely related to the prevalence of clinical and asymptomatic gallbladder disease independent of other gallbladder disease risk factors, such as age, race, diet, and body mass index. Each 0.5 mg/dL increase in serum ascorbic acid levels was associated with an ~13% lower prevalence of both clinical gallbladder disease ($P = 0.006$) and asymptomatic gallstones ($P < 0.05$) (45). We found no significant relation between serum ascorbic acid level and gallbladder disease among men although the OR estimates were consistent with a small protective effect (0.97 for clinical gallbladder disease and 0.91 for asymptomatic gallstones). The findings among men may reflect the lower prevalence of gallbladder disease among men and, as a consequence, the associated reduced statistical power to detect such an association.

The data on asymptomatic gallstones are particularly important. From the NHANES III cross-sectional data, we cannot exclude the possibility that participants changed their diets after being told they had gallbladder disease. Although it seems unlikely that women would consume less ascorbic acid–containing foods and supplements as a consequence of learning of gallbladder disease (thereby producing our findings), this possibility cannot be excluded. However, because NHANES III also collected abdominal ultrasound data on its participants, our findings relating to the presence of asymptomatic gallstones strengthen the hypothesis that ascorbic acid status may indeed be an important risk factor for gallstone formation, at least among women. Furthermore, the magnitude of the association between serum ascorbic acid levels and gallbladder disease prevalence was the same for both clinical and asymptomatic gallbladder disease among women, providing additional circumstantial evidence that the observed association reflects a true biological relationship. These NHANES III findings are the fourth study to report an association between ascorbic acid intake or blood levels and prevalence of gallbladder disease in women (45–48). It is the only one to have data on the presence of asymptomatic gallstones.

Serum Ascorbic Acid, Bone Mineral Density, and Fractures

Osteoporosis, affecting both women and men, is highly prevalent among older Americans, and is an important risk factor for clinical fractures (49). Ascorbic acid deficiency has been associated with decreased collagen synthesis and bone mineral density (BMD) in a few experimental animal studies (50,51). In animals, such deficiency affects vitamin D metabolism and binding and, in turn, the risk of osteoporosis (52). Ascorbic acid has also been reported to affect markers of osteoblast activity (53). In humans, some observational studies (54–61), but not all (62) have reported an association between ascorbic acid intake or blood levels and BMD. Although marked abnormalities in bone metabolism and growth have been described among children with scurvy (63), the relation of ascorbic acid to BMD over a wide range of intakes and blood levels has not been reported previously among a representative sample of the U.S. population. On the basis of the observation that ascorbic acid is a nutrient essential for collagen formation and normal bone development (63), we hypothesized that ascorbic acid status would be associated with BMD.

We analyzed data collected in NHANES III to examine whether dietary ascorbic acid intake and serum ascorbic acid levels were associated with BMD (measured at the proximal femur) and the prevalence of self-reported fractures of the hip, wrist, and spine (64). Because other investigators have described possible interactions between ascorbic acid and calcium intake (60), smoking (61), and postmenopausal estrogen therapy (59) on BMD or fracture, we also were interested in exploring whether these factors modified the association between ascorbic acid, BMD, and self-reported fractures. We identified three-way interactions between smoking, history of estrogen use, and dietary and serum ascorbic acid among postmenopausal women and therefore, analyzed the data for postmenopausal women stratified by smoking and estrogen use. We found that dietary ascorbic acid intake was independently associated with greater BMD among premenopausal women ($P = 0.002$). Among men, serum ascorbic acid was associated in a nonlinear fashion with BMD ($P < 0.05$), and dietary ascorbic acid intake was associated in a nonlinear fashion with self-reported fracture ($P = 0.05$) (see Fig. 6.1). Among postmenopausal women without a history of smoking or estrogen use, serum ascorbic acid was unexpectedly associated with lower BMD ($P = 0.01$). However, among postmenopausal women with a history of both smoking and estrogen use, serum ascorbic acid was associated with a 49% decrease in fracture prevalence ($P = 0.001$).

Ascorbic acid deficiency has been associated with osteoporosis in Black South African men in several older studies (54–56). More recent studies have examined dietary and supplement intake of ascorbic acid as a correlate of BMD (58–60, 65–68) or hip fracture (61). With the exception of the Honolulu Heart Study, which included men of Japanese ancestry (65), these other studies included only women (58–61,66–68) and none examined the relation of blood ascorbic acid levels to BMD or fracture. Our findings suggest that the relation between ascorbic acid, BMD, and prevalence of fractures differs between men and women and may be modified among postmenopausal women by use of tobacco and postmenopausal hormones.

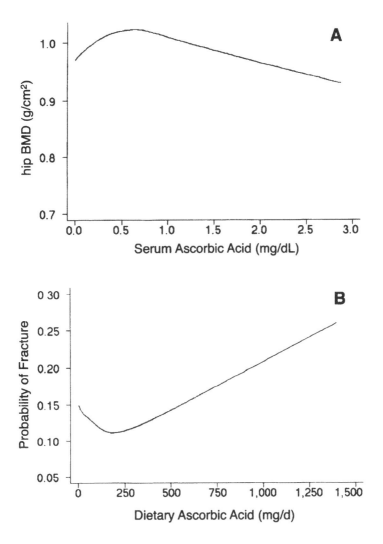

Fig. 6.1. Serum ascorbic acid and hip bone mineral density (A) and dietary ascorbic acid and self-reported fracture (B) among men. The relation of predicted hip bone mineral density (BMD) (gm/cm^2) among men enrolled in the Third National Health and Nutrition Examination Survey (NHANES III), 1988–1994 is expressed as a function of serum ascorbic acid (mg/dL) (A) and that of self-reported fracture as a function of dietary ascorbic acid intake (B). The figures are based on a multivariate model that adjusted for age, race, level of education, physical activity, body mass index, use of thiazide diuretics, dietary intake of calories, fat, protein, calcium, caffeine, and alcohol; history of smoking (never/past/current), history of diabetes, and serum levels of thyroid stimulating hormone, vitamin D, and vitamin E. Multivariate analyses revealed significant linear and quadratic terms for serum ascorbic acid and dietary ascorbic acid intake (both $P < 0.05$).

Serum Ascorbic Acid and Lead Poisoning

Lead pollution is an important public health problem (69). Because millions of American children are believed to have elevated blood lead levels, screening programs for childhood lead exposure have been established by the Centers for Disease Control (69). In 1984, as many as 3–4 million children were estimated to have blood lead levels >15 μg/dL (70) and even lower levels of lead exposure among children have been associated with adverse neuropsychological development (71). Work-related lead exposure has also been targeted as an area of concern by the Occupational Safety and Health Administration (69).

Calcium EDTA and other chelators are standard treatments for lead poisoning. Several animal studies, have examined the effect of ascorbic acid on lead toxicity. In rats fed a lead-containing diet, combined dietary supplementation with iron and ascorbic acid prevented growth depression and anemia, and lowered detectable lead levels in liver, kidney, and bone (72). These beneficial effects, however, were transient (73). Comparing the chelating effects of oral ascorbic acid and parenterally administered EDTA in lead-poisoned rats, Goyer and Cherian (74) reported that ascorbic acid and EDTA had equivalent chelating properties. The few published older studies among humans have yielded inconsistent results. Two case series reported significant clinical improvement among 337 workers with occupational lead exposure after daily administration of 100 mg of ascorbic acid (75,76). In an uncontrolled trial, the combined administration of zinc and ascorbic acid was reported to have reduced blood lead levels among 1000 psychiatric outpatients (77). In another study of 85 subjects who volunteered to consume a lead-containing drink, ascorbic acid supplementation produced small reductions in lead retention (78). Two other small clinical trials, however, concluded that ascorbic supplementation did not lower blood lead levels (79,80). In one of two these studies, 52 subjects were assigned to receive either ascorbic acid supplementation or placebo (79). Although reported to have no effect, 8 wk of ascorbic acid supplementation resulted in (nonsignificant) improvement. Compared with the placebo group, subjects treated with ascorbic acid were more likely to have a decrease of >5 μg/dL in blood lead level (relative risk = 0.64; 95% CI, 0.36–1.11; $P = 0.10$). In the second study of 45 men (80), 3 mo of ascorbic acid treatment resulted in (nonsignificant) 11–23% lower blood lead levels compared with levels in the placebo-treated group. It is possible that these two studies might have yielded significant results had larger numbers of participants been enrolled. Most importantly, a recent small randomized trial by Dawson and colleagues that studied 75 male smokers revealed that participants randomized to receive 1000 mg of ascorbic acid daily had an 81% decrease in their blood lead levels after 4 wk of treatment ($P < 0.001$) (81). Because there was no difference in urinary lead levels, the authors concluded that ascorbic acid supplementation likely decreased the absorption of lead from the gastrointestinal tract.

To ascertain whether ascorbic acid status was associated with blood lead levels and particularly with prevalence of elevated blood lead levels, we analyzed data collected in NHANES III that included serum ascorbic acid levels and blood lead levels for >19,000 Americans. To our knowledge, there have been no previous reports examining the relation between ascorbic acid and lead toxicity among a population-based sample of American adults and youth.

We found that a total of 57 adults (0.4%) and 22 youth (0.5%) (6–17 y old) had elevated lead levels (≥20 μg/dL for adults and ≥15 μg/dL for youth). Blood lead levels ranged from 0.5 to 56 μg/dL among adults and from 0.5 to 48.9 μg/dL among youth (82). Among youth, unadjusted, age-adjusted, and multivariate models revealed that serum ascorbic acid levels were inversely associated with prevalence of elevated blood lead levels. Youth in the highest serum ascorbic acid tertile were 89% less likely (95% CI, 65–96) to have elevated blood lead levels compared with youth in the lowest serum ascorbic acid tertile (multivariate P for trend < 0.002). Among adults, serum ascorbic acid level was also associated with prevalence of elevated blood lead levels; all models revealed that serum ascorbic acid levels were inversely associated with the prevalence of elevated blood lead levels. Compared with adults in the lowest serum ascorbic acid tertile, adults in the upper two tertiles were ~65–68% less likely to have elevated blood lead levels (multivariate P for trend = 0.03); ~4% of youth and 2% of adults with the lowest serum ascorbic acid levels had elevated blood lead levels (see Fig. 6.2).

In the context of the recent report by Dawson and colleagues (81) and the previous, albeit inconsistent, older reports regarding the effect of ascorbic acid on blood lead levels, we believe that the relation that we observed in NHANES III is likely real and not the result of residual confounding or bias. The increased consumption of ascorbic acid–containing foods and supplements represents a low-risk and potentially highly efficacious public health intervention with the potential to lower blood lead levels of the American population, including those segments of the population most at risk.

Serum Ascorbic Acid and Nontraditional CVD Risk Factors

The antioxidant theory of atherosclerosis hypothesizes, in part, that the oxidative modification of low density lipoprotein cholesterol over time results in vascular damage and ultimately, atherosclerosis (13,83). If this theory is correct, low blood levels of antioxidants, such as ascorbic acid, may be important modifiable risk factors for vascular disease (84). Traditional CVD risk factors, such as high blood cholesterol, hypertension, diabetes, and smoking, only partially account for CVD incidence, and additional risk factors continue to be identified. Some of these nontraditional CVD risk factors include markers of inflammation, such as C-reactive protein, plasma fibrinogen, and leukocyte count (85). The mechanisms linking some other factors, such as serum uric acid, homocysteine, creatinine, and albumin, to CVD have yet to be fully explained.

Because antioxidant status may be correlated with inflammation (86–88), and additionally, because several studies have indicated that ascorbic acid may have a uricosuric effect (89–91), we were interested in examining the relation of ascorbic acid to several nontraditional CVD risk factors. We postulated that if significant associations were identified, then controlling for differences in ascorbic acid status might be

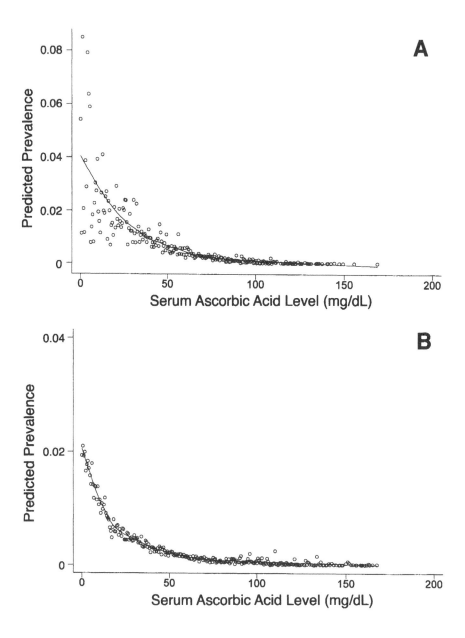

Fig. 6.2. The relation between serum ascorbic acid concentration in mg/dL and prevalence of elevated blood lead levels among (A) 4213 youth ages 6–16 y and (B) 15,365 adults age ≥ 17 y enrolled in the Third National Health and Nutrition Examination Survey (NHANES III), 1988–1994 based on a multivariate model.

important when considering whether these nontraditional risk factors were independent predictors of CVD. To ascertain whether serum ascorbic acid levels are associated with nontraditional CVD risk factors, we again analyzed data collected from NHANES III.

A total of 7345 women and 6390 men, aged 20–90 y enrolled in NHANES III with complete data were available for these analyses. Analyses of the relation of serum ascorbic acid levels to nontraditional CVD risk factors are presented in Table 6.1 for women and Table 6.2 for men. Among both women and men, serum ascorbic acid levels were inversely associated with serum levels of creatinine (P for trend < 0.001) and homocysteine (both P for trend ≤ 0.02) after adjustment for potential confounders. Among women, serum ascorbic acid levels were also inversely associated with plasma levels of fibrinogen (P for trend < 0.01), whereas among men, serum ascorbic acid levels were directly associated with higher levels of serum albumin (P for trend = 0.02). We found no association, however, between serum ascorbic acid levels and levels of serum uric acid or white blood cell count.

Most participants in NHANES III had nondetectable C-reactive protein levels. Using multivariable logistic regression models, we examined whether serum ascorbic acid levels were associated with detectable levels of C-reactive protein. Compared with women in the lowest serum ascorbic acid quartile, women in the highest quartile had a (nonsignificant) decreased odds of having detectable levels of C-reactive protein (OR = 0.81; 95% CI, 0.64–1.02; P = 0.08). Using multivariable linear regression models, serum ascorbic acid was inversely associated with C-reactive protein levels

TABLE 6.1

Relation of Serum Ascorbic Acid to Nontraditional Cardiovascular Disease Risk Factors Among 7345 Women Enrolled in the Third National Health and Nutrition Survey (NHANES III) 1988(1994

	Quartile of serum ascorbic acid				
	(n = 1894)	(n = 1835)	(n = 1791)	(n = 1825)	*P* for trend
Median (mg/dL)	0.24	0.68	0.97	1.30	
(Range)	(0.0–0.47)	(0.48–0.83)	(0.84–1.11)	(1.12–2.95)	
Adjusted mean levels[a]					
Log serum albumin (g/dL)	1.48	1.48	1.49	1.49	0.16
Log serum creatinine (mg/dL)[b]	–0.028	–0.034	–0.054	–0.060	<0.001
Plasma fibrinogen (mg/dL)[b]	302.7	298.2	293.5	286.5	<0.01
Log serum homocysteine (μmol/L)[b]	2.12	2.06	2.01	1.98	<0.001
Log serum uric acid (mg/dL)	1.73	1.74	1.74	1.71	0.11
Log leukocyte count (× 10⁹/L)	1.82	1.82	1.81	1.81	0.30

[a]Adjusted for age, race, body mass index, level of physical activity, current tobacco use, level of education, history of hypertension, history of diabetes, alcohol consumption, and aspirin use in the prior month. Uric acid analyses are additionally adjusted for diuretic use; homocysteine analyses are additionally adjusted for red blood cell folate, dietary intake of vitamin B₆, and serum vitamin B₁₂ levels.
[b]The number of women participants with complete data available for these analyses was 7329 for serum creatinine, 4085 for plasma fibrinogen, and 3605 for serum homocysteine.

TABLE 6.2
Relation of Serum Ascorbic Acid to Nontraditional Cardiovascular Disease Risk Factors Among 6390 Men Enrolled in the Third National Health and Nutrition Survey (NHANES III) 1988–1994

	Quartile of serum ascorbic acid				
	(n = 1638)	(n = 1587)	(n = 1614)	(n = 1551)	P for trend
Median (mg/dL)	0.16	0.52	0.81	1.12	
(Range)	(0.0–0.32)	(0.33–0.67)	(0.68–0.95)	(0.96–2.87)	
Adjusted mean levels[a]					
Log serum albumin (g/dL)	1.52	1.53	1.53	1.54	0.02
Log serum creatinine (mg/dL)[b]	0.231	0.229	0.214	0.198	<0.001
Plasma fibrinogen (mg/dL)[b]	287.9	286.5	287.4	286.8	0.90
Log serum homocysteine (μmol/L)[b]	2.38	2.37	2.31	2.30	0.02
Log serum uric acid (mg/dL)	1.95	1.98	1.97	1.94	0.29
Log leukocyte count (× 10^9/L)	1.82	1.79	1.80	1.79	0.10

[a]Adjusted for age, race, body mass index, level of physical activity, current tobacco use, level of education, history of hypertension, history of diabetes, alcohol consumption, and aspirin use in the prior month. Uric acid analyses are additionally adjusted for diuretic use; homocysteine analyses are additionally adjusted for red blood cell folate, dietary intake of vitamin B_6, and serum vitamin B_{12} levels.
[b]The number of men participants with complete data available for these analyses was 6371 for serum creatinine, 3738 for plasma fibrinogen, and 2721 for serum homocysteine.

among women with measurable levels ($P < 0.001$). Men in the highest serum ascorbic acid quartile had a significant decrease in the odds of having detectable levels of C-reactive protein compared with men in the lowest serum ascorbic acid quartile (OR = 0.67; 95% CI, 0.49–0.91; $P = 0.01$). Similar to the findings among women, serum ascorbic acid levels were independently and inversely associated with C-reactive protein levels among men with measurable levels ($P < 0.01$).

Serum Ascorbic and Helicobacter pylori

Chronic infection with *Helicobacter pylori* is an important risk factor for peptic ulcer disease (92) and possibly gastric cancer (93,94). Some epidemiologic studies have also linked lower dietary ascorbic acid consumption with an increased risk for gastric cancer (95,96). In one animal study, increased consumption of ascorbic acid inhibited the growth of *Helicobacter pylori* (97). To ascertain whether serum ascorbic acid is associated with serologic evidence of infection with *Helicobacter pylori* and particularly with strains expressing the *cagA* virulence factor, we recently analyzed data collected from NHANES III that included serum ascorbic acid levels and *Helicobacter pylori* serology for >6000 American adults (unpublished results).

Conclusions

Although ascorbic acid functions as an important water-soluble antioxidant, it has effects separate and apart from its antioxidant properties. We used the NHANES III data set to examine the relation of ascorbic acid to CVD and nontraditional CVD risk factors, gallbladder disease, lead toxicity, bone mineral density, and serologic evidence of infection with *Helicobacter pylori*.

These analyses have a number of strengths and limitations. First, NHANES III data are available for analysis to the general public for analysis, with the caveat that special statistical software and programming expertise are required. Because NHANES III surveyed a large probability sample of Americans using standardized questionnaire and laboratory protocols, findings derived from NHANES III should be generalizable to the U.S. population. Quantitative dietary information was limited by a single 24-h dietary recall. However, NHANES III measured serum ascorbic acid levels using modern HPLC on a very large sample of the U.S. population, thereby permitting a more reliable assessment of ascorbic acid status as a correlate of disease. However, as with other observational epidemiologic studies, we cannot exclude the possibility of residual confounding. Finally, the cross-sectional nature of survey data mandates that inferences regarding causality be made cautiously.

Data generated from analyses of observational epidemiologic studies, such as those from NHANES III, have to be integrated with evidence from bench research and animal studies. In the final analysis, hypotheses generated from such analyses will still require (when ethical and feasible) to be confirmed in humans using modern clinical trial methodology.

References

1. Wilson, J.D. (1987) *Harrison's Principles of Internal Medicine*, 11th edn. (Braunwald, E., Isselbacher, K.J., Petersdorf, R.G., Wilson, J.D., Martin, J.B., and Fauci, A.S., eds.) p. 414, McGraw-Hill Book Company, New York.
2. Mann, G.V. (1974) Hypothesis: The Role of Vitamin C in Diabetic Angiopathy, *Perspect. Biol. Med. 17*, 210–217.
3. Lind, J. (1757) *A Treatise of the Scurvy*, 2nd edn., Millar, London.
4. Frei, B., Stocker, R., England, L., and Ames, B.N. (1990) Ascorbate: The Most Effective Antioxidant in Human Plasma, *Adv. Exp. Med. Biol. 264*, 155–163.
5. Anonymous (1994) Plan and Operation of the Third National Health and Nutrition Examination Survey, (1988–94), National Center for Health Statistics, *Vital Health Stat. Series 1*, no. 32.
6. Gunter, E.W., Lewis, B.G., and Koncikowski, S.M. (1996) Laboratory Procedures Used for the Third Health and Nutrition Examination Survey (NHANES III), 1988–1994, U.S. Department of Health and Human Services, PHS, Centers for Disease Control and Prevention, Center for Environmental Health, Atlanta, GA and National Center for Health Statistics, Hyattsville, MD.
7. Gunter, E.W., Turner, W.E., Neese, J.W., and Bayse, D.D. (1985) Laboratory Procedures Used by the Clinical Chemistry Division, Centers for Disease Control, for the Second Health and Nutrition Examination Survey (HANES II), 1976–1980, U.S. Dept of Health and Human Services, PHS, Centers for Disease Control, Center for Environmental Health, Nutritional Biochemistry Branch, Atlanta, GA.
8. Jialal, I., Vega, G.L., and Grundy, S.M. (1990) Physiologic Levels of Ascorbate Inhibit the Oxidative Modification of Low Density Lipoprotein, *Atherosclerosis 82*, 185–191.

9. Harats, D., Ben-Naim, M., Dabach, Y., Hollander, G., Havivi, E., Stein, O., and Stein, Y. (1990) Effect of Vitamin C and E Supplementation on Susceptibility of Plasma Lipoproteins to Peroxidation Induced by Acute Smoking, *Atherosclerosis 85*, 47–54.

10. Frei, B. (1991) Ascorbic Acid Protects Lipids in Human Plasma and Low-Density Lipoprotein Against Oxidative Damage, *Am. J. Clin. Nutr. 54*, 1113S–1118S.

11. Frei, B., Forte, T.M., Ames, B.N., and Cross, C.E. (1991) Gas Phase Oxidants of Cigarette Smoke Induce Lipid Peroxidation and Changes in Lipoprotein Properties in Human Blood Plasma. Protective Effects of Ascorbic Acid, *Biochem. J. 277*, 133–138.

12. Retsky, K.L., and Frei, B. (1995) Vitamin C Prevents Metal Ion-Dependent Initiation and Propagation of Lipid Peroxidation in Human Low-Density Lipoprotein, *Biochim. Biophys. Acta 1257*, 279–287.

13. Gey, K.F. (1986) On the Antioxidant Hypothesis with Regard to Arteriosclerosis, *Bibl. Nutr. Dieta 37*, 53–91.

14. Steinberg, D. (1991) Antioxidants and Atherosclerosis: A Current Assessment, *Circulation 84*, 1420–1425.

15. Simon, J.A. (1992) Vitamin C and Cardiovascular Disease: A Review, *J. Am. Coll. Nutr. 11*, 107–125.

16. Simon, J.A., Schreiber, G.B., Crawford, P.B., Frederick, M.M., and Sabry, Z.I. (1993) Dietary Vitamin C and Serum Lipids in Black and White Girls, *Epidemiology 4*, 537–542.

17. Jacques, P.F., Sulsky, S.I., Perrone, G.A., and Schaefer, E.J. (1994) Ascorbic Acid and Plasma Lipids, *Epidemiology 5*, 19–26.

18. Hallfrisch, J., Singh, V.N., Muller, D.C., Baldwin, H., Bannon, M.E., and Andres, R. (1994) High Plasma Vitamin C Associated with High Plasma HDL- and HDL_2 Cholesterol, *Am. J. Clin. Nutr. 60*, 100–105.

19. Beetens, J.R., and Herman, A.G. (1983) Vitamin C Increases the Formation of Prostacyclin by Aortic Rings from Various Species and Neutralizes the Inhibitory Effect of 15-Hydroxy-Arachidonic Acid, *Br. J. Pharmacol. 80*, 249–254.

20. Srivastava, K.C. (1985) Ascorbic Acid Enhances the Formation of Prostaglandin E_1 in Washed Human Platelets and Prostacyclin in Rat Aortic Rings, *Prostaglandins Leukot. Med. 18*, 227–233.

21. Toivanen, J.L. (1987) Effects of Selenium, Vitamin E and Vitamin C on Human Prostacyclin and Thromboxane Synthesis in Vitro. *Prostaglandins Leukot. Med. 26*, 265–280.

22. Lefer, A.M. (1990) Prostacyclin, High Density Lipoproteins, and Myocardial Ischemia, *Circulation 81*, 2013–2015.

23. Ting, H.H., Timimi, F.K., Haley, E.A., Roddy, M.-A., Ganz, P., and Creager. M.A. (1997) Vitamin C Improves Endothelium-Dependent Vasodilation in Forearm Resistance Vessels of Humans with Hypercholesterolemia, *Circulation 95*, 2617–2622.

24. Timimi, F.K., Ting, H.H., Haley, E.A., Roddy, M.-A., Ganz, P., and Creager, M.A. (1998) Vitamin C Improves Endothelium-Dependent Vasodilation in Patients with Insulin-Dependent Diabetes Mellitus, *J. Am. Coll. Cardiol. 31*, 552–557.

25. Simon, J.A., Hudes, E.S., and Tice, J.A. (2001) Relation of Serum Ascorbic Acid to Mortality Among US Adults, *J. Am. Coll. Nutr. 20*, 255–263.

26. Simon, J.A., Hudes, E.S., Browner, W.S. (1998) Serum Ascorbic Acid and Cardiovascular Disease Prevalence in US Adults, *Epidemiology 9*, 316–321.

27. Simon, J.A., Hudes, and E.S. (1999) Serum Ascorbic Acid and Cardiovascular Disease Prevalence in US Adults: The Third National Health and Nutrition Examination Survey (NHANES III), *Ann. Epidemiol. 9*, 358–365.

28. Ginter, E., Zloch, Z., and Ondreicka, R. (1998) Influence of Vitamin C Status on Ethanol Metabolism in Guinea-Pigs, *Physiol. Res. 47*, 137–141.
29. Krasner, N., Dow, J., Moore, M.R., and Goldberg, A. (1974) Ascorbic-Acid Saturation and Ethanol Metabolism, *Lancet 2*, 693–695.
30. Ting, H.H., Timimi, F.K., Boles, K.S., Creager, S.J., Ganz, P., and Creager, M.A. (1996) Vitamin C Improves Endothelium-Dependent Vasodilation in Patients with Non-Insulin-Dependent Diabetes Mellitus, *J. Clin. Investig. 97*, 22–28.
31. Anonymous (1992) Gallstones and Laparoscopic Cholecystectomy, NIH Consensus Statement, September 14–16, 1992, Vol. 10, No. 3, pp. 1–26.
32. Trotman, B.W., Ostrow, J.D., and Soloway, R.D. (1974) Pigment vs Cholesterol Cholelithiasis: Comparison of Stone and Bile Composition, *Am. J. Dig. Dis. 19*, 585–590.
33. Trotman, B.W., and Soloway, R.D. (1975) Pigment vs Cholesterol Cholelithiasis: Clinical and Epidemiological Aspects, *Am. J. Dig. Dis. 20*, 735–740.
34. Paumgartner, G, and Sauerbruch, T. (1991) Gallstones: Pathogenesis, *Lancet 338*, 1117–1121.
35. Chatterjee, I.B., Majumder, A.K., Nandi, B.K., and Subramanian, N. (1975) Synthesis and Some Major Functions of Vitamin C in Animals, *Ann. N.Y. Acad. Sci. 258*, 24–46.
36. Ginter, E., Cerven, J., Nemec, R., and Mikul, L. (1971) Lowered Cholesterol Catabolism in Guinea Pigs with Chronic Ascorbic Acid Deficiency, *Am. J. Clin. Nutr. 24*, 1238–1245.
37. Ginter, E. (1975) Ascorbic Acid in Cholesterol and Bile Acid Metabolism, *Ann. N.Y. Acad. Sci. 258*, 410–421.
38. Hornig, D, and Weiser, H. (1976) Ascorbic Acid and Cholesterol: Effect of Graded Oral Intakes on Cholesterol Conversion to Bile Acids in Guinea Pigs, *Experientia 32*, 687–689.
39. Björkhem, I., and Kallner, A. (1976) Hepatic 7-Alpha Hydroxylation of Cholesterol in Ascorbate-Deficient and Ascorbate-Supplemented Guinea Pigs, *J. Lipid Res. 17*, 360–365.
40. Jenkins, S.A. (1977) Vitamin C and Gallstone Formation: A Preliminary Report, *Experentia 33*, 1616–1617.
41. Jenkins, S.A. (1977) Hypovitaminosis C and Cholelithiasis in Guinea Pigs, *Biochem. Biophys. Res. Commun. 77*, 1030–1035.
42. Jenkins, S.A. (1978) Biliary Lipids, Bile Acids and Gallstone Formation in Hypovitaminotic C Guinea-Pigs, *Br. J. Nutr. 40*, 317–322.
43. Ginter, E., Bobek, P., Kubec, F., Vozár, J., and Urbanová, D. (1982) Vitamin C in the Control of Hypercholesterolemia in Man, *Int. J. Vitam. Nutr. Res. Suppl. 23*, 137–152.
44. Simon, J.A. (1993) Ascorbic Acid and Cholesterol Gallstones, *Med. Hypotheses 40*, 81–84.
45. Simon, J.A., and Hudes, E.S. (2000) Serum Ascorbic Acid and Gallbladder Disease Prevalence Among U.S. Adults: The Third National Health and Nutrition Examination Survey (NHANES III), *Arch. Intern. Med. 160*, 931–936.
46. Ortega, R.M., Fernández-Azuela, M., Encinas-Sotillos, A., Andrés, P., and López-Sobaler, A.M. (1997) Differences in Diet and Food Habits Between Patients with Gallstones and Controls, *J. Am. Coll. Nutr. 16*, 88–95.
47. Simon, J.A., Grady, D., Snabes, M.C., Fong, J., and Hunninghake, D.B. (1998) Ascorbic Acid Supplement Use and the Prevalence of Gallbladder Disease, *J. Clin. Epidemiol. 51*, 257–265.
48. Simon, J.A., and Hudes, E.S. (1998) Serum Ascorbic Acid and Other Correlates of Gallbladder Disease Among U.S. Adults, *Am. J. Public Health 88*, 1208–1212.

49. Committee on Diet and Health, Food and Nutrition Board, National Research Council (1989) *Diet and Health: Implications for Reducing Chronic Disease Risk*, National Academy Press, Washington.

50. Kipp, D.E., Grey, C.E., McElvain, M.E., Kimmel, D.B., Robinson, R.G., and Lukert, B.P. (1996) Long-Term Low Ascorbic Acid Intake Reduces Bone Mass in Guinea Pigs, *J. Nutr. 126*, 2044–2049.

51. Kipp, D.E., McElvain, M., Kimmel, D.B., Akhter, M.P., Robinson, R.G., and Lukert, B.P. (1996) Scurvy Results in Decreased Collagen Synthesis and Bone Density in the Guinea Pig Animal Model, *Bone 18*, 281–288.

52. Sergeev, I.N., Arkhapchev, Y.P., and Spirichev, V.B. (1990) Ascorbic Acid Effects on Vitamin D Hormone Metabolism and Binding in Guinea Pigs, *J. Nutr.* 120, 1185–1190.

53. Franceschi, R.T. (1992) The Role of Ascorbic Acid in Mesenchymal Differentiation, *Nutr. Rev. 50*, 65–70.

54. Grusin, H., and Samuel, E. (1957) A Syndrome of Osteoporosis in Africans and Its Relationship to Scurvy, *Am. J. Clin. Nutr. 5*, 644–650.

55. Seftel, H.C., Malkin, C., Schmaman, A., Abrahams, C., Lynch, S.R., Charlton, R.W., and Bothwell, T.H. (1966) Osteoporosis, Scurvy, and Siderosis in Johannesburg Bantu, *Br. Med. J. 1*, 642–646.

56. Lynch, S.R., Berelowitz, I., Seftel, H.C., Miller, G.B., Krowitz, P., Charlton, R.W., and Bothwell, T.H. (1967) Osteoporosis in Johannesburg Bantu Males. Its Relationship to Siderosis and Ascorbic Acid Deficiency, *Am. J. Clin. Nutr. 20*, 799–807.

57. Odland, L.M., Mason, R.L., and Alexeff, A.I. (1972) Bone Density and Dietary Findings of 409 Tennessee Subjects. I. Bone Density Considerations, *Am. J. Clin. Nutr. 25*, 905–907.

58. Hernández-Avila, M., Stampfer, M.J., Ravnikar, V.A., Willett, W.C., Schiff, I., Francis, M., Longscope, C., and McKinlay, S. M. (1993) Caffeine and Other Predictors of Bone Density Among Pre- and Perimenopausal Women, *Epidemiology 4*, 128–134.

59. Leveille, S.G., LaCroix, A.Z., Koepsell, T.D., Beresford, S.A., Van Belle, G., and Buchner, D.M. (1997) Dietary Vitamin C and Bone Mineral Density in Postmenopausal Women in Washington State, USA, *J. Epidemiol. Community Health 51*, 479–485.

60. Hall, S.L., and Greendale, G.A. (1998) The Relation of Dietary Vitamin C Intake to Bone Mineral Density: Results from the PEPI Study, *Calcif. Tissue Int. 63*, 183–189.

61. Melhus, H., Michaëlsson, K., Holmberg, L., Wolk, A., and Ljunghall, S. (1999) Smoking, Antioxidant Vitamins, and the Risk of Hip Fracture, *J. Bone Miner. Res. 14*, 129–135.

62. Bhambhani, M.M., Bates, C.J., and Crisp, A.J. (1992) Plasma Ascorbic Acid Concentrations in Osteoporotic Outpatients, *Br. J. Rheumatol.* 31, 142–143.

63. Reid, M.E. (1954) Ascorbic Acid: Effects of Deficiency, in *The Vitamins: Chemistry, Physiology, Pathology* (Sebrell, W.H., Jr., and Harris, R.S., eds.) vol. I , Academic Press, New York.

64. Simon, J.A., and Hudes, E.S. (2001) Relation of Ascorbic Acid to Bone Mineral Density and Self-Reported Fractures Among US Adults, *Am. J. Epidemiol. 154*, 427–433.

65. Yano, K., Heilbrun, L.K., Wasnich, R.D., Hankin, J.H., and Vogel, J.M. (1985) The Relationship Between Diet and Bone Mineral Content of Multiple Skeletal Sites in Elderly Japanese-American Men and Women Living in Hawaii, *Am. J. Clin. Nutr. 42*, 877–888.

66. Sowers, M., Wallace, R.B., and Lemke, J.H. (1985) Correlates of Mid-Radius Bone Density Among Postmenopausal Women: A Community Study, *Am. J. Clin. Nutr. 41*, 1045–1053.

67. Freudenheim, J.L., Johnson, N.E., and Smith, E.L. (1986) Relationships Between Usual Nutrient Intake and Bone-Mineral Content of Women 35–65 Years of Age: Longitudinal and Cross-Sectional Analysis, *Am. J. Clin. Nutr. 44*, 863–876.
68. Michaëlsson, K., Holmberg, L., Mallmin, H., Wolk, A., Bergström, R., and Ljunghall, S. (1995) Diet, Bone Mass, and Osteocalcin: A Cross-Sectional Study, *Calcif. Tissue Int. 57*, 86–93.
69. US Preventive Services Task Force (1989) *Guide to Clinical Preventive Services: An Assessment of 169 Interventions,* Williams and Wilkins, Baltimore.
70. Anonymous (1988) *The Nature and Extent of Lead Poisoning in Children in the United States: A Report to Congress,* Agency for Toxic Substances and Disease Registry, Atlanta.
71. Baghurst, P.A., McMichael, A.J., Wigg, N.R., Vimpani, G.V., Roberson, E.F., Wigg, N.R., and Roberts, R.R. (1992) Environmental Exposure to Lead and Children's Intelligence at the Age of Seven Years: The Port Pirie Cohort Study, *N. Engl. J. Med. 327*, 1279–1284.
72. Suzuki, T., and Yoshida, A. (1979) Effect of Dietary Supplementation of Iron and Ascorbic Acid on Lead Toxicity in Rats, *J. Nutr. 109*, 983–988.
73. Suzuki, T, and Yoshida, A. (1979) Effectiveness of Dietary Iron and Ascorbic Acid in the Prevention and Cure of Moderately Long-Term Lead Toxicity in Rats, *J. Nutr. 109*, 1974–1978.
74. Goyer, R.A., and Cherian, M.G. (1979) Ascorbic Acid and EDTA Treatment of Lead Toxicity in Rats, *Life Sci. 24*, 433–438.
75. Holmes, H.N., Campbell, K., and Amberg, E.J. (1939) The Effect of Vitamin C on Lead Poisoning, *J. Lab. Clin. Med. 24*, 1119–1127.
76. Marchmont-Robinson, S.W. (1940) Effect of Vitamin C on Workers Exposed to Lead Dust, *J. Lab. Clin. Med. 26*, 1478–1481.
77. Sohler, A., Kruesi, M., and Pfeiffer, C.C. (1977) Blood Lead Levels in Psychiatric Outpatients Reduced by Zinc and Vitamin C, *J. Orthomol. Psychiatry 6*, 272–276.
78. Flanagan, P.R., Chamberlain, M.J., and Valberg, L.S. (1982) The Relationship Between Iron and Lead Absorption in Humans, *Am. J. Clin. Nutr. 36*, 823–829.
79. Lauwerys, R., Roels, H., Buchet, J.P., Bernard, A.A., Verhoeven, L., and Konings, J. (1983) The Influence of Orally-Administered Vitamin C or Zinc on the Absorption of and the Biological Response to Lead, *J. Occup. Med. 25*, 668–678.
80. Calabrese, E.J., Stoddard, A., Leonard, D.A., and Dinardi, S.R. (1987) The Effects of Vitamin C Supplementation on Blood and Hair Levels of Cadmium, Lead, and Mercury, in *Third Conference on Vitamin C* (Burns, J.J., Rivers, J.M., and Machlin, L.T., eds.) vol. 498, pp. 347–353, Annals of the New York Academy of Sciences, New York.
81. Dawson, E.B., Evans, D.R., Harris, W.A., Teter, M.C., and McGanity, W.J. (1999) The Effect of Ascorbic Acid Supplementation on the Blood Lead Levels of Smokers, *J. Am. Coll. Nutr. 18*, 166–170.
82. Simon, J.A., and Hudes, E.S. (1999) Relationship of Ascorbic Acid to Blood Lead Levels, *J. Am. Med. Assoc. 281*, 2289–2293.
83. Gey, K.F. (1994) Optimum Plasma Levels of Antioxidant Micronutrients: Ten Years of Antioxidant Hypothesis on Arteriosclerosis, *Bibl. Nutr. Dieta 51*, 84–99.
84. Gey, K.F. (1998) Vitamins E Plus C and Interacting Conutrients Required for Optimal Health, *Biofactors 7*, 113–174.
85. Danesh, J., Collins, R., Appleby, P., and Peto, R. (1998) Association of Fibrinogen, C-Reactive Protein, Albumin, or Leukocyte Count with Coronary Heart Disease: Meta-Analyses of Prospective Studies, *J. Am. Med. Assoc. 279*, 1477–1482.

86. Bayeta, E., and Lau, B.H.S. (2000) Pycnogenol Inhibits Generation of Inflammatory Mediators in Macrophages, *Nutr. Res. 20*, 249–259.
87. Beharka, A.A., and Han, S.N., Adolfsson, O., Wu, D., Smith, D., Lipman, R., Cao, G., Meydani, M., and Meydani, S.N. (2000) Long-Term Dietary Antioxidant Supplementation Reduces Production of Selected Inflammatory Mediators by Murine Macrophages, *Nutr. Res. 20*, 281–296.
88. Bowie, A.G., and O'Neill, L.A.J. (2000) Vitamin C Inhibits NF-(B Activation by TNF via the Activation of p38 Mitogen-Activated Protein Kinase, *J. Immunol. 165*, 7180–7188.
89. Stein, H.B., Hasan, A., and Fox, I.H. (1976) Ascorbic Acid-Induced Uricosuria: A Consequence of Megavitamin Therapy, *Ann. Intern. Med. 84*, 385–388.
90. Berger, L., Gerson, C.D., and Yü, T.-F. (1977) The Effect of Ascorbic Acid on Uric Acid Excretion with a Commentary on the Renal Handling of Ascorbic Acid, *Am. J. Med. 62*, 71–76.
91. Ersoy, A., Dilek, K., Yavuz, M., Güllülü, M., and Yurtkuran, M. (1999) The Effect of Vitamin C on Laboratory Tests in Haemodialysis Patients: Is There a Relationship Between the Administered Vitamin C Dose and Serum Uric Acid Levels? [letter], *Neprhol. Dial. Transplant. 14*, 2529–2530.
92. Anonymous (1994) NIH Consensus Conference on *Helicobacter pylori* in Peptic Ulcer Disease, *Helicobacter pylori* in Peptic Ulcer Disease, *J. Am. Med. Assoc. 272*, 65–69.
93. Parsonnet, J., Friedman, G.D., Vandersteen, D.P., Chang, Y., Vogelman, J.H., Orentreich, N., and Silbley, R.K. (1991) *Helicobacter pylori* Infection and the Risk of Gastric Carcinoma, *N. Engl. J. Med. 325*, 1127–1131.
94. Blaser, M.J., Perez-Perez, G.I., Kleanthous, H., Cover, T.L., Peek, R.M., Chyou, P.H., Stemmerman, G.N., and Nomura, A. (1995) Infection with *Helicobacter pylori* Strains Possessing *cagA* Is Associated with an Increased Risk for Developing Adenocarcinoma of the Stomach, *Cancer Res. 55*, 2111–2115.
95. Cohen, M., and Bhagavan, H.N. (1995) Ascorbic Acid and Gastrointestinal Cancer, *J. Am. Coll. Nutr. 14*, 565–578.
96. Flagg, E.W., Coates, R.J., and Greenberg, R.S. (1995) Epidemiologic Studies of Antioxidants and Cancer in Humans, *J. Am. Coll. Nutr. 14*, 419–427.
97. Zhang, H.-M., Wakisaka, N., Maeda, O., and Yamamoto, T. (1997) Vitamin C Inhibits the Growth of a Bacterial Risk Factor for Gastric Carcinoma. *Helicobacter pylori*, *Cancer 80*, 1897–1903.

Chapter 7

Vitamin C Status and Cardiovascular Disease: A Review of Prospective Studies

Catherine M. Loria

Division of Epidemiology and Clinical Applications, National Heart Lung and Blood Institute, Bethesda, MD 20892–7934

Introduction

Vitamin C is an essential nutrient for humans and its deficiency has long been known to cause scurvy (1). Less well understood is the role vitamin C plays in preventing cardiovascular disease (CVD), the leading cause of mortality in the United States for over 50 years (2). As an antioxidant, vitamin C may protect lipids, particularly low density lipoproteins (LDL), from oxidation. Oxidized LDL may contribute to the development of atherosclerotic lesions through several mechanisms (3,4). The recruitment of monocytes to the vascular intima is increased by oxidized LDL and monocytes can develop into macrophages. Oxidized LDL are more readily taken up by macrophages than nonoxidized LDL, a process in which macrophages can be converted to foam cells with later progression to fatty streaks and plaques. Oxidized LDL inhibit macrophage motility, preventing macrophages from leaving the intima and facilitating continued uptake of oxidized LDL. Oxidized LDL are also cytotoxic, leading to cell death, endothelial loss and denudation of the artery. Some antioxidants can neutralize oxygen-derived free radicals through hydrogen donation, preventing the chain reaction in which LDL are oxidized (4,5). Vitamin C is the only antioxidant that has been shown to prevent the initiation of the oxidation chain reaction in lipids (6), trapping free radicals in the aqueous phase before they can diffuse into lipids such as LDL.

Vitamin C can also regenerate oxidized vitamin E by reducing it back to its active form (7); vitamin E has been shown to slow the rate of LDL oxidation (4,6). Additionally, vitamin C may protect against both the initiation and progression of coronary heart disease (CHD) through other potential mechanisms. Because vitamin C is required for collagen synthesis (1), it may be important in maintaining vascular integrity (8). Vitamin C may also have antithrombotic and antiplatelet effects through its involvement in the synthesis of prostacyclin by the vascular wall endothelium (8). Vitamin C may affect serum cholesterol concentrations, possibly through its role in the production of enzymes involved in the biosynthesis or hydroxylation of cholesterol (9).

However, findings from epidemiologic studies examining the relationship between vitamin C and CVD have been inconsistent; consequently, the role that vitamin C plays in the etiology of CVD remains controversial. Previous prospective cohort studies have varied in methods for measuring vitamin C status, definitions used to classify morbidity and mortality, range of intake and serum levels within the study population, length of follow-up, cohort size and composition (age, gender, exclusions), and ability to control for potential confounders. Such variability may account for the inconsistent findings regarding vitamin C status and cardiovascular disease. The purpose of this chapter is to review the evidence from prospective cohort studies examining the relationship between vitamin C status and CVD mortality, taking into account the variability in study design.

Materials and Methods

We limited our review to prospective cohort studies that measured vitamin C status before ascertainment of morbidity and mortality and that used appropriate statistical techniques to estimate relative risk (RR). We also limited studies to those that excluded prevalent cases at study entry to avoid potential biases resulting from their inclusion; prevalent cases may be more likely to change their dietary habits as a result of their disease or their disease may affect blood ascorbate levels. Restriction of the cohort to those free of disease at baseline also allows examination of the benefits of vitamin C in primary prevention of CVD. We included studies assessing vitamin C status measured by both blood assay or dietary intake (including or excluding supplement use). We reviewed studies examining various end points, and studies are reviewed in descending order of inclusiveness of CVD end points. We also address whether adjustment was made for potentially confounding variables, particularly smoking, and how these may affect interpretation of findings.

Studies Using Blood Ascorbate Concentrations

Three studies have examined the relation of blood ascorbate levels to CVD, with two finding significant associations (Table 7.1). Using the broadest definition of CVD [9th revision of the International Classification of Disease (ICD-9), codes 390–459], Simon *et al.* (10) found that 30- to 75-y-old participants in the second National Health and Nutrition Examination Survey (NHANES II) with saturated (≥1.1 mg/dL) or normal (0.5–1.0 mg/dL) compared with marginal (<0.4 mg/dL) serum ascorbate levels had a 34 and 33% reduced risk, respectively, of dying from CVD over 12–16 y. In a cohort of roughly the same age but more than twice as large, Khaw *et al.* (11) found a significant association between plasma ascorbate quintile and CVD mortality (excluding rheumatic heart disease and diseases of arterioles and veins from previous definition) after adjustment for age during 4 y of follow-up. Levels identified as most protective were consistent with saturation but multivariate-adjusted RR for quintiles were not presented. However, the multivariate-adjusted RR was 0.64 (0.51–0.78) in men and 0.81 (0.62–1.06) in women per increase of 0.35 mg/dL serum ascorbate. In the same NHANES II cohort used in the study by Simon *et al.* (10), Loria *et al.* (12) did not find an association between serum ascorbate quartile and a narrower definition of

CVD, including end points likely to be affected by antioxidant mechanisms (e.g., ischemic heart disease, cerebrovascular disease, sudden death, and diseases of arterioles). Adjustment for potential confounders in multivariate analyses from all three studies were generally comparable; however, the first two studies used smoking status (i.e., current, past, never) whereas the last study used average number of cigarettes smoked per day.

Two of four studies found a significant relation between blood ascorbate and CHD mortality defined using ICD-9 codes 410–414 (Table 7.1). Plasma ascorbate quintiles were associated with CHD mortality during 4 y of follow-up; however, these data were adjusted only for age; multivariate-adjusted RR in men was 0.63 (0.42–0.94) and women 0.56 (0.36–0.87) per increase of 0.35 mg/dL serum ascorbate (11). In a smaller study of free-living men and women ≥60 y old, Sahyoun *et al.* (13) found that participants with plasma ascorbate 1.0–1.5 mg/dL vs. 0.9 mg/dL had a decreased risk of heart disease mortality; participants with levels ≥1.6 mg/dL also had an increased risk although it was not significant. Although this study adjusted for a number of confounders, adjustment was not made for smoking. Two other studies examining CHD mortality in middle-aged men (14) and elderly men and women (15) found no significant association with plasma ascorbate. Another study examining the relation of plasma ascorbate with nonfatal and fatal CHD in middle-aged men failed to find a significant association (16). Only one study examined the relation between plasma ascorbate and stroke; elderly men and women in the highest (>0.49 mg/dL) compared with the lowest (≤0.21 mg/dL) tertile had a 30% decreased risk of dying from stroke although the RR risk was borderline significant and was adjusted only for age and sex (15).

Studies Using Vitamin C Intake

Two studies examined the relation between vitamin C intake and broad definitions of CVD in representative samples of U.S. adults (Table 7.2). The first study, using data from NHANES I with 13–16 y of follow-up, found that vitamin C intakes >50 mg/d in combination with vitamin C supplement use had a standardized mortality ratio (SMR) of 0.7 (17). However, SMR, unlike RR estimates, are estimated using an external comparison group and do not permit adjustment for potential confounding factors. In NHANES II, vitamin C intakes of 32–73 mg/d among men were associated with an increased (60%) risk of CVD mortality after adjustment for CVD risk factors, supplement use, and total energy intake, but there was no increased risk in the middle two dietary intake quartiles (Loria, unpublished data). There was no association between vitamin C intake and CVD mortality among women in NHANES II. This study also used a narrower definition of CVD than the NHANES I study, excluding rheumatic, hypertensive and pulmonary heart disease, and diseases of veins.

Four studies examined the relation between vitamin C intake and fatal and nonfatal CHD; however, none yielded significant associations. In a study of U.S. male health professionals who were 40–75 y of age and followed for 4 y, total vitamin C intakes (diet and supplements) were not associated with the risk of CHD after adjust-

TABLE 7.1

Prospective Cohort Studies Examining Serum or Plasma Ascorbate Concentrations and Cardiovascular Disease (CVD)[a]

Reference	Cohort	Gender	Age range (y)	Cohort size	Follow-up (y)	Endpoints (ICD-9)	Exposure (mg/dL)	RR	95% CI
CVD									
(10)	U.S.	Both	30–75	6035	12–16	390–459	<0.4	1.0	
							0.5–1	0.67	0.5–0.9
							1.1–2.7	0.66	0.4–0.99
(11)	UK	Men	45–79	8860	2–6	400–438	0.4	1.0	
							0.7	0.9	0.6–1.5
							0.8	0.67	0.4–1.2
							1.0	0.29	0.2–0.6
							1.3	0.29	0.2–0.6
		Women		10,636			0.5	1.0	
							0.9	0.41	0.2–0.9
							1.0	0.36	0.2–0.9
							1.2	0.6	0.3–1.3
							1.5	0.41	0.2–1.0
(12)	U.S.	Men	30–75	3347	12–16	410–414,427.5, 428.9,429.2, 430–438, 440–444	<0.5	1.4	0.9–2.3
							0.5–0.9	1.00	0.6–1.5
							0.9–1.3	1.30	0.9–2.0
							>1.3	1.0	
		Women		3724		(same)	<0.7	0.9	0.6–1.5
							0.7–1.2	0.9	0.6–1.4
							1.2–1.5	0.9	0.6–1.4
							>1.5	1	

CHD								
(11) UK	Men	45–79	8860	2–6	410–414	0.4	1	
						0.7	1.18	0.7–2.1
						0.8	0.92	0.5–1.7
						1.0	0.35	0.2–0.8
						1.3	0.32	0.2–0.8
	Women		10,636		(same)	0.5	1	
						0.9	0.23	0.7–0.9
						1.0	0.39	0.2–1.2
						1.2	0.32	0.1–1.0
						1.5	0.07	0.1–0.7
(14) Basel	Men	Middle-age	2974	12	410–414	<0.4	1.25	NS
(15) Britain	Both	65+	730	20	410–414	<0.21	1	
						0.21–0.49	0.9	0.6–1.3
						>0.49	0.9	0.6–1.3
(13) Boston	Both	60+	725	12	Heart disease mortality	0.9	1	
						1.0–1.5	0.51	0.3–0.9
						1.6	0.53	0.3–1.1
Fatal or nonfatal MI								
(16) Eastern Finland	Men	42–60	1605	3–8	Fatal or nonfatal MI	<0.20	2.08	0.8–5.3
						0.2–0.58	0.87	0.4–1.9
						0.59–0.88	0.62	0.3–1.5
						0.89–1.14	0.92	0.4–2.1
						>1.14	1.0	
Cerebrovascular diseases								
(15) Britain	Both	65+	730	20	433–438	<0.21	1	
						0.21–0.49	1.1	0.7–1.7
						>0.49	0.7	0.4–1.1

[a]Abbreviations: ICD-9, International Classification of Diseases, 9th Revision; RR, relative risk; CI, confidence interval; CHD, coronary heart disease; NS, nonsignificant; MI, myocardial infarction.

TABLE 7.2
Prospective Cohort Studies Examining Vitamin C Intake and Cardiovascular Disease (CVD)[a]

Reference	Cohort	Gender	Age range (y)	Cohort size	Follow-up (y)	Endpoints (ICD-9)	Exposure (mg/d)	RR	95% CI
CVD mortality									
(17)	U.S.	Men	25–74	4479	13–16	390–459	<50 mg/d	1.00	0.9–1.1
							50 + nosup	1.00	0.9–1.1
							50 + sup	0.70	0.5–0.8
		Women		6869		(same)	<50 mg/d	1.00	0.9–1.1
							50 + nosup	1.00	0.9–1.1
							50 + sup	0.80	0.6–0.96
Unpublished[b]	U.S.	Men	30–75	3347	12–16	410–414,427.5, 428.9,429.2, 430–438, 440–444	<32	1.30	0.8–2.2
							32–73	1.60	1.0–2.6
							74–139	1.20	0.7–2.11
							140+	1.00	
		Women		3724		(same)	<29	0.80	0.5–1.4
							29–73	0.80	0.5–1.2
							74–134	0.90	0.6–1.4
							135+	1.00	
Fatal and nonfatal CHD									
(18)	U.S.	Men	40–75	39,910	4	Fatal CHD, nonfatal MI, CABG, angioplasty, stroke	92	1.0	0.8–1.3
							149	1.0	0.9–1.5
							218	1.2	0.8–1.3
							392	1.0	
							1162	0.9	0.7–1.2
(19)	U.S.	Women	30–55	87,245	8	CHD	Q1	1.0	
							Q5	0.8	NS
(20)	Wales	Men	45–59	2423	5	410–414, MI, angina	<34.7	1.50	NS
							34.8–43.5	1.30	
							43.6–52.3	1.40	
							52.4–66.4	1.30	
							≥66.5	1.00	
(21)	Rotterdam	Both	55–95	4802	4	121–124 (ICD-10)	<87	1.00	
							87–126	0.94	0.58–1.37
							>126	0.88	0.56–1.38

CHD mortality

(22)	Finland	Men	30–69	2748	12–16	410–414 (ICD-8)	≤60	1.00	
							61–85	0.90	0.6–1.2
							>85	1.00	0.7–1.4
		Women		2385			≤61	1.00	
							62–91	0.50	0.3–0.99
							>91	0.50	0.2–0.98
(23)	Chicago	Men	40–55	5397	24	410–412	21–82	1.00	
							83–112	1.03	
							113–393	0.75	
							100	0.63	0.45–1.02
(15)	Britain	Both	65+	730	20	410–414	<28	1.00	
							28–44.9	0.90	0.7–1.4
							≥45	0.80	0.6–1.2
(13)	Boston	Both	60+	725	12	Heart disease mortality	90	1.00	
							91–387	0.50	0.3–0.9
							388	0.50	0.3–1.1
(24)	Iowa	Women	55–69	34,486	8	410,412, 429.2	<112.3	1.00	
							112.4–161.3	1.08	0.69–1.69
							161.4–226.7	0.85	0.53–1.37
							226.8–391.2	0.99	0.61–1.59
							>391.3	1.49	0.96–2.30

Cerebrovascular disease

(15)	Britain	Both	65+	730	20	433–438	<28	1.00	
							28–44.9	0.80	0.5–1.2
							≥45	0.40	0.2–0.6
(25)	Chicago	Men	40–55	1843	30	430–434 436–438	22–74	1.00	
							75–97	0.87	0.60–1.26
							98–122	0.87	0.60–12.6
							123–393	0.71	0.47–1.05
(26)	Iowa	Women	55–69	34,492	11	430–438	82.4	1.00	
							138.3	0.69	0.43–1.10
							190.6	0.77	0.48–1.21
							280.9	0.70	0.42–1.13
							678.7	1.24	0.81–1.89

[a]Abbreviations: ICD-9, International Classification of Diseases, 9th revision; RR, relative risk; CI, confidence interval; sup, supplement use; CHD, coronary heart disease; MI, myocardial infarction; CABG, coronary artery bypass grafting; Q, quintile; NS, nonsignificant.
[b]Data from Loria, unpublished.

ment for CHD risk factors and intakes of vitamin E and carotene (18). Among 30- to 55-y-old female nurses in the United States followed for 8 y, the risk of CHD was lower for women in the highest compared with the lowest quintile of total intake although this association was not significant (19). A study of middle-aged men living in Wales failed to find significant associations (20). Another study examining morbidity and mortality from acute MI did not yield significant associations in Dutch adults followed for 4 y (21).

Five studies examined the relation between vitamin C intake and fatal CHD; only one found a significant association although another tended to be significant. Dietary vitamin C intake was significantly associated with CHD mortality (ICD-8, codes 410–414) among Finnish women but not Finnish men (22). Women in the middle (≤61 mg/d) and highest (61–85 mg/d) tertile had a 50% lower risk of CHD mortality than women in the lowest tertile (<60-61 mg/d) (22). Middle-aged men in Chicago followed for 24 y who had intakes between 113 and 393 mg/d compared with 21–82 mg/d had a decreased risk of dying from acute MI (ICD-9, codes 410–412) although it only was borderline significant (23). The RR was significant, however, in nonsmokers when analyses were stratified (23). Another study examining CHD mortality failed to find a significant association in elderly adults (15). In a smaller study of free-living men and women (60 y old, Sahyoun *et al.* (13) did not find a significant relation between total vitamin C intake and heart disease mortality, presumably CHD. Another study examining mortality from acute MI failed to find a significant association in Iowa women followed for 6 y (24).

One of three studies that examined the relation between vitamin C intake and stroke found a significant association, whereas another reported only a borderline significant finding. Adults ≥65 y old followed for 20 y who consumed >45 mg vitamin C/d had a 60% reduced risk of dying from stroke compared with those consuming <28 mg/d (15). However, RR estimates were adjusted for age and sex only and did not account for other potential confounders. Middle-aged men in Chicago followed for 30 y who had intakes between 123 and 393 mg/d compared with 22–74 mg/d had a decreased risk of stroke although it only was borderline significant (25). In a study of Iowa women 55–69 y of age, there was no increased risk of dying from stroke after 11 y (26).

Discussion

The evidence regarding vitamin C status and CVD from prospective cohort studies was reviewed, taking into account the variability in study design. Findings were more likely to be positive when serum ascorbate concentration vs. intake was used to assess vitamin C status. This observation is not surprising given that these indicators may measure quite different aspects of vitamin C status, i.e., estimated body ascorbic acid pool vs. vitamin C intake from foods and supplements. The relationship between vitamin C intakes and serum ascorbate level is complex because serum levels are saturable (1,27), absorption is dose dependent (1), and may be modified by other factors, such as cigarette smoking, disease state, and medications (1,4,28–30). On the basis of this review of prospective studies, available stores may be the more important aspect of vitamin C status with respect to CVD than intake level, although the potential for misclassification associated with dietary assessment methods may also account for greater inconsistency in studies examining intake levels.

Findings also tended to be positive when a wider definition of CVD was used; this pattern was observed in studies using both serum concentrations and intake data. Studies that examined CHD, either fatal or nonfatal, as an end point yielded inconsistent findings, although those using serum ascorbate concentrations tended to be positive compared with those using vitamin C intake. The evidence for a relationship between stroke and vitamin C status was also weak. No studies examined the association of vitamin C on morbidity alone; it is possible that vitamin C may affect morbidity differently than mortality. A tendency to positive associations with wider definitions including end points both likely and less likely to be affected by antioxidant mechanisms suggests that vitamin C may be protective of CVD through multiple mechanisms, including but not limited to oxidation.

Most of the studies reviewed controlled for important potential confounders; however, some analyses were adjusted only minimally for age and sex and these results should be interpreted with caution, especially those that did not adjust for cigarette smoking. In one of the two most recent studies examining serum ascorbate in a large cohort, Khaw *et al.* (11) adjusted only for age and sex in analyses by ascorbate quintile but presented multivariate-adjusted models for continuous ascorbate level. In two different analyses of the other recently published large cohort, NHANES II, one study controlled for smoking status (i.e., current, past, never) and found a significant association (10), whereas the other adjusted for number of cigarettes smoked and failed to find a significant association (12). This discrepancy raises the question of possible residual confounding by smoking although these studies also used differing end points, vitamin C status classification, inclusion criteria, and classification of other adjustment variables, all of which may have contributed to the discrepant findings. More research is required on appropriate adjustment for cigarette smoking in statistical models, given the important association of smoking with serum ascorbate levels. Cigarette smoking decreases serum ascorbate levels (28,31) and increases CVD risk possibly by generating oxygen-derived free radicals (1,32). Further consideration of whether smoking should be treated as a potential confounder, effect modifier, part of the causal mechanism, or some combination of these is warranted.

Summary

Findings from prospective cohort studies regarding vitamin C status and CVD were slightly more consistent when stratifying by vitamin C assessment method and CVD end points. On the other hand, data from such studies remain inconclusive possibly because of limitations of cohort studies, including limited ability to adjust adequately for potential confounders or effect modifiers.

References

1. Jacob, R.A. (1994) Vitamin C, in *Modern Nutrition in Health and Disease* (Shils, M.E., Olson, J.A., and Shike, M., eds.) Lea and Febiger, Philadelphia.
2. Peters, K.D., Kochanek, K.D., and Murphy, S.L. (1998) Report of Final Mortality Statistics, 1996, National Center for Health Statistics, Hyattsville, MD.
3. Luc, G., and Fruchart, J.C. (1991) Oxidation of Lipoproteins and Atherosclerosis, *Am. J. Clin. Nutr. 53*, 206S–209S.
4. Abbey, M., Nestel, P.J., and Baghurst, P.A. (1993) Antioxidant Vitamins and Low-Density-Lipoprotein Oxidation, *Am. J. Clin. Nutr. 58*, 525–532.
5. Farrell, P.M., and Roberts, R.J. (1994) Vitamin E, in *Modern Nutrition in Health and Disease* (Shils, M.E., Olson, J.A., and Shike, M., eds.) Lea and Febiger, Philadelphia.
6. Frei, B. (1991) Ascorbic Acid Protects Lipids in Human Plasma and Low-Density Lipoprotein Against Oxidative Damage, *Am. J. Clin. Nutr. 54*, 1113S–1118S.
7. Esterbauer, H., Dieber-Rotheneder, M., Striegl, G., and Waeg, G. (1991) Role of Vitamin E in Preventing the Oxidation of Low-Density Lipoprotein, *Am. J. Clin. Nutr. 53*, 314S–321S.
8. Simon, J.A. (1992) Vitamin C and Cardiovascular Disease: A Review, *J. Am. Coll. Nutr. 11*, 107–125.
9. Carr, A.C., and Frei, B. (1999) Toward a New Recommended Dietary Allowance for Vitamin C Based on Antioxidant and Health Effects in Humans, *Am. J. Clin. Nutr. 69*, 1086–1107.
10. Simon, J.A., Hudes, E.S., and Tice, J.A. (2001) Relation of Serum Ascorbic Acid to Mortality Among US Adults, *J. Am. Coll. Nutr. 20*, 255–263.
11. Khaw, K.T., Bingham, S., Welch, A., Luben, R., Wareham, N., Oakes, S., and Day, N. (2001) Relation Between Plasma Ascorbic Acid and Mortality in Men and Women in EPIC-Norfolk Prospective Study: A Prospective Population Study. European Prospective Investigation into Cancer and Nutrition, *Lancet 357*, 657–663.
12. Loria, C.M., Klag, M.J., Caulfield, L.E., and Whelton, P.K. (2000) Vitamin C Status and Mortality in US Adults, *Am. J. Clin. Nutr. 72*, 139–145.
13. Sahyoun, N.R., Jacques, P.F., and Russell, R.M. (1996) Carotenoids, Vitamins C and E, and Mortality in an Elderly Population, *Am. J. Epidemiol. 144*, 501–511.
14. Gey, K.F., Stahelin, H.B., and Eichholzer, M. (1993) Poor Plasma Status of Carotene and Vitamin C Is Associated with Higher Mortality from Ischemic Heart Disease and Stroke: Basel Prospective Study, *Clin. Investig. 71*, 3–6.
15. Gale, C.R., Martyn, C., Winter, P.D., and Cooper, C. (1995) Vitamin C and Risk of Death from Stroke and Coronary Heart Disease in Cohort of Elderly People, *Br. Med. J. 310*, 1563–1566.
16. Nyyssonen, K., Porkkala-Sarataho, E., Kaikkonen, J., and Salonen, J.T. (1997) Ascorbate and Urate Are the Strongest Determinants of Plasma Antioxidative Capacity and Serum Lipid Resistance to Oxidation in Finnish Men, *Atherosclerosis 130*, 223–233.
17. Enstrom, J.E. (1994) *Vitamin C Intake and Mortality Among a Sample of United States Population: New Results*, Hippokrates Verlag, Stuttgart.
18. Rimm, E.B., Stampfer, M.J., Ascherio, A., Giovannucci, E., Colditz, G.A., and Willett, W.C. (1993) Vitamin E Consumption and the Risk of Coronary Heart Disease in Men, *N. Engl. J. Med. 328*, 1450–1456.
19. Manson, J.E., Gaziano, J.M., Jonas, M.A., and Hennekens, C.H. (1993) Antioxidants and Cardiovascular Disease: A Review, *J. Am. Coll. Nutr. 12*, 426–432.

20. Fehily, A.M., Yarnell, J.W., Sweetnam, P.M., and Elwood, P.C. (1993) Diet and Incident Ischaemic Heart Disease: The Caerphilly Study, *Br. J. Nutr. 69*, 303–314.
21. Klipstein-Grobusch, K., den Breeijen, J.H., Grobbee, D.E., Boeing, H., Hofman, A., and Witteman, J.C. (2001) Dietary Antioxidants and Peripheral Arterial Disease: The Rotterdam Study, *Am. J. Epidemiol. 154*, 145–149.
22. Knekt, P., Reunanen, A., Jarvinen, R., Seppanen, R., Heliovaara, M., and Aromaa, A. (1994) Antioxidant Vitamin Intake and Coronary Mortality in a Longitudinal Population Study, *Am. J. Epidemiol. 139*, 1180–1189.
23. Pandey, D.K., Shekelle, R., Selwyn, B.J., Tangney, C., and Stamler, J. (1995) Dietary Vitamin C and Beta-Carotene and Risk of Death in Middle-Aged Men. The Western Electric Study, *Am. J. Epidemiol. 142*, 1269–1278.
24. Kushi, L.H., Folsom, A.R., Prineas, R.J., Mink, P.J., Wu, Y., and Bostick, R.M. (1996) Dietary Antioxidant Vitamins and Death from Coronary Heart Disease in Postmenopausal Women, *N. Engl. J. Med. 334*, 1156–1162.
25. Daviglus, M.L., Orencia, A.J., Dyer, A.R., Liu, K., Morris, D.K., Persky, V., Chavez, N., Goldberg, J., Drum, M., Shekelle, R.B., and Stamler, J. (1997) Dietary Vitamin C, Beta-Carotene and 30-Year Risk of Stroke: Results from the Western Electric Study, *Neuroepidemiology 16*, 69–77.
26. Yochum, L.A., Folsom, A.R., and Kushi, L.H. (2000) Intake of Antioxidant Vitamins and Risk of Death from Stroke in Postmenopausal Women, *Am. J. Clin. Nutr. 72*, 476–483.
27. Gibson, R.S. (1990) *Principles of Nutritional Assessment*, 1st edn., Oxford University Press, New York.
28. Pelletier, O. (1968) Smoking and Vitamin C Levels in Humans, *Am. J. Clin. Nutr. 21*, 1259–1267.
29. Pelletier, O. (1970) Vitamin C Status of Cigarette Smokers and Non-Smokers, *Am. J. Clin. Nutr. 23*, 520–524.
30. Lee, W., Davis, K.A., Rettmer, R.L., and Labbe, R.F. (1988) Ascorbic Acid Status: Biochemical and Clinical Considerations, *Am. J. Clin. Nutr. 48*, 286–290.
31. Kallner, A., Hartmann, D., and Hornig, D. (1979) Steady-State Turnover and Body Pool of Ascorbic Acid in Man, *Am. J. Clin Nutr. 32*, 530–539.
32. Machlin, L.J., and Bendich, A. (1987) Free Radical Tissue Damage: Protective Role of Antioxidant Nutrients, *FASEB J. 1*, 441–445.

Chapter 8

Potential Adverse Effects of Vitamins C and E

Carol S. Johnston

Department of Nutrition, Arizona State University East, Mesa, AZ 85212

Introduction

Vitamins C and E are effective free radical scavengers. Water-soluble vitamin C is an excellent antioxidant in plasma, whereas lipid-soluble vitamin E protects cell membranes from peroxidation. One third of U.S. adults ingest a vitamin and/or mineral supplement daily. A majority of supplement users (65%) consume broad-spectrum multivitamin/mineral products. However, 4 and 8% of U.S. adults specifically ingest vitamin E and vitamin C supplements, respectively, on a daily basis (1). These two vitamins are the most commonly supplemented micronutrients in the United States (2). Of all supplement users, the average vitamin C intake in the form of supplements is ~135 mg; but 10% of supplement users (3% of U.S. adults) consume >1000 mg of vitamin C daily (1,3). This level of intake is 10-fold greater than the recommended dietary allowance (RDA) for vitamin C, 75–90 mg/d. The average intake of vitamin E among supplement users is 10 mg daily[1] and 10% of supplement users consume ~140 mg vitamin E daily (3), nearly 1000% of the RDA, 15 mg/d. The safety of such doses has been questioned, yet there is little evidence directly indicating that these levels of intake are harmful for healthy individuals in the general population. Rather, high dietary intakes of vitamin C have been associated with reduced risk of degenerative diseases, and high-dose vitamin E decreases platelet adhesion, inhibits oxidation of low density lipoproteins (LDL), and is promoted for the prevention and treatment of coronary artery disease.

---

[1]Vitamin E is defined herein as the natural α-tocopherol as presented by the Food and Nutrition Board of the Institute of Medicine in the year 2000 to establish the Dietary Reference Intakes for vitamin E. Thus, vitamin E derived from supplements was assumed to be *all rac*-α-tocopherol (*RRR*-, *RSR*-, *RRS*-, and *RSS*-α-tocopherol) and a factor of 0.50 was applied to convert mg of this synthetic form of vitamin E to mg natural α-tocopherol. To convert IU to mg natural α-tocopherol, a factor of 0.45 was applied to the 2*R*-stereoisomers of tocopherol and a factor of 0.67 was applied to the natural *RRR*-α-tocopherol.

Supplementation and Mortality

Vitamin C Supplement Use

All-cause mortality in a sample of the U.S. population [National Health and Nutrition Examination Survey (NHANES) I participants, n = 10,550; 749 deaths] was inversely related to vitamin C intake [standardized mortality ratio as a function of the dietary vitamin C index, 0.58; 95% confidence interval (CI), 0.36–0.80] (4). The vitamin C index ranged from persons consuming <50 mg vitamin C daily to those consuming ≥50 mg vitamin C daily from diet alone and those consuming ≥50 mg vitamin C daily from diet plus an average of ~800 mg daily from supplementation. After adjustment for age, sex, and 10 potentially confounding variables (race, history of serious disease, education, cigarette smoking, recreational exercise, alcohol consumption, energy consumed, fat, serum cholesterol, and dietary vitamin A), the inverse relationship of vitamin C intake to mortality remained strong (standardized mortality ratio, 0.62; 95% CI, 0.36–0.88). This inverse relation of total mortality to vitamin C intake was stronger and more consistent in this population than the relation of total mortality to serum cholesterol and dietary fat intake (4). In a separate population, there were no differences in mortality between individuals consuming above and below 250 mg vitamin C/d; however, those consuming >750 mg vitamin C/d experienced only 40% of the total cohort death rate during the 10-y study period (3).

In the NHANES II Mortality Study (n = 8453; 1327 total deaths), all-cause mortality was reduced 29% (*P* < 0.001) in individuals with high serum ascorbate (serum values ≥1.1 mg/dL) after multivariate adjustment for potential confounding variables, but vitamin C supplement use was not significantly associated with mortality (*P* = 0.19) (5). The risk of cardiovascular disease death was reduced 25% (*P* = 0.09) in individuals with high serum ascorbate, but this risk reduction was not related to the use of vitamin C supplements (5). However, the risk of cancer death in men specifically using vitamin C supplements was reduced 65% (*P* = 0.046) compared with a 31% reduction in men with high serum ascorbate. Fatal cancer risk was not associated with vitamin C supplement use in women. In an American Cancer Society cohort (n = 711,891; 4404 deaths from colorectal cancer), long-term vitamin C supplementation (≥10 y) lowered risk of rectal cancer mortality (–60%) but did not affect risk of colon cancer mortality (6).

Losconzy *et al.* (7) reported no significant effect of vitamin C supplementation on all-cause mortality, coronary disease mortality, or cancer mortality, yet vitamin supplement use may have been underreported because subjects completed questionnaires for nonprescription medications only, not vitamin supplementation specifically. Two other large, prospective population studies did not specifically address vitamin C supplementation use, yet men in the highest stratum for vitamin C status, also representing the greatest number of vitamin C supplement users, had significantly reduced risk of all-cause mortality, cardiovascular disease mortality, and cancer mortality (8,9). Women in the highest stratum for vitamin C status had reduced risk of all-cause mortality and cardiovascular disease in one of these studies (8) but not the other (9). These data indicate the general safety of vitamin C in that mortality end points for the major causes of death in the United States are either unaltered or perhaps favorably affected by high-dose vitamin C ingestion.

Vitamin E Supplement Use

In an elderly cohort (n = 11,178), vitamin E supplementation alone reduced all-cause mortality risk 20% independently of other factors (7), and risk for coronary disease mortality and cancer mortality were reduced 36 and 28%, respectively, in vitamin E users (7). In the American Cancer Society cohort, short- or long-term vitamin E supplement use was not associated with reduced risk of colorectal cancer; however, long-term vitamin E use was associated with decreased risk of colorectal cancer mortality in participants who were current cigarette smokers (–40%) (6). Two large, prospective populations studies showed a reduced risk for all-cause mortality in men ingesting ≥113 mg vitamin E daily (–28%, P = 0.06) (10) and trends toward a reduction in the risk of all-cause mortality (–13%) and cardiovascular mortality (–42%) in women ingesting >45 mg vitamin E/d (11).

The prospective, randomized, placebo-controlled Alpha-Tocopherol, Beta Carotene Cancer Prevention Study (n = 29,133; 876 new lung cancer diagnosis and 1415 other cancer diagnoses) showed no effect of vitamin E supplementation (25 mg/d) on the incidence of lung cancer in male smokers (12). However, in this trial, there was a beneficial effect of vitamin E use against prostate cancer (34% lower incidence) and colorectal cancer (17% lower incidence). Importantly, a potential adverse effect of vitamin E use was observed in these male smokers, i.e., a 5% increased incidence of hemorrhagic stroke (12). In this same cohort, there was a nonsignificant increased risk of fatal coronary heart disease in smokers ingesting vitamin E supplements (+33%) (13). In the Heart Outcomes Prevention Evaluation Study (n = 9541; 1511 primary cardiovascular events), a high dosage of vitamin E (180 mg/d for 5 y) had no effect on risk of cardiovascular mortality or all-cause mortality in high-risk patients (14). There was a nonsignificant increased risk of stroke mortality among the vitamin E users (+17%, P = 0.13), but there was no increase in hemorrhagic stroke among the vitamin E users.

In 196 hemodialysis patients with preexisting cardiovascular disease who were randomized to receive 360 mg vitamin E daily or placebo, risk of cardiovascular disease mortality was not significantly reduced, but the difference in the survival curves for cardiovascular disease mortality by treatment tended to be significant (P = 0.06) (15). Despite a slight reduction in cardiovascular disease mortality risk, risk for all-cause mortality did not differ by treatment due to a nonsignificant increase in noncardiac mortality in patients ingesting vitamin E, including two cases of death associated with hemorrhage (15). Ascherio *et al.* (16) specifically examined the effect of regular vitamin E supplement ingestion (113 mg) on risk of ischemic and hemorrhagic strokes in mostly nonsmoking men (n = 43,738) participating in the Health Professionals Follow-up Study. After multivariate adjustments, there was a slight but nonsignificant risk of hemorrhagic stroke (+3%) and ischemic stroke (+16%) in users of vitamin E.

Hence, intervention trials to date do not support a regimen of high-dose vitamin E ingestion for reducing risk of cardiovascular and/or cancer death in high-risk populations. Furthermore, there is some evidence indicating a potential adverse effect on risk for hemorrhagic stroke in users of vitamin E supplements. Hemorrhagic stroke is classified as subarachnoid hemorrhage or intracerebral hemorrhage, and the risk profiles of these two subtypes are distinct. Subarachnoid hemorrhage is typically caused by rupture of an arterial aneurysm and has been related to a smoking-induced elastase/α-antitrypsin imbalance (17), whereas intracerebral hemorrhage may be related more to hypertension and necrosis of small arterioles (18). The number of cigarettes smoked increased the risk of subarachnoid hemorrhage but not intracerebral hemorrhage (18). In smokers, vitamin E supplementation raised the risk of fatal subarachnoid hemorrhage (+181%, $P = 0.005$) but did not alter risk of fatal intracerebral hemorrhage (19); furthermore, a high serum vitamin E concentration was related to a 53% lower incidence of intracerebral hemorrhage in smokers (18). Thus, the antiplatelet and anticlotting actions of vitamin E may, to some degree, exacerbate an arterial rupture, particularly in smokers, but these actions, as well as the antioxidant action of vitamin E, might lesson the development of fibrinoid necrosis (18). The antiplatelet and anticlotting actions of vitamin E are discussed in a later section.

Adverse Reactions to Supplemental Vitamin C

In the general population, the data overwhelmingly indicate that regular ingestion of vitamin C supplements is unlikely to have major adverse effects. The low toxicity of supplemental vitamin C can probably be attributed to decreased bioavailability and increased urinary excretion as the dose is increased. In one study, ~70% of a 500-mg dose was absorbed and >50% of the absorbed dose was excreted in urine unmetabolized. At the 1250-mg dosage level, only 50% of the dose was absorbed and nearly all of the absorbed dose was excreted (20). Furthermore, the oxidized form of vitamin C, dehydroascorbic acid, is not highly reactive and is effectively cleared from fluids and rapidly reduced to ascorbic acid (21,22). Nonetheless, concerns have been raised regarding adverse effects of high doses of vitamin C.

Systemic Conditioning (Rebound Scurvy)

There is some evidence that systemic conditioning (the accelerated metabolism or disposal of ascorbic acid) may occur after prolonged supplementation of high doses of vitamin C. The physiologic relevance of accelerated vitamin C metabolism in otherwise healthy individuals, however, has not been addressed adequately. In human subjects consuming a vitamin C–deficient diet (5 mg/d) in a live-in metabolic unit, abrupt withdraw of vitamin C supplementation (600 mg/d for 3 wk) was associated with a significant reduction in the mean leukocyte vitamin C concentration (16.8 ± 4.37 and 9.4 ± 5.33 μg vitamin C/10^8 cells, presupplementation and at 4 wk after withdrawal of supplementation, respectively) (23). Both of these leukocyte vitamin C concentrations were indicative of vitamin C depletion. Yet simply removing vitamin C–rich fruits and vegetables from the diets of healthy free-living adults can lower mean plasma vitamin C concentrations into the depleted or deficient range within 1–3 wk (24). Thus, even in adults not supplementing their diets with vitamin C, removal of vitamin C from the diet (in the form of supplements or foods), results in rapid vitamin C depletion, reflecting how poorly vitamin C status is maintained in the absence of a dietary source.

Well-nourished subjects receiving vitamin C supplements may also display accelerated vitamin C metabolism; however, because indices of vitamin C nutrition remain within normal ranges, the term "rebound scurvy" is misleading. In guinea pigs consuming a standard guinea pig diet containing 0.1% vitamin C (w/w), chronic vitamin C administration [1 g/(kg body.d) intraperitoneally for 4 wk] appeared to be associated with an increased rate of vitamin C turnover. Mean plasma vitamin C concentrations fell significantly below control values at wk 2 and 5 after the abrupt withdrawal of the administered vitamin C (25); however, all mean plasma vitamin C concentrations measured postwithdrawal were well within normal ranges. In human subjects consuming normal diets, the mean plasma vitamin C concentration did not fall significantly below the presupplement value at 7 or 23 d after the abrupt withdrawal of vitamin C supplements (2 g/d for 9 d) (26).

Kidney Stone Formation

Almost 75% of kidney stones are calcium oxalate, and a lesser number are uric acid (5%). High doses of vitamin C may moderately increase urinary oxalic acid and urinary uric acid levels; yet, an association between high intakes of vitamin C and the actual formation of kidney stones has never been demonstrated.

In human subjects with hyperuricemia, an acute dose of vitamin C, 0.5 or 2.0 g, did not significantly alter urinary clearance of uric acid. However, a single 4.0-g dose of vitamin C significantly increased mean uric acid clearance by 71% in these subjects (27); it was unclear, however, whether this rise in uric acid clearance was associated with urinary uric acid concentrations above the normal range (1.5–4.4 mmol/d). In three subjects ingesting 8 g of vitamin C daily for up to 7 d, urinary uric acid was elevated over the normal range in one subject (27). In subjects with and without gout, vitamin C infusion (2.5-10 mg/min to achieve plasma levels from 3 to 12 mg/100 mL) raised urinary urate clearance only moderately (28), and there were no differences in response between subjects with and without gout. In a carefully controlled inpatient trial, mean urinary uric acid was not elevated in seven young men after chronic daily ingestion of 200 or 400 mg vitamin C (20). Chronic daily ingestion of 1000 mg of vitamin C did significantly raise mean urinary uric acid to a concentration outside the normal range (~5.5 mmol/d).

Vitamin C intake has been related to urinary oxalate ($r = 0.33$, $P < 0.001$) (29). Mean urinary oxalate was not significantly altered in seven men consuming 200 or 400 mg vitamin C daily for 4–5 wk (20); although chronic daily ingestion of 1000 mg vitamin C raised urinary oxalate nearly 50% ($P < 0.05$), urinary oxalate values remained within the normal range (228–684 μmol/d). In six healthy subjects, vitamin C supplementation, 10 g/d, did not alter urinary oxalate in five of the subjects, and the elevated level noted in the remaining subject was within the normal range (30).

These data indicate that high doses of vitamin C alter urinary excretion of uric and oxalate acid; however, diets high in purines, organ meats and fish, or chronic low-dose aspirin, can cause hyperuricosuria, and foods rich in oxalates (e.g., tea, coffee, nuts, beans, chocolates) can increase oxalate concentrations in urine to levels above the normal range (31). Hence vitamin C is one of many diet constituents associated with alterations in urinary constituents of renal stones. Furthermore, diets high in animal protein or low in fluids, bacterial products, and heavy physical exertion have also been implicated in renal stone disease. Because uric acid can precipitate in the absence of hyperuricosuria, and urinary saturation with calcium oxalate is common in the general population, factors other than urinary levels of uric acid and oxalic acid are crucial for stone formation, including renal epithelial cell responses, acid urine, and proteins of renal tubular origin (31).

In the Harvard Prospective Health Professional Follow-Up Study (45,251 men; 751 incident cases of kidney stones after 6 y), the relative risk of developing kidney stones in users of vitamin C supplements ($\geq$1500 mg vitamin C daily vs. <250 mg daily) was 0.89 after multivariate adjustment (32). Using data from NHANES II, Simon and Hudes (33) found no associations between serum vitamin C and prevalence of kidney stones in women or men. In the Nurses' Health Study cohort (85,557 women; 1078 incident cases after 14 y), supplemental vitamin C was not associated with risk of kidney stones (multivariate relative risk, 1.06) (34). Nonetheless, high doses of vitamin C could invalidate urinary measures of uric acid and oxalic acid, and probably should not be taken by individuals predisposed to renal calculi or with renal failure (35).

Prooxidant Activity

Ascorbic acid nonenzymatically detoxifies a variety of reactive radicals and oxygen species, including hydroxyl, superoxide, peroxyl and nitrogen dioxide radicals, and in the interstitial fluids, plasma, or the cytosol, ascorbic acid acts as a primary antioxidant to scavenge free radicals that are generated by biological interactions or by cellular metabolism. In the scavenging process, ascorbic acid loses a single electron to form the ascorbate free radical, which is only weakly reactive and which may be quickly reduced back to ascorbic acid, or, with the loss of a second electron, is converted to dehydroascorbic acid. *In vivo*, dehydroascorbic acid may be recycled back to ascorbic acid in erythrocytes or numerous other cell types, or may incur irreversible ring opening to form 2,3-diketo-1-gulonic acid, a nonreactive metabolite. Given these metabolic pathways, vitamin C is an efficient protective antioxidant and unlikely to demonstrate prooxidant activity *in vivo* (36).

In vitro, however, vitamin C reduces unbound transition metals, which then participate in free radical generation, most notably the reduction of ferric iron to the ferrous form, which is active in the Fenton reaction and the decomposition of hydrogen peroxide to form hydroxyl radicals (37). The extremely reactive hydroxyl radical quickly reacts with most molecules including DNA, and is implicated in mutagenesis and the initiation of cancer. *In vivo*, most transition metals are bound to protein, thus preventing metal-dependent free radical reactions. Experiments in iron-overloaded plasma demonstrated that iron, even in excess of the amount bound by transferrin, was unable to participate in a Fenton reaction in the presence of ascorbic acid (38). The complexity of plasma, including the presence of catalase or ferroxidase activity, may provide an environment inhibiting a Fenton reaction when transition metals are unbound (38). In a separate study, when excess iron was added to adult plasma, endogenous and exogenous ascorbic acid delayed the onset of iron-induced lipid peroxidation in a dose-dependent manner (39). Furthermore, LDL incubated in buffer were protected from metal ion–dependent oxidative modification by dehydroascorbic acid and from metal ion–independent oxidative modification by ascorbic acid (40). Thus, in biological systems, ascorbic acid acts as an antioxidant, even in the presence of iron overload.

In iron-loaded guinea pigs, significantly less oxidative damage, as indicated by reduced levels of liver or plasma F_2-isoprostanes, was recorded in animals fed high-dose vitamin C vs. animals fed low-dose vitamin C (41). In humans, vitamin C supplementation (260 mg/d for 6 wk) was associated with significant reductions in several DNA adducts that are markers of oxidized DNA base damage, i.e., 8-oxo-guanine and 5-OH methyl uracil (42). Total base damage (compiled data for 12 different DNA adducts), however, was unaffected by vitamin C supplementation. Vitamin C supplementation (260 mg/d) with iron (14 mg/d) was associated with a significant reduction in the 5-OH methyl uracil DNA adduct but with significant increases in 5-OH methyl hydantoin and 5-OH cytosine. However, total base damage was unaltered by the vitamin C plus iron supplementation regimen.

In a controversial report, prooxidant effects of supplemental vitamin C were reported in healthy volunteers consuming 500 mg of vitamin C/d for 6 wk (43). Lymphocytes extracted from these individuals displayed an increase in oxidative damage to DNA as indicated by a significant rise in mean levels of 8-oxo-adenine during the supplementation period. Yet, the same report clearly depicted a significant, 50% decrease in levels of 8-oxo-guanine over the same time period. This is noteworthy because the oxidation of guanine has been highly correlated with mortality from heart disease ($r = 0.95$), colorectal cancer ($r = 0.91$), pancreatic cancer ($r = 0.82$), and lung cancer ($r = 0.32$) (44). Furthermore, hydroxyl radicals and singlet oxygen directly target guanine residues, and the resulting G→T transversion mutations have been observed in ras protooncogenes and p53 tumor-suppresser genes, cancer-related genes that are implicated in 40–50% of human cancers (45,46). There are few or no data in the literature implicating 8-oxo-adenine in mutations of cancer-related genes. Other reports have also demonstrated that vitamin C supplementation (250 or 500 mg/d for 28 d) significantly reduced oxidative damage to DNA *in vivo* as indicated by reduced tissue levels of 8-oxo-guanine (47,48).

Recently, Lee *et al*. (49) demonstrated that ascorbic acid in the absence of transitional metals reduced lipid peroxides *in vitro*, producing the genotoxin 4,5-epoxy-2(*E*)-decenal which reacts with DNA bases to form etheno-DNA adducts. *In vivo*, etheno-adducts in white blood cell DNA were inversely correlated with vegetable or vitamin E consumption (50) and positively associated with linoleic acid intake (51). Furthermore, citrus fruits, berries, or vitamin C intake were reportedly inversely related to *hprt* mutation frequency in peripheral blood lymphocytes, a marker of genotoxic effects due to lipid peroxidation (52).

Iron Overload and Iron Toxicity

Vitamin C promotes the absorption of dietary nonheme iron and contributes to the prevention of iron deficiency anemia (53,54). Mealtime vitamin C in the range of 50–100 mg has the most marked effect on iron absorption, and supplemental vitamin C beyond these levels had minimal additional effects on iron status in healthy individuals (55). Whether vitamin C supplementation might enhance the uptake and storage of nonheme iron in individuals with iron-overload pathologies (e.g., hereditary hemochromatosis and thalassemia major) is unclear. Moreover, these patients have measurable nontransferrin-bound iron in serum, which is associated with oxidative stress and products of lipid peroxidation (56,57). Theoretically, vitamin C could act as a prooxidant in the presence of unbound iron and fuel the production of lipid peroxides.

A recent case report of a 36-y-old woman with hereditary hemochromatosis resulting in end-stage cardiomyopathy, suggests that high-dose vitamin C supplement use (>1000 mg/d) may have predisposed her to a more fulminant disease course (58). In fact, vitamin C deficiency may have a protective role in some patients with severe iron overload (59). Vitamin E supplementation (300–600 mg/d), however, improved measures of oxidative stress in β-thalassemia patients, significantly reducing levels of LDL-conjugated dienes and red blood cell malondialdehyde, and increasing the resistance of red blood cells to osmotic lysis (60,61). In a rat model of hemochromatosis, vitamin E supplementation reduced lipid peroxidation in iron-loaded livers by 50% (62). This therapeutic dichotomy between vitamin C and vitamin E supplementation is noteworthy.

Gastrointestinal Effects

Gastrointestinal symptoms such as nausea, abdominal cramps, and diarrhea may occur in ~10% of subjects when vitamin C doses exceed 2 or 3 g/d. These symptoms have no long-term consequences, and buffered ascorbate salts consumed with a meal would alleviate these effects. However, osmotic diarrhea and gastrointestinal disturbances were the only selected critical end points for the establishment of a Tolerable Upper Intake Level (UL) for vitamin C, 2000 mg/d (63). That is, the *only* adverse effect of high-dose vitamin C considered by the Food and Nutrition Board in the formulation of the UL for vitamin C was osmotic diarrhea.

Other Adverse Effects

Several other commonly cited adverse effects of high-dose vitamin C, hemolysis related to glucose-6-phosphate dehydrogenase deficiency (64) and destruction of vitamin B_{12} (65), have never been demonstrated under controlled, experimental conditions. Conversely, serum vitamin B_{12} and serum vitamin C were directly related (33,65); each 57 pmol/L rise in vitamin C was associated with a 27 pmol/L increase in vitamin B_{12} ($P < 0.001$) (33).

Adverse Reactions to Supplemental Vitamin E

Meydani *et al.* (66) demonstrated the safety of vitamin E supplementation (27–360 mg/d for 4 mo) using numerous toxicity indices including nutrient status, liver enzyme function, serum autoantibodies, plasma lipoprotein concentrations, bleeding time, and neutrophil cytotoxicity. Other large intervention trials reported no significant adverse events related to vitamin E supplementation (180–360 mg/d) compared with placebo, providing reassurance for the conduct of large, long-term trials to address the efficacy of vitamin E supplementation (14,15). It is clear, however, that there are a few populations in which caution should be employed regarding vitamin E supplementation, mainly populations at risk of clinically important bleeding (67). Regular cigarette smoking has been associated with aneurysm formation and hemorrhage (17,68), possibly due to reduced levels of serum α1-antitrypsin with increased arterial collagen catabolism, a situation exacerbated by reduced platelet activity (68,69). Because there was a slight increased risk of fatal subarachnoid hemorrhage in smokers supplementing with vitamin E (see above), vitamin E supplementation should be carefully considered for these individuals.

Antiplatelet Effects

Vitamin E supplementation in adults (150 mg/d for 8 wk) increased platelet vitamin E content 70% and decreased platelet aggregation induced by ADP or arachidonic acid (70). Using increasing dosages of vitamin E (50–268 mg/d), Mabile *et al.* (71) reported that the inhibition of platelet aggregation occurred at the 50 mg dosage and was not further affected by the higher dosages. Thus low-dose vitamin E is sufficient to reduce platelet activity. Vitamin E also reduced platelet adhesion to human endothelial cells *in vitro* (72). The antiplatelet effects of vitamin E may be associated with decreased cyclooxygenase activation due to the antioxidant properties of vitamin E, or vitamin E may function to inhibit protein kinase C independent of antioxidant properties (73,74). Although this inhibition of platelet activation may potentially benefit patients at risk of developing thrombotic complications (75), supplementary vitamin E is not currently recommended for either primary or secondary prevention of myocardial infarction (76). Anecdotal reports indicate that the likelihood of postoperative hemorrhage may increase after the preoperative use of vitamin E (67,77). Moreover, patients receiving antithrombotic therapy (e.g., warfarin or aspirin) should be carefully monitored if using vitamin E supplements.

In a study of male smokers, vitamin E use (25 mg/d), alone or in combination with aspirin use (>100 mg/d), increased the risk of clinically important bleeding (78). The extent of bleeding was greater in smokers supplementing with vitamin E than in those who took neither vitamin E nor aspirin (27.1 and 25.8%, respectively, $P = 0.05$). Bleeding tendency was similar in users of aspirin and nonusers of vitamin E or aspirin. The extent of bleeding was greatest in the smokers using both vitamin E and aspirin (33.4%). Bleeding times, however, were not prolonged by vitamin E supplementation (360 mg/d for 1–4 mo) in healthy adults (66,79).

Prooxidant Activity

When protecting against the propagation of fatty acid peroxidation, vitamin E is oxidized to the α-tocopherol radical. *In vivo*, vitamin C and ubiquinol (coenzyme Q) reduce the α-tocopherol radical and regenerate vitamin E (80,81). Dihydrolipoic acid also plays a prominent role in vitamin E recycling; however, its role is indirect in that dihydrolipoic acid reduces dehydroascorbic acid, oxidized vitamin C, and maintains high levels of ascorbic acid to provide for effective recycling of vitamin E (80). Vitamin C levels in plasma are variable and dependent on dietary intake, and dietary vitamin C depletion has been associated with reduced vitamin E concentrations *in vivo* (82,83). About 30% of Americans have marginal vitamin C status, reflecting a diet low in fruits and vegetables or conditions of high oxidative stress, such as smoking or diabetes (84,85). Vitamin E supplementation under these conditions may lead to a redox imbalance by further depleting vitamin C causing prooxidative effects (85,86).

Brown *et al.* (85) examined the effect of a wide range of supplemental vitamin E (0, 70, 140, 560, or 1050 mg/d for 20 wk) on lipid peroxidation in male smokers and in men who had never smoked. Red blood cell vitamin E concentrations were significantly elevated in all groups ingesting vitamin E. The susceptibility of red blood cells to hydrogen peroxide–induced lipid peroxidation *in vitro* was reduced substantially (–63 to –73%, $P < 0.05$) in smokers ingesting 70–1050 mg vitamin E. In the nonsmokers, lipid peroxidation was decreased in those ingesting 70, 140, or 560 mg vitamin E; however, lipid peroxidation was significantly elevated (+36%) in nonsmokers consuming the highest dosage of vitamin E, 1050 mg/d. Moreover, the erythrocyte vitamin E:plasma vitamin C ratio was highest in this group, indicating the highest degree of antioxidant imbalance of all groups. Thus, the increased susceptibility to red blood cell peroxidation in this group may reflect a decrease in vitamin C regeneration of tocopherol and, hence, a prooxidative effect of vitamin E (85). A redox imbalance was also cited as contributing to a 44% increase in platelet-leukocyte coaggregates *ex vivo* in type 2 diabetics ingesting 536 mg vitamin E daily for 6 mo (86).

It seems likely that individuals consuming high-fat diets and who have higher blood lipid concentrations are more likely to respond to vitamin E supplementation because vitamin E absorption and transport would be optimized under these circumstances (87). Thus, a prooxidant effect of vitamin E (or perhaps more appropriately termed, a redox imbalance) might occur only when certain conditions overlap, i.e., inadequate vitamin C nutrition, high blood lipids, and high dietary fat (85).

Necrotizing Enterocolitis

Newborn infants, especially premature infants, often have low plasma vitamin E concentrations and were once considered to suffer from a vitamin E deficiency manifesting as hemolytic anemia, the retinopathy of prematurity, and/or bronchopulmonary dysplasia. Prophylactic use of high-dose vitamin E in these populations, however, did not prevent anemia (88) or retinopathy (89). Furthermore, vitamin E use by neonates has been associated with an increased incidence of necrotizing enterocolitis as well as hemorrhagic complications of prematurity (89,90). The adequate level of vitamin E in infant formulas is set at 0.5 mg/100 kcal and an upper limit has been proposed at 10 mg/100 kcal (63,91).

Supplementation and Drug Interactions

The anticoagulants aspirin and warfarin (Coumadin) are commonly prescribed to patients at risk of heart attacks or stroke. High-dose vitamin E can magnify the anticlotting effects of these drugs and increase the possibility of clinically significant bleeding (78). Bleeding with anticoagulant drug use occurs more frequently in older patients and in patients at risk of vitamin K deficiency due to antibiotic therapy or fat malabsorption syndromes (92). Thus, high-dose vitamin E therapy might be contradicted in these patient populations.

β-Hydroxyl-β-methyl glutarate-CoA reductase inhibitors (statins) in combination with niacin are often the initial therapy prescribed for dyslipidemia to reduce LDL cholesterol and raise high density lipoprotein (HDL) cholesterol. A recent clinical trial examining the efficacy of different lipid-altering and/or antioxidant strategies on coronary artery disease indices demonstrated that when statin/niacin therapy was combined with antioxidant use (β-carotene, 12.5 mg; vitamin C, 500 mg; vitamin E, 268 mg) the beneficial response of HDL to the statin/niacin therapy was markedly attenuated ($P = 0.057$) (93). Although these results must be verified, they demonstrate the importance of investigating interactive effects of popular drug therapies and high-dose vitamin supplementation.

Concern has been expressed regarding high-dose vitamin C use by cancer patients because high intracellular concentrations of vitamin C may enhance the tumor's ability to resist the oxidative stress associated with radiation therapy and/or chemotherapy (94). However, clinical trials have not addressed this theoretical concern. In fact, vitamin C has been shown to contribute to the antitumor effects of mitomycin C in cultured leukemia cells (HL 60) by acting as an electron donor to mitomycin C (95). Finally, it has been documented that high urinary vitamin C can interfere with routine urinalysis for hemoglobin or glucose if dipstick tests rely on redox indicator systems (96) or with the measurement of oxalate in urine and plasma (97). Protocols to avoid vitamin C interference in these measurements should be followed.

References

1. Subar, A.F., and Block, G. (1990) Use of Vitamin and Mineral Supplements: Demographics and Amounts of Nutrients Consumed, *Am. J. Epidemiol. 132*, 1091–1101.
2. Block, G., Cox, C., Madans, J., Schreiber, G.B., Licitra, L., and Melia, N. (1988) Vitamin Supplement Use, by Demographic Characteristics, *Am. J. Epidemiol. 127*, 297–309.
3. Moss, A.J., Levy, A.S., Kim, I., and Park, Y.K. (1989) Use of Vitamin and Mineral Supplements in the United States: Current Users, Types of Products, and Nutrients. U.S. Department of Health and Human Services, Public Health Service, CDC, National Center for Health Statistics, No. 174, Hyattsville, MD.
4. Enstrom, J.E., Kanim, L.E., and Klein, M.A. (1992) Vitamin C Intake and Mortality Among a Sample of the United States Population, *Epidemiology 3*, 194–202.
5. Simon, J.A., Hudes, E.S., and Tice, J.A. (2001) Relation of Serum Ascorbic Acid to Mortality Among US Adults, *J. Am. Coll. Nutr. 20*, 255–263.
6. Jacobs, E.J, Connell, C.J., Patel, A.V., Chao, A., Rodriguez, C., Seymour, J., McCullough, M.L., Calle, E.E., and Thun, M.J. (2001) Vitamin C and Vitamin E Supplement Use and Colorectal Cancer Mortality in a Large American Cancer Society Cohort, *Cancer Epidemiol. Biomark. Prev. 10*, 17–23.
7. Losonczy, K.G., Harris, T.B., and Havlik, R.J. (1996) Vitamin E and Vitamin C Supplement Use and Risk of All-Cause and Coronary Heart Disease Mortality in Older Persons: The Established Populations for Epidemiologic Studies of the Elderly, *Am. J. Clin. Nutr. 64*, 190–196.
8. Khaw, K.T., Bingham, S., Welch, A., Luben, R., Wareham, N., Oakes, S., and Day, N. (2001) Relation Between Plasma Ascorbic Acid and Mortality in Men and Women in EPIC-Norfolk Prospective Study: A Prospective Population Study, *Lancet 357*, 657–663.
9. Loria, C.M, Klag, M.J., Caulfield, L.E., and Whelton, P.K. (2000) Vitamin C Status and Mortality in U.S. Adults, *Am. J. Clin. Nutr. 72*, 139–145.
10. Rimm, E.B., Stampfer, M.J., Ascherio, A., Giovannucci, E., Colditz, G.A., and Willett, W.C. (1993) Vitamin E Consumption and the Risk of Coronary Heart Disease in Men, *N. Engl. J. Med. 328*, 1450–1456.
11. Stampfer, M.J., Hennekens, C.H., Manson, J.E., Colditz, G.A., Rosner, B., and Willett, W.C. (1993) Vitamin E Consumption and the Risk of Coronary Disease in Women, *N. Engl. J. Med. 328*, 1444–1449.
12. The Alpha-Tocopherol, Beta Carotene Cancer Prevention Study Group. (1994) The Effect of Vitamin E and Beta Carotene on the Incidence of Lung Cancer and Other Cancers in Male Smokers, *N. Engl. J. Med. 330*, 1029–1035.
13. Rapola, J.M., Virtamo, J., Ripatti, S., Huttunen, J.K., Albanes, D., Taylor, P.R., and Heinonen, O.P. (1997) Randomised Trial of α-Tocopherol and β-Carotene Supplements on Incidence of Major Coronary Events in Men with Previous Myocardial Infarction, *Lancet 349*, 1715–1720.
14. The Heart Outcomes Prevention Evaluation Study Investigators (2000) Vitamin E Supplementation and Cardiovascular Events in High-Risk Patients, *N. Engl. J. Med. 342*, 154–160.
15. Boaz, M., Smetana, S., Weinstein, T., Matas, Z., Gafter, U., Iaina, A., Knecht, A., Weissgarten, Y., Brunner, D., Fainaru, M., and Green, M.S. (2000) Secondary Prevention with Antioxidants of Cardiovascular Disease in Endstage Renal Disease (SPACE): Randomized Placebo-Controlled Trial, *Lancet 356*, 1213–1218.
16. Ascherio, A., Rimm, E.B., Hernan, M.A., Giovannucci, E., Lawachi, I., Stampfer, M.J.,

and Willett, W.C. (1999) Relation of Consumption of Vitamin E, Vitamin C, and Carotenoids to Risk for Stroke Among Men in the United States, *Ann. Intern. Med. 130*, 963–970.

17. Juvela, S., Poussa, K., and Porras, M. (2001) Factors Affecting Formation and Growth of Intracranial Aneurysms, *Stroke 32*, 485–491.
18. Leppala, J.M., Virtamo, J., Fogelholm, R., Albanes, D., and Heinonen, O.P. (1999) Different Risk Factors for Different Stroke Subtypes, *Stroke 30*, 2535–2540.
19. Leppala, J.M., Virtamo, J., Fogelholm, R., Huttunen, J.K., Albanes, D., Taylor, P.R., and Heinonen, O.P. (2000) Controlled Trial of α-Tocopherol and β-Carotene Supplements on Stroke Incidence and Mortality in Male Smokers, *Arterioscler. Thromb. Vasc. Biol. 20*, 230–235.
20. Levine, M., Conry-Cantilena, C., Wang. Y., Welch, R.W., Washko, P.W., Dhariwal, K.R., Park, J.B., Lazarev, A., Graumlich, J.F., and King, J. (1996) Vitamin C Pharmacokinetics in Healthy Volunteers: Evidence for a Recommended Dietary Allowance, *Proc. Natl. Acad. Sci. USA. 93*, 3704–3709.
21. May, J.M., Qu, Z., and Li, X. (2001) Requirement for GSH in Recycling of Ascorbic Acid in Endothelial Cells, *Biochem. Pharmacol. 62*, 873–881.
22. Mendiratta, S., Qu, Z., and May J.M. (1998) Erythrocyte Ascorbate Recycling: Antioxidant Effects in Blood, *Free Radic. Biol. Med. 24*, 789–797.
23. Omaye, S.T., Skala, J.H., and Jacob, R.A. (1986) Plasma Ascorbic Acid in Adult Males: Effects of Depletion and Supplementation, *Am. J. Clin. Nutr. 44*, 257–264.
24. Johnston, C.S., and Corte, C. (1999) Individuals with Marginal Vitamin C Status Are at High Risk of Developing Vitamin C Deficiency, *J. Am. Diet. Assoc. 99*, 854–856.
25. Tsao, C.S., and Leung, P.Y. (1988) Urinary Ascorbic Acid Levels Following the Withdrawal of Large Doses of Ascorbic Acid in Guinea Pigs, *J Nutr. 118*, 895–900.
26. Schrauzer, G.N., and Rhead, W.J. (1973) Ascorbic Acid Abuse: Effects of Long Term Ingestion of Excessive Amounts on Blood Levels and Urinary Excretion, *Int. J. Vitam. Nutr. Res. 43*, 201–211.
27. Stein, H.B., Hasan, A., and Fox, I.H. (1976) Ascorbic Acid-Induced Uricosuria, *Ann. Intern. Med. 84*, 385–388.
28. Berger, L., and Gerson, C.D. (1977) The Effect of Ascorbic Acid on Uric Acid Excretion with a Commentary on the Renal Handling of Ascorbic Acid, *Am. J. Med. 62*, 71–76.
29. Trinchieri, A., Ostini, F., and Nespoli, R. (1998) Hyperoxaluria in Patients with Idiopathic Calcium Nephrolithiasis, *J. Nephrol. 11 (Suppl. 1)*, 70–72.
30. Tsao, C.S., and Salimi, S.L. (1984) Effect of Large Intake of Ascorbic Acid on Urinary and Plasma Oxalic Acid Levels, *Int. J. Vitam. Nutr. Res. 54*, 245–249.
31. Bihi, G., and Meyers, A. (2001) Recurrent Renal Stone Disease — Advances in Pathogenesis and Clinical Management, *Lancet 358*, 651–656.
32. Curhan, G.C., Willett, W.C., Rimm, E.B., and Stampfer, M.J. (1996) A Prospective Study of the Intake of Vitamins C and B6 and the Risk of Kidney Stones in Men, *J. Urol. 155*, 1847–1851.
33. Simon, J.A., and Hudes, E.S. (1999) Relation of Serum Ascorbic Acid to Serum Vitamin B12, Serum Ferritin, and Kidney Stones in US Adults, *Arch. Intern. Med. 159*, 619–624.
34. Curan, G.C., Willett, W.C., Speizer, F.E., and Stampfer, M.J. (1999) Intake of Vitamins B6 and C and the Risk of Kidney Stones in Women, *J. Am. Soc. Nephrol. 10*, 840–845.
35. Ono. K. (1986) Secondary Hyperoxalemia Caused by Vitamin C Supplementation in Regular Hemodialysis Patients, *Clin. Nephrol. 26*, 239–243.

36. Rose, R.C., and Bode, A.M. (1993) Biology of Free Radical Scavengers: An Evaluation of Ascorbate, *FASEB J. 7*, 1135–1142.

37. Cai, L., Koropatnick, J., and Cherian, M.G. (2001) Roles of Vitamin C in Radiation-Induced DNA Damage in Presence and Absence of Copper, *Chem.-Biol. Interact. 137*, 75–88.

38. Minetti, M., Forte, T., Soriani, M., Quaresima, V., Menditto, A., and Ferrari, M. (1992) Iron-Induced Ascorbate Oxidation in Plasma as Monitored by Ascorbate Free Radical Formation, *Biochem. J. 282*, 459–465.

39. Berger, R.M., Polidori, M.C., Dabbagh, A., Evans, P.J., Halliwell, B., Morrow, J.K., Roberts, L.J., and Frei, B. (1997) Antioxidant Activity of Vitamin C in Iron-Overloaded Human Plasma, *J. Biol. Chem. 272*, 15656–15660.

40. Retsky, K.L., Freeman, M.W., and Frei, B. (1993) Ascorbic Acid Oxidation Product(s) Protect Human Low Density Lipoprotein Against Atherogenic Modification, *J. Biol. Chem. 268*, 1304–1309.

41. Chen, K., Suh, J., Carr, A.C., Morrow, J.D., Zeind, J., and Frei, B. (2000) Vitamin C Suppresses Oxidative Lipid Damage In Vivo, Even in the Presence of Iron Overload, *Am. J. Physiol. 279*, E1406–E1412.

42. Proteggente, A.R., Rehman, A., Halliwell, B., and Rice-Evans, C.A. (2000) Potential Problems of Ascorbate and Iron Supplementation: Pro-Oxidant Effect In Vivo? *Biochem. Biophys. Res. Commun. 277*, 535–540.

43. Podmore, I.D., Griffiths. H.R., Herbert, K.E., Mistry, N., Mistry, P., and Lunec, J. (1998) Vitamin C Exhibits Pro-Oxidant Properties, *Nature 392*, 559.

44. Collins, A.R., Gedik, C.M., Olmedilla, B., Southon, S., and Bellizzi, M. (1998) Oxidative DNA Damage Measured in Human Lymphocytes: Large Differences Between Sexes and Between Countries, and Correlations with Heart Disease Mortality, *FASEB J. 12*, 1397–1400.

45. LePage, F., Margot, A., Grollman, A.P., Sarasin, A., and Gentil, A. (1995) Mutagenicity of a Unique 8-Oxoguanine in a Human Ha-ras Sequence in Mammalian Cells, *Carcinogenesis 16*, 2779–2784.

46. Wiseman, H., and Halliwell, B. (1996) Damage to DNA by Reactive Oxygen and Nitrogen Species: Role in Inflammatory Disease and Progression to Cancer, *Biochem. J. 313*, 17–29.

47. Fraga, C.G., Motchnik, P.A., Shigenaga, M.D., Helbock, H.J., Jacob, R.A., and Ames, B.N. (1991) Ascorbic Acid Protects Against Endogenous Oxidative DNA Damage in Human Sperm, *Proc. Natl. Acad. Sci. USA 88*, 11003–11006.

48. Lee, B.M., Lee, S.K., and Kim, H.S. (1998) Inhibition of Oxidative DNA Damage, 8-OHdG, and Carbonyl Contents in Smokers Treated with Antioxidants (Vitamin E, Vitamin C, β-Carotene and Red Ginseng), *Cancer Lett. 132*, 219–227.

49. Lee, S.H., Oe, T., and Blair, I.A. (2001) Vitamin C-Induced Decomposition of Lipid Hydroperoxides to Endogenous Genotoxins, *Science 292*, 2083–2086.

50. Hagenlocher, T., Nair, J., Becker, N., Korfmann, A., and Bartsch, H. (2001) Influence of Dietary Fatty Acid, Vegetable, and Vitamin Intake on Etheno-DNA Adducts in White Blood Cells of Healthy Female Volunteers: A Pilot Study, *Cancer Epidemiol. Biomark. Prev. 10*, 1187–1191.

51. Bartsch, H., Nair, J., and Velic, I. (1997) Etheno-DNA Base Adducts as Tools in Human Cancer Aetiology and Chemoprevention, *Eur. J. Cancer Prev. 6*, 529–534.

52. Mayer, C., Schmezer, P., Freese, R., Mutanen, M., Hietanen, E., Obe, G., Basu, S., and Bartsch, H. (2000) Lipid Peroxidation Status, Somatic Mutations and Micronuclei in

Peripheral Lymphocytes: A Case Observation on a Possible Interrelationship, *Cancer Lett.* *152*, 169–173.

53. Mao, X., and Yao, G. (1992) Effect of Vitamin C Supplementations on Iron Deficiency Anemia in Chinese Children, *Biomed. Environ. Sci. 5*, 125–129.

54. Sharma, D.C., and Mathur, R. (1995) Correction of Anemia and Iron Deficiency in Vegetarians by Administration of Ascorbic Acid, *Indian J. Physiol. Pharmacol. 39*, 403–406.

55. Fleming, D.J., Jacques, P.F., Dallal, G.E., Tucker, K.L., Wilson, P.W., and Wood, R.J. (1998) Dietary Determinants of Iron Stores in a Free-Living Elderly Population: The Framingham Heart Study, *Am. J. Clin. Nutr. 67*, 722–733.

56. Reller, K., Dresow, B., Collell, M., Fischer, R., Engelhardt, R., Nielsen, P., Durken, M., Politis, C., and Piga, A. (1998) Iron Overload and Antioxidant Status in Patients with β-Thalassemia Major, *Ann. N.Y. Acad. Sci. 850*, 463–465.

57. Cappellini, M.D., Tavazzi, D., Duca, L., Marelli, S., and Fiorelli, G. (2000) Non-Transferrin-Bound Iron, Iron-Related Oxidative Stress and Lipid Peroxidation in β-Thalassemia Intermedia. *Transfus. Sci. 23*, 245–246.

58. Schofield, R.S., Aranda, J.M., Hill, J.A., and Streiff, R. (2001) Cardiac Transplantation in a Patient with Hereditary Hemochromatosis: Role of Adjunctive Phlebotomy and Erythropoietin, *J. Heart Lung Transplant. 20*, 696–698.

59. Cohen, A., Cohen, I.J., and Schwartz, E. (1981) Scurvy and Altered Iron Stores in Thalassemia Major, *N. Engl. J. Med. 304*, 158–160.

60. Giardini, O., Cantani, A., Donfrancesco, A., Martino, F., Mannarino, O., D'Eufemia, P., Miano, C., Ruberto, U., and Lubrano, R. (1985) Biochemical and Clinical Effects of Vitamin E Administration in Homozygous Beta-Thalassemia, *Acta Vitaminol. Enzymol. 7*, 55–60.

61. Tesoriere, L., D'Arpa, D., Butera, D., Allergra, M., Renda, D., Maggio, A., Bongiorno, A., and Livrea, M.A. (2001) Oral Supplements of Vitamin E Improve Measures of Oxidative Stress in Plasma and Reduce Oxidative Damage to LDL and Erythrocytes in Beta-Thalassemia Intermedia Patients, *Free Radic. Res. 34*, 529–540.

62. Brown, K.E., Poulos, J.E., Li, L., Soweid, A.M., Ramm, G.A., O'Neill, R., Britton, R.S., and Bacon, B.R. (1997) Effect of Vitamin E Supplementation on Hepatic Fibrogenesis in Chronic Dietary Iron Overload, *Am. J. Physiol. 272*, G116–G23.

63. Food and Nutrition Board. (2000) *Dietary Reference Intakes for Vitamin C, Vitamin E, Selenium, and Carotenoids*, National Academy Press, Washington.

64. Rees, D.C., Kelsey, H., and Richards, J.D.M. (1993) Acute Haemolysis Induced by High Dose Ascorbic Acid in Glucose-6-Phosphate Dehydrogenase Deficiency, *Br. Med. J. 306*, 841–842.

65. Herbet, V., Jacob, E., and Wong, K.T.J. (1978) Low Serum Vitamin B12 Levels in Patients Receiving Ascorbic Acid in Megadose: Studies Concerning the Effect of Ascorbate on Radioisotope Vitamin B12 Assay, *Clin. Nutr. 31*, 253–258.

66. Meydani, S.N., Meydani ,M., Blumberg, J.B., Leka, L.S., Pedrosa, M., Diamond, R., Schaefer, E.J. (1998) Assessment of the Safety of Supplementation with Different Amounts of Vitamin E in Healthy Older Adults, *Am. J. Clin. Nutr. 68*, 311–318.

67. Spencer, A.P. (2000) Vitamin E: Cautionary Issues, *Curr. Treat. Options Cardiovasc. Med. 2*, 193–195.

68. Singh, K., Bonaa, K.H., Jacobsen, B.K, Bjork, L., and Solberg, S. (2001) Prevalence of and Risk Factors for Abdominal Aortic Aneurysms in a Population-Based Study, *Am. J. Epidemiol. 154*, 236–244.

69. Gaetani, P., Tartara, F., Tancioni, F., Klersy, C., Forlino, A., and Baena, R.R. (1996) Activity of Alpha 1-Antitrypsin and Cigarette Smoking in Subarachnoid Haemorrhage from Ruptured Aneurysm, *J. Neurol. Sci. 141*, 33–38.

70. Calzada, C., Bruckdorfer, K.R., and Rice-Evans, C.A. (1997) The Influence of Antioxidant Nutrients on Platelet Function in Healthy Volunteers, *Atherosclerosis 128*, 97–105.

71. Mabile, L., Bruckdorfer, K.R., and Rice-Evans, C. (1999) Moderate Supplementation with Natural α-Tocopherol Decreases Platelet Aggregation and Low-Density Lipoprotein Oxidation, *Atherosclerosis 147*, 177–185.

72. Szuwart, T., Brzoska, T., Luger, T.A., Filler, T., Peuker, E., and Dierichs, R. (2000) Vitamin E Reduces Platelet Adhesion to Human Endothelial Cells In Vitro, *Am. J. Hematol. 65*, 1–4.

73. Freedman, J.E., Farhat, J.H., Loscalzo, J., and Keaney, JF. (1996) α-Tocopherol Inhibits Aggregation of Human Platelets by a Protein Kinase C-Dependent Mechanism, *Circulation 94*, 2434–2440.

74. Freedman, J.E., Li, L., Sauter, R., and Keaney, J.F. (2000) α-Tocopherol and Protein Kinase C Inhibition Enhance Platelet-Derived Nitric Oxide Release, *FASEB J. 14*, 2377–2379.

75. Janicki, K., Dmoszynska, A., Janicka, L., Stettner, S., and Jesipowicz, J. (1992) Influence of Antiplatelet Drugs on Occlusion of Arteriovenous Fistula in Uraemic Patients, *Int. Urol. Nephrol. 24*, 83–89.

76. Spinler, S.A., Hilleman, D.E., Cheng, J.W., Howard, P.A., Mauro, V.F., Lopez, L.M., Munger, M.A., Gardner, S.F., and Nappi, J.M. (2001) New Recommendations from the 1999 American College or Cardiology/American Heart Association Acute Myocardial Infarction Guidelines, *Ann. Pharmacother. 35*, 589–617.

77. Norred, C.L., and Finlayson, C.A. (2000) Hemorrhage After the Preoperative Use of Complementary and Alternative Medicines, *AANA J. 68*, 217–220.

78. Liede, K.E., Haukka, J.K., Saxen, L.M., and Heinonen, O.P. (1998) Increased Tendency Towards Gingival Bleeding Caused by Joint Effect of α-Tocopherol Supplementation and Acetylsalicylic Acid, *Ann. Med. 30*, 542–546.

79. Stampfer, M.J., Jakubowski, J.A., Faigel, D., Vaillancourt, R., and Deykin, D. (1988) Vitamin E Supplementation Effect on Human Platelet Function, Arachidonic Acid Metabolism, and Plasma Prostacyclin Levels, *Am. J. Clin. Nutr. 47*, 700–706.

80. Kagan, V.E., and Tyurina, Y.Y. (1998) Recycling and Redox Cycling of Phenolic Antioxidants, *Ann. N.Y. Acad. Sci. 854*, 425–434.

81. Lass, A., and Sohal, R.S. (1998) Electron Transport-Linked Ubiquinone Dependent Recycling of α-Tocopherol Inhibits Autooxidation of Mitochondrial Membranes, *Arch. Biochem. Biophys. 352*, 229–236.

82. Jacob, R.A., Kutnink, M.A., Csallany, A.S., Daroszewska, M., and Burton, G.W. (1996) Vitamin C Nutriture Has Little Short-Term Effect on Vitamin E Concentrations in Healthy Women, *J. Nutr. 126*, 2268–2277.

83. Tanaka, K., Hashimoto, T., Tokumaru, S., Iguchi, H., and Kojo, S. (1997) Interactions Between Vitamin C and Vitamin E Are Observed in Tissues of Inherently Scorbutic Rats, *J. Nutr. 127*, 2060–2064.

84. Seghieri, G., Martinoli, L., Felice, M.D., Anichini, R., Fazzinig, A., Ciuti, M., Miceli, M., Gaspag, L., and Franconig, F. (1998) Plasma and Platelet Ascorbate Pools and Lipid Peroxidation in Insulin-Dependent Diabetes Mellitus, *Eur. J. Clin. Investig. 28*, 659–663.

85. Brown, K.M., Morrice, P.C., and Duthie, G.G. (1997) Erythrocyte Vitamin E and Plasma Ascorbate Concentrations in Relation to Erythrocyte Peroxidation in Smokers and Nonsmokers: Dose Response to Vitamin E Supplementation, *Am. J. Clin. Nutr. 65*, 496–502.

86. Ferber, P., Moll, K., Koschinsky, T., Rosen, P., Susanto, F., Schwippert, B., and Tschope, D. (1999) High Dose Supplementation of RRR-α-Tocopherol Decreases Cellular Hemostatis but Accelerates Plasmatic Coagulation in Type 2 Diabetes Mellitus, *Horm. Metab. Res. 31*, 665–671.

87. Keenoy, B.M., Shen, H., Engelen, W., Vertommen, J., Van Dessel, G., Lagrou, A., and DeLeeuw, I. (2001) Long-Term Pharmacologic Doses of Vitamin E Only Moderately Affect the Erythrocytes of Patients with Type 1 Diabetes Mellitus, *J. Nutr. 131*, 1723–1730.

88. Zipursky, A., Brown, E.J., Watts, J., Milner, R., Rand, C., Blanchette, V.S., Bell, E.F., Paes, B., and Ling, E. (1987) Oral Vitamin E Supplementation for the Prevention of Anemia in Premature Infants: A Controlled Trial, *Pediatrics 79*, 61–68.

89. Phelps, D.L., Rosenbaum, A.L., Isenberg, S.J., Leake, R.D., Dorey, F.J. (1987) Tocopherol Efficacy and Safety for Preventing Retinopathy of Prematurity: A Randomized, Controlled, Double-Masked Trial, *Pediatrics 79*, 489–500.

90. Finer, N.N., Peters, K.L., Hayek, Z., and Merkel, C.L. (1984) Vitamin E and Necrotizing Enterocolitis, *Pediatrics 73*, 387–393.

91. Bell, E.F. (1989) Upper Limit of Vitamin E in Infant Formulas, *J. Nutr. 119*, 1829–1831.

92. Sebastian, J.L., and Tresch, D.D. (2000) Use of Oral Anticoagulants in Older Patients, *Drugs Aging 16*, 409–435.

93. Cheung, M.C., Zhao, X.Q., Chait, A., Albers, J.J., and Brown, B.G. (2001) Antioxidant Supplements Block the Response of HDL to Simvastatin-Niacin Therapy in Patients with Coronary Artery Disease and Low HDL, *Arterioscler. Thromb. Vasc. Biol. 21*, 1320–1326.

94. Agus, D.B., Vera, J.C., and Golde, D.W. (1999) Stromal Cell Oxidation: A Mechanism by Which Tumors Obtain Vitamin C, *Cancer Res. 59*, 4555–4558.

95. Kammerer, C., Czermak, I., Getoff, N., and Kodym, R. (1999) Enhancement of Mitomycin C Efficiency by Vitamin C, E-Acetate and Beta-Carotene Under Irradiation. A Study In Vitro, *Anticancer Res. 19*, 5319–5321.

96. Brigden, M.L., Edgell, D., McPherson, M., Leadbeater, A., and Hoag, G. (1991) High Incidence of Significant Urinary Ascorbic Acid Concentrations in a West Coast Population—Implications for Routine Urinalysis, *Clin. Chem. 38*, 426–431.

97. Wilson, D.M., and Liedtke, R.R. (1991) Modified Enzyme-Based Colorimetric Assay of Urinary and Plasma Oxalate with Improved Sensitivity and No Ascorbate Interference: Reference Values and Sample Handling Procedures, *Clin. Chem. 37*, 1229–1235.

Chapter 9

Vitamin E: An Introduction

Lester Packer[a] and Ute C. Obermüller-Jevic[b]

[a]University of Southern California, School of Pharmacy, Department of Molecular Pharmacology and Toxicology, Los Angeles, CA 90089–9121
[b]University of California, Davis, Department of Internal Medicine, Division of Pulmonary and Critical Care Medicine, Davis, CA 95616

Introduction

Eighty years ago vitamin E was discovered to be an essential micronutrient by Evans and Bishop (1). With the later recognition that this vitamin is not only necessary for successful reproduction but also the major lipophilic antioxidant in the human body, and subsequently, the causative role of oxidative stress in the pathogenesis of age-related and common chronic diseases, intense research has been conducted to evaluate the possible protective and curative potential of vitamin E. In the meantime, several other specific biological functions of vitamin E have been revealed apart from its antioxidant function including modulation of cell signaling [e.g., protein kinase C (PKC) inhibition], gene expression, and cell proliferation. Nonetheless, there are still many gaps in understanding the physiologic role of vitamin E in humans. Except in the case of nutrient deficiency, not much is clearly proven about the functional importance of vitamin E for human health and disease prevention. The following introduction provides a brief overview of the known biological functions of vitamin E, focusing on α-tocopherol as the major form of vitamin E in the human body, and highlights some milestones in the past eight decades of vitamin E research.

Molecular Structure of Vitamin E

Vitamin E is the generic name for a group of eight plant-derived, lipid-soluble substances ("tocols") including four tocopherols and four tocotrienols (Fig. 9.1). The molecular structure of vitamin E is comprised of a chromanol ring with a side chain located at the C2 position. Tocopherols have a saturated phytyl side chain and tocotrienols have an unsaturated isoprenoid side chain. The number and position of methyl groups located around the chromanol ring vary among the different tocopherols and tocotrienols, and account for the designation as α-, β-, γ-, or δ-forms. The α-forms of tocopherol and tocotrienol contain three methyl groups, whereas the β- and γ-forms have two, and the δ-forms have one methyl group. Natural and natural-source tocopherols occur in *RRR*-configuration only (formerly designated as *d*-α-tocopherol). In contrast, synthetic α-tocopherol consists of an equal racemic mixture of eight stereoisomers (*RRR, RSR, RRS, RSS, SRR, SSR, SRS, SSS*) arising from the

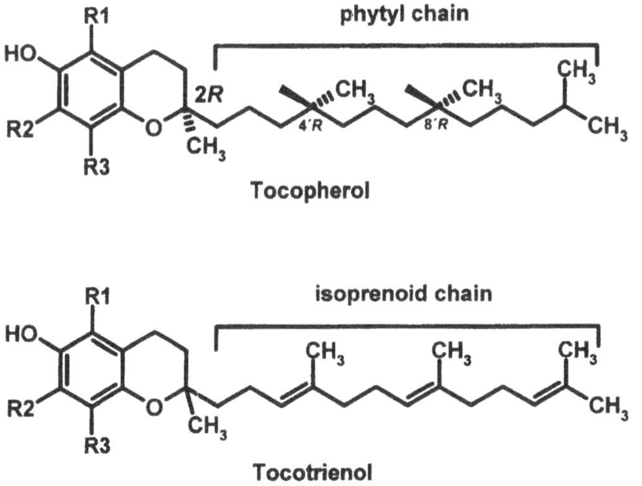

The four isoforms of tocopherols and tocotrienols:

α-:	R_1=CH_3	R_2=CH_3	R_3=CH_3
β-:	R_1=CH_3	R_2=H	R_3=CH_3
γ-:	R_1=H	R_2=CH_3	R_3=CH_3
δ-:	R_1=H	R_2=H	R_3=CH_3

Fig. 9.1. The molecular structure of vitamin E. Tocopherols and tocotrienols exist in four different isoforms (α-, β-, γ-, and δ-forms). Tocopherol is shown in its naturally occurring *RRR*-configuration. *Source*: Adapted from Food and Nutrition Board, 2000 (Ref. 6).

three chiral centers of the molecule at positions C2, C4′ and C8′ and designated as all-*rac*-α-tocopherol (or *dl*-α-tocopherol) (2).

Biological Activity of Vitamin E

The various forms of vitamin E differ in their biological activities. α-Tocopherol is the most common form of vitamin E occurring in human blood and tissues, and it has the highest biological activity among all tocopherols and tocotrienols. Moreover, the human body preferentially accumulates the *RRR*- or 2*R*-forms of vitamin E (3–5). Thus natural *RRR*-α-tocopherol has a higher bioavailability and "biological activity" than synthetic all-*rac*-α-tocopherol. This was considered in the new guidelines on vitamin E intake as published in 2000 by the Food and Nutrition Board (FNB), U.S. National Academy of Sciences (6). According to the FNB, only *RRR*-α-tocopherol itself, or the *RRR*- and 2*R*-stereoisomers of all-*rac*-α-tocopherol meet the vitamin E requirements in humans, resulting in a twofold higher potency of natural vs. synthetic α-tocopherol sources. This value has been challenged by Hoppe and Krennrich (7).

The biological activity of vitamin E in dietary supplements is usually expressed as international units (IU). As published in the United States Pharmacopeia (USP), 1 IU is defined as the biological activity of 1 mg all-*rac*-α-tocopheryl acetate. Other biological activities of vitamin E are 1 mg all-*rac*-α-tocopherol = 1.1 IU; 1 mg *RRR*-α-tocopherol = 1.49 IU; 1 mg γ-tocopherol = 0.15 IU (8). Formerly, the biological activities of different vitamin E forms were often reported as α-tocopherol equivalents (α-TE). One α-TE was defined, for example, as 1 mg α-tocopherol, or 2 mg β-tocopherol, or 10 mg γ-tocopherol, or 3.3 mg α-tocotrienol (9).

Dietary Sources, Supplements, and Recommended Intake of Vitamin E

Vegetable oils and lipid-rich plant products (e.g. nuts, seeds, grains) are the main dietary sources of vitamin E (10,11). In Western diets, vitamin E intake derives mainly from fats and oils contained in margarine, mayonnaise, salad dressing, and desserts, and increasingly also from fortified food (e.g., breakfast cereals, milk, fruit juices) (12–14). It is noteworthy that the U.S. diet contains large amounts of γ-tocopherol compared with populations in other Western countries, which is a result of the high consumption of soybean and corn oils containing more γ- than α-tocopherol (15). Vitamin E used for food fortification or dietary supplements consists mainly of α-tocopherol, derived either from natural sources (i.e., methylated γ-tocopherol from vegetable oil) or from synthetic production; it is usually esterified to increase stability.

The most recent data available on vitamin E intake in the United States were reported in 1999 from the Third National Health and Nutrition Examination Survey (NHANES III, 1988–1994) among >16,000 U.S. adults showing a mean vitamin E intake of 10.4 α-TE in men and 8.0 α-TE in women (16). With the new definition of vitamin E activity from the FNB (6) distinguishing among α-tocopherol stereoisomers, these intake values of vitamin E are likely not as high. New studies are required to update these data on vitamin E intake because dietary patterns have changed during recent years and the consumption of fortified food and vitamin E supplements has increased.

The FNB indicates a recommended daily allowance (RDA) for adults of 15 mg vitamin E including food, fortified food, and supplements (6). This RDA refers to α-tocopherol because it is the only form of vitamin E that has been shown to reverse deficiency symptoms in humans. The recommendations are based largely on *in vitro* studies of M. Horwitt dating back to 1960; these studies are still considered to offer the most adequate data for defining the physiologic status and health benefits of vitamin E in humans. In these studies, prevention of H_2O_2-induced erythrocyte hemolysis was used as test system (17). However, except when vitamin E is clearly deficient the erythrocyte hemolysis test is not a useful indicator of vitamin E functional status. Thus, the guidelines of the FNB were discussed critically in the literature and it was strongly agreed that other assays and new biomarkers were required in the future to define the physiologic role and beneficial potential of vitamin E in humans (18,19).

The recommended daily intake of 15 mg vitamin E is considered rather unlikely to be achieved by North Americans and other Western countries through diet alone.

Although universal dietary supplementation has not been not recommended, a dose of 200 mg/d vitamin E (*RRR*-α-tocopherol) has been suggested to saturate plasma levels and elevate tissue levels; this may have possible health effects in the long term (20). In the future, subpopulations at risk for vitamin E deficiency in well-nourished populations such as the United States and other Western countries have to be defined. These specific groups, e.g., the elderly or individuals suffering from chronic diseases, may benefit in particular from regular intake of vitamin E supplements.

As an upper limit for supplemental vitamin E intake, the FNB published a dose for adults of 1 g/d α-tocopherol (i.e., 1500 IU *RRR*- or 1100 IU all-*rac*-α-tocopherol); this dose is considered safe, showing no apparent side effects (6). In human intervention studies, various doses of vitamin E up to 3600 IU/d have been used (21). Nevertheless, conclusive evidence from long-term studies regarding biological effects and safety of chronic intake of pharmacologic doses of vitamin E are lacking. In the Alpha-Tocopherol, Beta Carotene (ATBC) Cancer Prevention Study (22) with Finnish smokers consuming 50 IU/d vitamin E for 6 y, an increase in mortality from hemorrhagic stroke was observed; however, other intervention studies did not report such an adverse effect (23,24). It was suggested that pharmacologic doses of vitamin E are contraindicated in persons with blood coagulation disorders because vitamin E might exacerbate defects in the blood coagulation system due to its inhibitory effects on platelet aggregation (19,25).

Uptake, Distribution and Metabolism of Vitamin E

Together with dietary fat, α-tocopherol and all other forms of vitamin E are absorbed in the digestive tract, incorporated into chylomicrons and transported in the lymphatic system. Part of the absorbed stereoisomers is taken up into extrahepatic tissues by the action of lipoprotein lipase, and the remainder is delivered in chylomicron remnants to the liver (26). The distribution of vitamin E into the circulation is regulated by a cytosolic α-tocopherol transfer protein (α-TTP) in the liver, which is selective for α-tocopherol in its *RRR*- or 2*R*-forms (27). Other tocopherols and tocotrienols exert much less affinity for α-TTP as assessed by ligand specificity, e.g., 38% for β-tocopherol, 9% for γ-tocopherol, 2% for δ-tocopherol, and 12% for α-tocotrienol (28). α-TTP plays an important role in maintaining plasma levels of vitamin E (29). The function of α-TTP and the mechanism of α-tocopherol regulation were reviewed recently (30,31).

In the liver, vitamin E is incorporated into very low density lipoproteins (VLDL) and released to the systemic blood circulation. Excess amounts of α-tocopherol along with the other absorbed forms of tocopherols and tocotrienols are metabolized or eliminated by the biliary tract. VLDL are converted into low density lipoproteins (LDL), and excess surface components including α-tocopherol are transferred to high density lipoproteins (HDL). Delivery of α-tocopherol to peripheral tissues takes place *via* binding of LDL to LDL receptors and subsequent cellular uptake. Vitamin E tends to accumulate in adipose tissues. The metabolism and kinetics of vitamin E as well as the function of vitamin E regulatory proteins have been reviewed (32,33). Cytosolic tocopherol-associated proteins (TAP) have been reported, showing α-tocopherol–specific binding characteristics and also nuclear translocation and transcriptional activation in various mammalian cell types and organs (34,35). TAP seems to play an important role in vitamin E–induced gene expression (36); however, its biological functions are not widely known. A further cytosolic vitamin E regulatory protein, i.e., tocopherol-binding protein (TBP) has been reported with yet unknown functions (37,38).

Data on serum levels of vitamin E from population-based studies are scarce. Analysis of the data obtained from the NHANES III study shows mean serum levels of α-tocopherol of 26.8 (0.65–232.18) μmol/L and a mean α-tocopherol:cholesterol ratio of 5.1 (16). Although no differences in gender have been reported, significant differences were found among races, with the highest serum levels in Caucasians and the lowest in African-Americans. Approximately one third of the NHANES III study population had low vitamin E concentrations (<20 μmol/L). A recent study in five European countries among 350 healthy adults showed mean α-tocopherol serum levels of 26.4 (13.23–46.40) μmol/L and a mean α-tocopherol:cholesterol ratio of 6.5 μmol/mmol (39). The lowest acceptable vitamin E level in plasma has been determined to be 11.6 μmol/L (0.5 mg/dL) and a ratio of vitamin E:cholesterol of 2.25 μmol/mmol (40).

It should be mentioned that serum levels of vitamin E correlate with cholesterol levels, and hence do not necessarily correlate with vitamin E intake (16). However, except for non- or poorly responding subjects, serum levels of vitamin E can usually be increased up to threefold by intake of dietary supplements reaching a saturation level (20).

Milestones in Vitamin E Research

Four milestones of vitamin E research have been recognized, namely, (i) its biological importance in reproduction and essentiality as a micronutrient; (ii) its unique role as lipophilic antioxidant in lipoproteins and cell membranes; (iii) its effects on cell signaling and gene expression; and (iv) the identification of vitamin E deficiency diseases and the recognition of beneficial effects of vitamin E in human health and disease prevention.

Milestone I: Biological importance of vitamin E in reproduction and essentiality. Initially, vitamin E was recognized as a nutritional factor required to ensure normal reproduction in rats (1) and was named according to a consecutive alphabetical order preceded by the discovery of vitamins A–D (41). Later, vitamin E was called α-tocopherol from the Greek term "tokos" (childbirth), "phero" (to bear), and –ol, indicating an alcohol (42).

Rats fed a diet low in vitamin E had reduced fertility and a high rate of fetal resorption. However, when animals were fed a lipophilic fraction from lettuce or, as later shown, wheat germ oil, their fertility was retained and a successful implantation of the embryo was observed (42,43).

The biological activity of vitamin E is based on this rat resorption-gestation assay. The family of natural vitamin E molecules as well as the stereoisomers of all-

rac-α-tocopherol all exhibit vitamin E activity to varying degrees in this bioassay. Unfortunately, this assay of reproductive activity in pregnant rats may have limited relevance to human health.

Milestone II: Unique role of vitamin E as a lipophilic antioxidant in lipoproteins and cell membranes. In the 1950s, it was recognized under the leadership of A.L. Tappel's group that vitamin E is the body's major lipid-soluble antioxidant protecting the lipoproteins and membranes in which it resides against free radical–mediated lipid peroxidation which, if not prevented or interrupted by vitamin E, causes widespread oxidative molecular damage and pathology (44,45). All natural isoforms and synthetic stereoisomers of vitamin E exhibit to varying degrees the ability to inhibit lipid peroxidation as a "chain-breaking" antioxidant (46–48).

Vitamin E destroys primarily peroxyl radicals and thus protects polyunsaturated fatty acids (PUFA) from oxidation (49). Additionally, vitamin E scavenges a variety of oxygen-derived free radicals including alkoxyl radicals, superoxide, and other reactive oxygen species (ROS) such as singlet oxygen and ozone, and it reacts with nitrogen species (50). The antioxidant action of vitamin E has been documented in numerous human and animal studies, including protection of lipids (51–55) and DNA (56–58). Only one study investigated in an animal model of diabetic pregnancy the effect of vitamin E on the formation of oxidatively modified proteins showing reduction of protein carbonyls by combined treatment with vitamins E and C (59).

Vitamin E participates in an *antioxidant network* (a concept advanced by L. Packer's group); thus, vitamin E radicals can be recycled or regenerated back to their native form, e.g., by vitamin C (60–62). When vitamin C radicals form as a result, they may be regenerated by thiol or polyphenol antioxidants in the body. These interactions between redox antioxidant substances and enzymes form the basis for the body's underlying antioxidant defense system (Fig. 9.2). All naturally occurring analogs and synthetic stereoisomers of vitamin E exhibit redox cycling activity to varying degrees in the system of antioxidant defense against oxidative stress (63–66).

When vitamin E interacts with lipid peroxyl and other lipid radicals in cell membranes or lipoproteins, a tocopheroxyl or tocotrienoxyl radical is formed (67,68). However, like other natural antioxidants such as vitamin C or polyphenols, vitamin E radicals are not as dangerous a species as the ones they have destroyed because the free electron is delocalized around their chemical ring structure. This is the unique feature of such biological antioxidants.

"Oxidative stress" is the term particularly designated by H. Sies to call attention to the imbalance that arises when exposure to oxidants changes the normal redox status of tissue antioxidants (69,70). Oxidative stress occurs from an increased formation of ROS and/or from a diminished ability to inactivate these species. So-called "free radicals" or reactive oxygen and nitrogen species are continuously generated under physiologic conditions from metabolism or arise from strenuous and traumatic exercise, and from exposure to environmental stress (e.g., ultraviolet light, cigarette smoke, pollution, and chemicals), infectious microorganisms, viruses, parasites, and during the aging process. These highly reactive species continuously interact in bio-

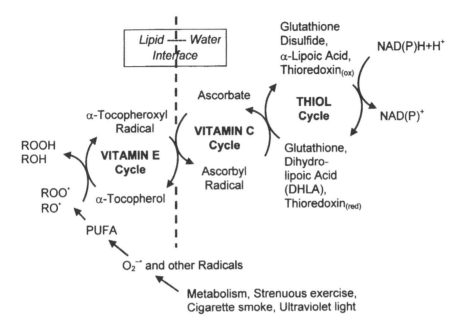

Fig. 9.2. The antioxidant network concept.

logical systems, and there is evidence that both oxidants and antioxidant defense must be kept in balance to minimize molecular, cellular, and tissue damage (71,72).

Direct cellular defenses against oxidative stress are comprised of antioxidant enzymes such as superoxide dismutase, catalase, glutathione (GSH), and thioredoxin reductase/peroxidase on the one hand, and of antioxidants on the other hand. Secondary defenses include proteolytic and repair processes that degrade and eliminate damaged molecules and thus reduce damage. If antioxidant defenses are overwhelmed due to a huge amount of oxidative stress, or if defense systems are impaired due to direct damage, age-dependent degradation processes, or lack of micronutrient supply, accumulation of oxidative damage occurs. All of the major *aqueous* low-molecular-weight antioxidants such as GSH (or its oxidation product glutathione disulphide, GSSG), vitamin C (ascorbate or its 1- or 2-electron oxidized form, the ascorbate radical or dehydroascorbate), protein thiols such as thioredoxin and glutaredoxin, and the *lipophilic* antioxidants ubiquinol (and its 1-electron oxidation product, the ubiquinone semiquinone radical, or 2-electron oxidation product, ubiquinone), and vitamin E as tocopherols or tocotrienols (and their 1-electron oxidation product, tocopheroxyl, or tocotrienoxyl radicals or 2-electron oxidation product, tocopherylquinones) are in a redox balance equilibrium, which can be shifted to also reflect "oxidative stress." Recovery of the oxidative imbalance seen in oxidative stress requires NADPH- and NADH-dependent reactions of reducing metabolism. Indeed, all of the redox-based antioxidants appear to interact with one another through an antioxidant network of nonenzymatic and enzymatic reactions.

Mild oxidative stress affects cell signaling pathways and gene expression and is involved in the regulation of cellular functions (73). However, if an excess of oxidative stress occurs and the rate of oxidation exceeds the rate of repair, the strength of the entire antioxidant system will be weakened and concentrations of antioxidants will decrease (74). It is hypothesized that people with low amounts of antioxidants may be more prone to oxidative tissue damage, and thus to disease. The concept of oxidative stress has been implicated in many diseases including infectious diseases, age-related degenerative diseases, and chronic diseases (75–78).

Oxidative stress is recognized as an important factor in aging and age-related diseases. The free radical theory of aging, first postulated by Harman in 1956 (79), suggests that degeneration processes occur due to the accumulation of oxidative damage over a life span. This damage results physiologically from oxidant by-products of normal metabolism and may be increased by environmental factors resulting in accelerated aging processes and disease. Molecular markers of tissue damage to lipids (e.g., isoprostanes, age pigment, or lipofuscin), protein (carbonyl and nitrotyrosine derivatives of protein) and DNA (products of DNA fragmentation and oxidized bases, e.g., 8-hydroxy-2-deoxyguanosine) accumulate during aging, and tissues of aged individuals are more susceptible to oxidative damage (75). This can result from an impaired antioxidant defense system because antioxidant enzymes are less active in aging persons and dietary antioxidants are often insufficient.

Milestone III: Effects of vitamin E on cell signaling and gene expression. In the early 1990s, inhibition of PKC activity by vitamin E was suggested by Azzi's group as the crucial factor for inhibition of cell proliferation in smooth muscle cells (80–83). PKC plays a major role in cell signaling and modulates gene expression during cell growth, proliferation, and differentiation. Thus, PKC activity is an important factor contributing to disorders such as vascular disease, cancer, diabetes, and other age-related degenerative diseases (84–87).

Vitamin E was found to inhibit PKC activity in many cell types including smooth muscle cells, monocytes, macrophages, neutrophils, fibroblasts, and mesangial cells, and the effects were confirmed repeatedly in animal studies (88–90). Inhibition of PKC activity by vitamin E occurs indirectly *via* activation of a phosphatase that cleaves the active, phosphorylated form of PKC, or by modulating diacylglycerol kinase activity (91,92). What is novel is that inhibition of PKC is apparently independent of the antioxidant activity of vitamin E (93). Hence, the biological role of vitamin E goes beyond its antioxidant function.

Several other mechanisms have been proposed suggesting that vitamin E modulates cell signaling. Recent advances in molecular biology and the availability of microarray techniques for studying effects of vitamin E on gene expression have revealed novel vitamin E–sensitive genes and signal transduction pathways as recently reviewed by Rimbach and co-workers (94).

Vitamin E regulates at the transcriptional level the expression of several genes including collagen $\alpha 1$ and α-TTP in liver (95,96), collagenase in skin (97), adhesion molecules and chemokines such as vascular cell adhesion molecule 1 (VCAM-1) and monocyte chemoattractant protein 1 (MCP-1) in endothelial cells (98), different integrins in erythroleukemia cells (99), α-tropomyosin in smooth muscle cells (100), and scavenger receptors class A (SR-A) and CD 36 in macrophages and smooth muscle cells (101,102). At the post-translational level, vitamin E regulates the expression of cyclooxygenase in monocytes leading to a decrease in prostaglandin E_2 levels (103). Because this latter effect was also observed using other vitamin E homologues compatible with antioxidant activity, this function of vitamin E may involve redox-signaling.

These newly discovered actions of vitamin E may help to explain the observed beneficial effects of vitamin E in the chronic and degenerative diseases of aging. Future experimental studies are required to achieve the following: (i) to clarify the role of PKC inhibition in the described activities of vitamin E, and the effects of TAP on gene expression and biological activity; (ii) to determine to what extent antioxidant function, i.e., proton and/or electron transfer, plays a role in the effects of vitamin E on cell signaling and gene expression; and (iii) to investigate potentially different effects on cell signaling and biological activities among isoforms of vitamin E as well as stereoisomers of synthetic α-tocopherol.

Milestone IV: Recognition of vitamin E deficiency diseases and the beneficial effects of vitamin E supplements in human health and disease prevention. Clinically manifest vitamin E deficiency is rare. Clear indications of vitamin E deficiency have been found in premature infants and in children suffering from an abnormal ability to absorb vitamin E as in abetalipoproteinemia, chronic cholestatic liver disease, cystic fibrosis, and short-bowel syndrome (104,105).

A mutation in the α-TTP gene as found in "familial isolated vitamin E deficiency" results in a failure to deliver α-tocopherol to the systemic circulation (106,107). In these cases, clinical vitamin E deficiency is recognized in infants showing symptoms of retrolental fiblasia and intraventricular hemorrhage, and in children suffering from muscular dystrophy, ataxia, and other disorders (108,109). High-dose vitamin E supplementation can reduce or eliminate clinical symptoms in these patients (110); however, to achieve amelioration of neurological symptoms, early diagnosis and early start of treatment are crucial.

More frequently, a chronic suboptimal supply of vitamin E (i.e., theoretically an intake below RDA levels) occurs population wide and in all age groups (16,111). This may cause impaired defense against oxidative stress and increased susceptibility to oxidative injury and adverse health effects. In most diseases that have been examined with any degree of scrutiny, evidence for "oxidative stress" and oxidative damage has been observed in some stage of disease initiation or progression. Therefore it seems obvious that vitamin E, like other antioxidants, may prevent or delay disease progression. However, can vitamin E at dietary levels or as supplements given during a short period of our life span effectively prevent chronic diseases and delay age-related degenerative disorders? Indeed, much evidence from observational, clinical, and

TABLE 9.1
Ongoing Clinical Trials Involving Vitamin E (July 2002)[a]

Study[b]	Institution	End Points	Participants	Phase	Type and design	Treatment	Follow-up	Expected completion
Vitamin E to treat uveitis-associated macular edema[1]	National Eye Institute (NEI), Bethesda, MD	Cystoid macular edema; uveitis	80	I	Treatment, randomized, double-blind	1600 IU/d vitamin E, or placebo, for 4 mo	mo 4 and 5	N/A
Donepezil and vitamin E to prevent CD after cranial radiation therapy[2]	Multicenter	Small cell lung cancer; depression; delirium	104	III	Supportive care randomized, double-blind, two-arm	Vitamin E twice daily; and donepezil once daily, or placebo, for at least 1 mo	Every 6 mo	2004
Memory impairment study[3]	Multicenter	MCI	720, ≥ 55 y old	III	Prevention, treatment, randomized, double-blind, three-arm	2000 IU/d vitamin E or donepezil, or placebo for 3 y	mo 3 and 6, then every 6 mo	2002
SELECT[4]	Multicenter	Prostate cancer	32,400 men, ≥ 55 y old	III	Prevention, randomized, double-blind, four-arm	400 IU/d vitamin E and/or 200 µg/d selenium, or placebo	Every 6 mo for 7 y	2012
Lymphedema study[5]	Royal Marsden Hospital, London, UK	Lymphedema	100 women	II	Supportive care; randomized, double-blind, two-arm	Vitamin E and pentoxifylline, or placebo twice a day for 6 mo	At 6 mo and 1 y	N/A
Lung cancer study[6]	Multicenter	Lung cancer	60 former smokers	II	Supportive care; prevention, randomized, double-blind, three-arm	Vitamin E and/or isotretinoin	Annually for 2 y	N/A
Vitamin E as add-on therapy for children with epilepsy[7]	The Children's Hospital, Denver, CO	Epilepsy	50, 1–18 y old	IV	Treatment, randomized, double-blind, two-arm	Vitamin E or placebo	6 mo	N/A
Carotid atherosclerosis trial[8]	University of Texas, Dallas, TX	Cardiovascular	120	II	Treatment, randomized, double-blind	1200 IU/d vitamin E, or placebo, for 2 y	Every 6 mo for 2 y	N/A

Title	Center	Disease	Sample	Phase	Design	Intervention	Follow-up	Year
Diet and PSA levels in patients with prostate cancer[9]	Memorial Sloan-Kettering Cancer Center, New York, NY	Prostate cancer	154	III	Prevention, treatment, randomized, double-blind, two-arm	"Intensive nutritional intervention or general nutritional instruction" for 18 mo	mo 1 and 3, then every 3 mo	N/A
Laser and medical treatment of diabetic macular edema[10]	NEI, Bethesda, MD	Diabetes mellitus; macular edema	60	I	Treatment, randomized	1600 IU vitamin E, or placebo, pre- and postphotocoagulation	N/A	N/A
AREDS[11]	Multicenter	Cataracts; age-related macular degeneration	4757 ≥ 55 y old	III	Treatment, randomized, double-blind, four-arm	Antioxidants (500 mg vitamin C, 400 IU vitamin E, 15 mg β-carotene) and/or 80 mg zinc daily, or placebo	Every 6 mo for at least 5 y	2005
WHS[12]	Multicenter	Cardiovascular; cancer	39,876 women, ≥ 45 y old	III	Prevention, randomized, double-blind, 2 × 2 factorial	600 IU vitamin E and/or aspirin on alternate days, or placebo	At least 5 y	2004
PHS II[13]	Multicenter	Cardiovascular; cancer; eye disease	15,000 male physicians, ≥ 55 y old	III	Prevention, randomized, double-blind, 2 × 2 × 2 factorial	400 IU vitamin E and/or 50 mg β-carotene on alternate days, and/or 500 mg/d vitamin C and/or multivitamin daily, or placebo	Annually for at least 5 y	2002 (extended until 2007)
WACS[14]	Multicenter	Cardiovascular	8000 women, ≥ 40 y old	III	Prevention, randomized, double blind, 2 × 2 × 2 factorial	600 IU vitamin E and/or 50 mg β-carotene on alternate days, and/or 500 mg/d vitamin C, or placebo	Annually for at least 5 y	2002 (extended until 2006)
SUVIMAX[15]	Multicenter	Cardiovascular	13,000 women ≥ 35 y old, men ≥ 45 y old	III	Prevention, randomized, double-blind	30 mg/d vitamin E, 120 mg/d vitamin C, 6 mg β-carotene, 100 μg/d selenium and 20 mg/d zinc, or placebo	Annually for 8 y	2003

(continued)

TABLE 9.1 *(continued)*
Ongoing Clinical Trials Involving Vitamin E (July 2002)[a]

Study[b]	Institution	End Points	Participants	Phase	Type and design	Treatment	Follow-up	Expected completion
WAVE[16]	Multicenter	Cardiovascular	420 women, ≥ 38 y old	III	Prevention, treatment, randomized, double-blind, 2 × 2 factorial	800 IU/d vitamin E and 1000 mg/d vitamin C, or placebo, and/or hormone replacement therapy	3 y	2002
CLIPS[17]	Multicenter	Leg ischemia; cardiovascular	350	III	Prevention, treatment, randomized, double-blind, 2 × 2 factorial	Antioxidants (600 mg/d vitamin E, 250 mg/d vitamin C, 20 mg/d β-carotene) and/or aspirin, or placebo	Up to 4 y	2001

[a]Data for studies 1–12, 14, and16 were obtained from www.clinicaltrials.gov; for study 13, from Ref. 112); for study 15, www.suvimax.org; and for study 17, from www.vas-int.org.
[b]The official titles of the studies are as follows:

[1]Randomized Masked Study to Evaluate the Use of Vitamin E in the Treatment of Uveitis-Associated Macular Edema.
[2]Phase III Randomized Study of Donepezil and Vitamin E in the Prevention of Cognitive Dysfunction Following Cancer Treatment That Included Prophylactic Cranial Irradiation in Patients with Small Cell Lung Cancer.
[3]A Randomized, Double-Blind, Placebo-Controlled Trial to Evaluate the Safety and Efficacy of Vitamin E and Donepezil HCl (Aricept) to Delay Clinical Progression from Mild Cognitive Impairment (MCI) to Alzheimer's Disease (AD).
[4]Phase III Randomized Study of Selenium and Vitamin E for the Prevention of Prostate Cancer.
[5]Phase II Randomized Study of Vitamin E and Pentoxifylline in Women with Lymphedema After Radiotherapy for Breast Cancer.
[6]Phase II Randomized Study of Isotretinoin with or Without Vitamin E for the Chemoprevention of Lung Cancer.
[7]Double-Blind, Placebo-Controlled Trial of Vitamin E as Add-On Therapy for Children with Epilepsy.
[8]Effect of High Dose Vitamin E on Carotid Atherosclerosis.
[9]Phase III Randomized Study of the Effect of a Diet Low in Fat and High in Soy, Fruits, Vegetables, Green Tea, Vitamin E, and Fiber on PSA [prostate-specific antigen] Levels in Patients with Prostate Cancer.
[10]Preliminary Assessment of Laser and Medical Treatment of Diabetic Macular Edema.
[11]Age-Related Eye Disease Study.
[12]Women's Health Study—Trial of Aspirin and Vitamin E in Women.
[13]Physicians' Health Study.
[14]Women's Antioxidant and Cardiovascular Study.
[15]Supplementation en Vitamines et Minéraux Antioxidants.
[16]Women's Angiographic Vitamin and Estrogen Trial (completed but the results have not yet been published).
[17]Critical Leg Ischemia Prevention Study (completed but the results have not yet been published).

experimental studies documents a beneficial role of vitamin E, and several small clinical trials and some large human intervention studies are ongoing (Table 9.1). The following chapters on vitamin E in this book highlight recent findings and discuss biological functions, health benefits, and the potential therapeutic role of vitamin E in humans.

Acknowledgments

The authors are very grateful to Dr. Estibaliz Olano-Martin, Department of Respiratory and Critical Care Medicine, University of California, Davis for helpful discussions and for summarizing ongoing clinical trials.

References

1. Evans, H.M., and Bishop, K.S. (1922) On the Existence of a Hitherto Unrecognized Dietary Factor Essential for Reproduction, *Science 56*, 650.
2. Horwitt, M. (1976) Vitamin E: A Reexamination, *Am. J. Clin. Nutr. 29*, 569–578.
3. Burton, G.W., Traber, M.G., Acuff, R.V., Walters, D.N., Kayden, H., Hughes, L., and Ingold, K.U. (1998) Human Plasma and Tissue Alpha-Tocopherol Concentrations in Response to Supplementation with Deuterated Natural and Synthetic Vitamin E, *Am. J. Clin. Nutr. 67*, 669–684.
4. Weiser, H., Riss, G., and Kormann, A.W. (1996) Biodiscrimination of the Eight Alpha-Tocopherol Stereoisomers Results in Preferential Accumulation of the Four 2*R* Forms in Tissues and Plasma of Rats, *J. Nutr. 126*, 2539–2549.
5. Machlin, L. (1990) Vitamin E, in *Handbook of Vitamins* (Machlin, L., ed.) pp. 99–145, Marcel Dekker, New York.
6. Food and Nutrition Board (2000) *Dietary Reference Intakes for Vitamin C, Vitamin E, Selenium, and Carotenoids*, pp. 186–283, National Academy Press, Washington, DC.
7. Hoppe, P., and Krennrich, G. (2000) Bioavailability and Potency of Natural-Source and All-Racemic Alpha-Tocopherol in the Human: A Dispute, *Eur. J. Nutr. 39*, 183–193.
8. The United States Pharmacopeia (1999) *The United States Pharmacopeia 24. National Formulary 19*, USP, Rockville, MD.
9. National Research Council (1989) *Recommended Dietary Allowances*, National Academy Press, Washington.
10. Dial, S., and Eitenmiller, R. (1995) Tocopherols and Tocotrienols in Key Foods in the U.S. Diet, in *Nutrition, Lipids, Health, and Disease* (Ong, A., Niki, E., and Packer, L., eds.) pp. 327–342, AOCS Press, Champaign, IL.
11. Sheppard, A., Pennington, J., and Weihrauch, J. (1993) Analysis and Distribution of Vitamin E in Vegetable Oils and Foods, in *Vitamin E in Health and Disease* (Packer, L., and Fuchs, J., eds.) pp. 9–31, Marcel Dekker, New York.
12. Ma, J., Hampl, J.S., and Betts, N.M. (2000) Antioxidant Intakes and Smoking Status: Data from the Continuing Survey of Food Intakes by Individuals 1994–1996, *Am. J. Clin. Nutr. 71*, 774–780.
13. Sichert-Hellert, W., Kersting, M., Alexy, U., and Manz, F. (2000) Ten-Year Trends in Vitamin and Mineral Intake from Fortified Food in German Children and Adolescents, *Eur. J. Clin. Nutr. 54*, 81–86.
14. Murphy, S.P., Subar, A.F., and Block, G. (1990) Vitamin E Intakes and Sources in the United States, *Am. J. Clin. Nutr. 52*, 361–367.

15. Jiang, Q., Christen, S., Shigenaga, M., and Ames, B. (2001) γ-Tocopherol, the Major Form of Vitamin E in the US Diet, Deserves More Attention, *Am. J. Clin. Nutr. 74*, 714–722.

16. Ford, E.S., and Sowell, A. (1999) Serum Alpha-Tocopherol Status in the United States Population: Findings from the Third National Health and Nutrition Examination Survey, *Am. J. Epidemiol. 150*, 290–300.

17. Horwitt, M. (1960) Vitamin E and Lipid Metabolism in Man, *Am. J. Clin. Nutr. 8*, 451–461.

18. Traber, M.G. (2001) Vitamin E: Too Much or Not Enough? *Am. J. Clin. Nutr. 73*, 997–998.

19. Horwitt, M.K. (2001) Critique of the Requirement for Vitamin E, *Am. J. Clin. Nutr. 73*, 1003–1005.

20. Frei, B., and Traber, M.G. (2001) The New US Dietary Reference Intakes for Vitamins C and E, *Redox Rep. 6*, 5–9.

21. Tanyel, M.C., and Mancano, L.D. (1997) Neurologic Findings in Vitamin E Deficiency, *Am. Fam. Physician 55*, 197–201.

22. The ATBC Cancer Prevention Study Group (1994) The Alpha-Tocopherol, Beta-Carotene Lung Cancer Prevention Study: Design, Methods, Participant Characteristics, and Compliance, *Ann. Epidemiol. 4*, 1–10.

23. GISSI-Prevenzione Investigators (1999) Dietary Supplementation with n-3 Polyunsaturated Fatty Acids and Vitamin E After Myocardial Infarction: Results of the GISSI-Prevenzione Trial, *Lancet 354*, 447–455.

24. HOPE Study Investigators (2000) Vitamin E Supplementation and Cardiovascular Events in High-Risk Patients, *N. Engl. J. Med. 342*, 154–160.

25. Diplock, A.T. (1995) Safety of Antioxidant Vitamins and Beta-Carotene, *Am. J. Clin. Nutr. 62*, 1510S–1516S.

26. Kayden, H.J., and Traber, M.G. (1993) Absorption, Lipoprotein Transport, and Regulation of Plasma Concentrations of Vitamin E in Humans, *J. Lipid Res. 34*, 343–358.

27. Arita, M., Sato, Y., Miyata, A., Tanabe, T., Takahashi, E., Kayden, H.J., Arai, H., and Inoue, K. (1995) Human Alpha-Tocopherol Transfer Protein: cDNA Cloning, Expression and Chromosomal Localization, *Biochem. J. 306*, 437–443.

28. Hosomi, A., Arita, M., Sato, Y., Kiyose, C., Ueda, T., Igarashi, O., Arai, H., and Inoue, K. (1997) Affinity for Alpha-Tocopherol Transfer Protein as a Determinant of the Biological Activities of Vitamin E Analogs, *FEBS Lett. 409*, 105–108.

29. Jishage, K., Arita, M., Igarashi, K., Iwata, T., Watanabe, M., Ogawa, M., Ueda, O., Kamada, N., Inoue, K., Arai, H., and Suzuki, H. (2001) Alpha-Tocopherol Transfer Protein Is Important for the Normal Development of Placental Labyrinthine Trophoblasts in Mice, *J. Biol. Chem. 276*, 1669–1672.

30. Stocker, A., and Azzi, A. (2000) Tocopherol-Binding Proteins: Their Function and Physiological Significance, *Antioxid. Redox Signal. 2*, 397–404.

31. Traber, M., and Arai, H. (1999) Molecular Mechanisms of Vitamin E Transport, *Annu. Rev. Nutr. 19*, 343–355.

32. Blatt, D., Leonard, S., and Traber, M. (2001) Vitamin E Kinetics and the Function of Tocopherol Regulatory Proteins, *Nutrition 17*, 799–805.

33. Brigelius-Flohé, R., and Traber, M. (1999) Vitamin E: Function and Metabolism, *FASEB J. 13*, 1145–1155.

34. Zimmer, S., Stocker, A., Sarbolouki, M.N., Spycher, S.E., Sassoon, J., and Azzi, A. (2000) A Novel Human Tocopherol-Associated Protein: Cloning, In Vitro Expression, and Characterization, *J. Biol. Chem. 275*, 25672–25680.

35. Stocker, A., Zimmer, S., Spycher, S.E., and Azzi, A. (1999) Identification of a Novel Cytosolic Tocopherol-Binding Protein: Structure, Specificity, and Tissue Distribution, *IUBMB Life 48*, 49–55.

36. Azzi, A., Breyer, I., Feher, M., Pastori, M., Ricciarelli, R., Spycher, S., Staffieri, M., Stocker, A., Zimmer, S., and Zingg, J.-M. (2000) Specific Cellular Responses to α-Tocopherol, *J. Nutr. 130*, 1649–1652.

37. Gordon, M., Campbell, F., and Dutta-Roy, A. (1996) Alpha-Tocopherol-Binding Protein in the Cytosol of the Human Placenta, *Biochem. Soc. Trans. 24*, 202S.

38. Dutta-Roy, A., Leishman, D., Gordon, M., Campbell, F., and Duthie, G. (1993) Identification of a Low Molecular Mass (14.2 kDa) Alpha-Tocopherol-Binding Protein in the Cytosol of Rat Liver and Heart, *Biochem. Biophys. Res. Commun. 196*, 1108–1111.

39. Olmedilla, B., Granado, F., Southon, S., Wright, A.J., Blanco, I., Gil-Martinez, E., Berg, H., Corridan, B., Roussel, A.M., Chopra, M., and Thurnham, D.I. (2001) Serum Concentrations of Carotenoids and Vitamins A, E, and C in Control Subjects from Five European Countries, *Br. J. Nutr. 85*, 227–238.

40. Commission of the European Communities (1993) *Report of the Scientific Committee for Food Nutrient and Energy Intakes for the European Community, 31st series*, Luxembourg.

41. Sharman, I. (1977) Vitamins: Essential Dietary Constituents Discovered, *Endeavour 1*, 97–102.

42. Evans, H., Emerson, O., and Emerson, G. (1936) The Isolation from Wheat Germ Oil of an Alcohol, Alpha-Tocopherol, Having the Properties of Vitamin E, *J. Biol. Chem. 113*, 319–332.

43. Evans, H.M., and Burr, G.O. (1925) The Anti-Sterility Vitamin Fat Soluble E, *Proc. Natl. Acad. Sci. USA 11*, 334–342.

44. Tappel, A.L. (1965) Free-Radical Lipid Peroxidation Damage and Its Inhibition by Vitamin E and Selenium, *Fed. Proc. 24*, 73.

45. Tappel, A.L. (1962) Vitamin E as the Biological Lipid Antioxidant, *Vitam. Horm. 20*, 493–510.

46. Kamal-Eldin, A., and Appelqvist, L. (1996) The Chemistry and Antioxidant Properties of Tocopherols and Tocotrienols, *Lipids 31*, 671–701.

47. Packer, L. (1994) Vitamin E Is Nature's Master Antioxidant, *Sci. Am. Sci. Med. 1*, 54–63.

48. Witting, L. (1980) Vitamin E and Lipid Antioxidants in Free-Radical-Initiated Reactions, in *Free Radicals in Biology* (Pryor, W., ed.) p. 295, Academic Press, New York.

49. Burton, G., and Ingold, K. (1983) Is Vitamin E the Only Lipid-Soluble, Chain-Breaking Antioxidant in Human Blood Plasma and Erythrocyte Membranes? *Arch. Biochem. Biophys. 221*, 281–290.

50. Wang, X., and Quinn, P. (1999) Vitamin E and Its Function in Membranes, *Prog. Lipid Res. 38*, 309–336.

51. Maccarrone, M., Meloni, C., Manca-di-Villahermosa, S., Cococcetta, N., Casciani, C.U., Finazzi-Agro, A., and Taccone-Gallucci, M. (2001) Vitamin E Suppresses 5-Lipoxygenase-Mediated Oxidative Stress in Peripheral Blood Mononuclear Cells of Hemodialysis Patients Regardless of Administration Route, *Am. J. Kidney Dis. 37*, 964–969.

52. Rhoden, E.L., Pereira-Lima, L., Teloken, C., Lucas, M.L., Bello-Klein, A., and Rhoden, C.R. (2001) Beneficial Effect of Alpha-Tocopherol in Renal Ischemia-Reperfusion in Rats, *Jpn. J. Pharmacol. 87*, 164–166.

53. Engelen, W., Keenoy, B.M., Vertommen, J., and De Leeuw, I. (2000) Effects of Long-Term Supplementation with Moderate Pharmacologic Doses of Vitamin E Are Saturable and Reversible in Patients with Type 1 Diabetes, *Am. J. Clin. Nutr. 72*, 1142–1149.

54. Ciabattoni, G., Davi, G., Collura, M., Iapichino, L., Pardo, F., Ganci, A., Romagnoli, R., Maclouf, J., and Patrono, C. (2000) In Vivo Lipid Peroxidation and Platelet Activation in Cystic Fibrosis, *Am. J. Respir. Crit. Care Med. 162*, 1195–1201.

55. Pratico, D., Tangirala, R.K., Rader, D.J., Rokach, J., and FitzGerald, G.A. (1998) Vitamin E Suppresses Isoprostane Generation in Vivo and Reduces Atherosclerosis in ApoE-Deficient Mice, *Nat. Med. 4*, 1189–1192.

56. Sardas, S., Yilmaz, M., Oztok, U., Cakir, N., and Karakaya, A.E. (2001) Assessment of DNA Strand Breakage by Comet Assay in Diabetic Patients and the Role of Antioxidant Supplementation, *Mutat. Res. 490*, 123–129.

57. Factor, V.M., Laskowska, D., Jensen, M.R., Woitach, J.T., Popescu, N.C., and Thorgeirsson, S.S. (2000) Vitamin E Reduces Chromosomal Damage and Inhibits Hepatic Tumor Formation in a Transgenic Mouse Model, *Proc. Natl. Acad. Sci. USA 97*, 2196–2201.

58. Pincheira, J., Navarrete, M.H., de la Torre, C., Tapia, G., and Santos, M.J. (1999) Effect of Vitamin E on Chromosomal Aberrations in Lymphocytes from Patients with Down's Syndrome, *Clin. Genet. 55*, 192–197.

59. Cederberg, J., Siman, C.M., and Eriksson, U.J. (2001) Combined Treatment with Vitamin E and Vitamin C Decreases Oxidative Stress and Improves Fetal Outcome in Experimental Diabetic Pregnancy, *Pediatr. Res. 49*, 755–762.

60. Constantinescu, A., Han, D., and Packer, L. (1993) Vitamin E Recycling in Human Erythrocyte Membranes, *J. Biol. Chem. 268*, 10906–10913.

61. Maguire, J.J., Wilson, D.S., and Packer, L. (1989) Mitochondrial Electron Transport-Linked Tocopheroxyl Radical Reduction, *J. Biol. Chem. 264*, 21462–21465.

62. Thiele, J.J., Schroeter, C., Hsieh, S.N., Podda, M., and Packer, L. (2001) The Antioxidant Network of the Stratum Corneum, *Curr. Probl. Dermatol. 29*, 26–42.

63. Packer, L., Weber, S.U., and Rimbach, G. (2001) Molecular Aspects of Alpha-Tocotrienol Antioxidant Action and Cell Signaling, *J. Nutr. 131*, 369S–373S.

64. Guo, Q., and Packer, L. (2000) Ascorbate-Dependent Recycling of the Vitamin E Homologue Trolox by Dihydrolipoate and Glutathione in Murine Skin Homogenates, *Free Radic. Biol. Med. 29*, 368–374.

65. Lopez-Torres, M., Thiele, J.J., Shindo, Y., Han, D., and Packer, L. (1998) Topical Application of Alpha-Tocopherol Modulates the Antioxidant Network and Diminishes Ultraviolet-Induced Oxidative Damage in Murine Skin, *Br. J. Dermatol. 138*, 207–215.

66. Haramaki, N., Stewart, D.B., Aggarwal, S., Ikeda, H., Reznick, A.Z., and Packer, L. (1998) Networking Antioxidants in the Isolated Rat Heart Are Selectively Depleted by Ischemia-Reperfusion, *Free Radic. Biol. Med. 25*, 329–339.

67. Liebler, D.C., and Burr, J.A. (1992) Oxidation of Vitamin E During Iron-Catalyzed Lipid Peroxidation: Evidence for Electron-Transfer Reactions of the Tocopheroxyl Radical, *Biochemistry 31*, 8278–8284.

68. Scarpa, M., Rigo, A., Maiorino, M., Ursini, F., and Gregolin, C. (1984) Formation of Alpha-Tocopherol Radical and Recycling of Alpha-Tocopherol by Ascorbate During

Peroxidation of Phosphatidylcholine Liposomes. An Electron Paramagnetic Resonance Study, *Biochim. Biophys. Acta 801*, 215–219.

69. Sies, H. (1997) Oxidative Stress: Oxidants and Antioxidants, *Exp. Physiol.* 82, 291–295.
70. Sies, H., and Cadenas, E. (1985) Oxidative Stress: Damage to Intact Cells and Organs, *Philos. Trans. R. Soc. Lond. B. Biol. Sci. 311*, 617–631.
71. Elsayed, N.M. (2001) Antioxidant Mobilization in Response to Oxidative Stress: a Dynamic Environmental-Nutritional Interaction, *Nutrition 17*, 828–834.
72. Evans, P., and Halliwell, B. (2001) Micronutrients: Oxidant/Antioxidant Status, *Br. J. Nutr. 85 (Suppl. 2)*, S67–S74.
73. Droge, W. (2002) Free Radicals in the Physiological Control of Cell Function, *Physiol. Rev. 82*, 47–95.
74. Davies, K.J. (2000) Oxidative Stress, Antioxidant Defenses, and Damage Removal, Repair, and Replacement Systems, *IUBMB Life 50*, 279–289.
75. Halliwell, B. (2001) Free Radical Reactions in Human Disease, in *Environmental Stressors in Health and Disease* (Fuchs, J., and Packer, L., eds.) pp. 1–16, Marcel Dekker, New York.
76. Hensley, K., and Floyd, R.A. (2002) Reactive Oxygen Species and Protein Oxidation in Aging: A Look Back, a Look Ahead, *Arch. Biochem. Biophys. 397*, 377–383.
77. Terman, A. (2001) Garbage Catastrophe Theory of Aging: Imperfect Removal of Oxidative Damage? *Redox Rep. 6*, 15–26.
78. Sastre, J., Pallardo, F.V., and Vina, J. (2000) Mitochondrial Oxidative Stress Plays a Key Role in Aging and Apoptosis, *IUBMB Life 49*, 427–435.
79. Harman, D. (1956) Aging: A Theory Based on Free Radical and Radiation Chemistry, *J. Gerontol. 11*, 298–300.
80. Chatelain, E., Boscoboinik, D.O., Bartoli, G.M., Kagan, V.E., Gey, F.K., Packer, L., and Azzi, A. (1993) Inhibition of Smooth Muscle Cell Proliferation and Protein Kinase C Activity by Tocopherols and Tocotrienols, *Biochim. Biophys. Acta 1176*, 83–89.
81. Boscoboinik, D., Szewczyk, A., Hensey, C., and Azzi, A. (1991) Inhibition of Cell Proliferation by Alpha-Tocopherol. Role of Protein Kinase C, *J. Biol. Chem. 266*, 6188–6194.
82. Boscoboinik, D., Szewczyk, A., and Azzi, A. (1991) Alpha-Tocopherol (Vitamin E) Regulates Vascular Smooth Muscle Cell Proliferation and Protein Kinase C Activity, *Arch. Biochem. Biophys. 286*, 264–269.
83. Mahoney, C.W., and Azzi, A. (1988) Vitamin E Inhibits Protein Kinase C Activity, *Biochem. Biophys. Res. Commun. 154*, 694–697.
84. Nishizuka, Y. (2001) The Protein Kinase C Family and Lipid Mediators for Transmembrane Signaling and Cell Regulation, *Alcohol Clin. Exp. Res. 25*, 3s–7s.
85. Meier, M., and King, G.L. (2000) Protein Kinase C Activation and Its Pharmacological Inhibition in Vascular Disease, *Vasc. Med. 5*, 173–185.
86. Ways, D.K., and Sheetz, M.J. (2000) The Role of Protein Kinase C in the Development of the Complications of Diabetes, *Vitam. Horm. 60*, 149–193.
87. Carter, C.A. (2000) Protein Kinase C as a Drug Target: Implications for Drug or Diet Prevention and Treatment of Cancer, *Curr. Drug Targets 1*, 163–183.
88. Freedman, J.E., Farhat, J.H., Loscalzo, J., and Keaney, J.F., Jr. (1996) α-Tocopherol Inhibits Aggregation of Human Platelets by a Protein Kinase C-Dependent Mechanism, *Circulation 94*, 2434–2440.
89. Koya, D., Lee, I.K., Ishii, H., Kanoh, H., and King, G.L. (1997) Prevention of

Glomerular Dysfunction in Diabetic Rats by Treatment with D-Alpha-Tocopherol, *J. Am. Soc. Nephrol. 8*, 426–435.

90. Studer, R.K., Craven, P.A., and DeRubertis, F.R. (1997) Antioxidant Inhibition of Protein Kinase C-Signaled Increases in Transforming Growth Factor-Beta in Mesangial Cells, *Metabolism 46*, 918–925.

91. Ricciarelli, R., Tasinato, A., Clement, S., Ozer, N.K., Boscoboinik, D., and Azzi, A. (1998) α-Tocopherol Specifically Inactivates Cellular Protein Kinase C Alpha by Changing Its Phosphorylation State, *Biochem. J. 334*, 243–249.

92. Clement, S., Tasinato, A., Boscoboinik, D., and Azzi, A. (1997) The Effect of Alpha-Tocopherol on the Synthesis, Phosphorylation and Activity of Protein Kinase C in Smooth Muscle Cells After Phorbol 12-Myristate 13-Acetate Down-Regulation, *Eur. J. Biochem. 246*, 745–749.

93. Tasinato, A., Boscoboinik, D., Bartoli, G.M., Maroni, P., and Azzi, A. (1995) d-α-Tocopherol Inhibition of Vascular Smooth Muscle Cell Proliferation Occurs at Physiological Concentrations, Correlates with Protein Kinase C Inhibition, and Is Independent of Its Antioxidant Properties, *Proc. Natl. Acad. Sci. USA 92*, 12190–12194.

94. Rimbach, G., Minihane, A., Majewicz, J., Fischer, A., Pallauf, J., Virgli, F., and Weinberg, P. (2002) Regulation of Cell Signaling by Vitamin E, *Prog. Nutr. Soc.*, in press.

95. Chojkier, M., Houglum, K., Lee, K.S., and Buck, M. (1998) Long- and Short-Term D-α-Tocopherol Supplementation Inhibits Liver Collagen Alpha1(I) Gene Expression, *Am. J. Physiol. 275*, G1480–G1485.

96. Shaw, H.M., and Huang, C. (1998) Liver Alpha-Tocopherol Transfer Protein and Its mRNA Are Differentially Altered by Dietary Vitamin E Deficiency and Protein Insufficiency in Rats, *J. Nutr. 128*, 2348–2354.

97. Ricciarelli, R., Maroni, P., Ozer, N., Zingg, J., and Azzi, A. (1999) Age-Dependent Increase of Collagenase Expression Can Be Reduced by Alpha-Tocopherol Via Protein Kinase C Inhibition, *Free Radic. Biol. Med. 27*, 729–737.

98. Zapolska-Downar, D., Zapolski-Downar, A., Markiewski, M., Ciechanowicz, A., Kaczmarczyk, M., and Naruszewicz, M. (2000) Selective Inhibition by Alpha-Tocopherol of Vascular Cell Adhesion Molecule-1 Expression in Human Vascular Endothelial Cells, *Biochem. Biophys. Res. Commun. 274*, 609–615.

99. Breyer, I., and Azzi, A. (2001) Differential Inhibition by Alpha- and Beta-Tocopherol of Human Erythroleukemia Cell Adhesion: Role of Integrins, *Free Radic. Biol. Med. 30*, 1381–1389.

100. Aratri, E., Spycher, S.E., Breyer, I., and Azzi, A. (1999) Modulation of Alpha-Tropomyosin Expression by Alpha-Tocopherol in Rat Vascular Smooth Muscle Cells, *FEBS Lett. 447*, 91–94.

101. Ricciarelli, R., Zingg, J.M., and Azzi, A. (2000) Vitamin E Reduces the Uptake of Oxidized LDL by Inhibiting CD36 Scavenger Receptor Expression in Cultured Aortic Smooth Muscle Cells, *Circulation 102*, 82–87.

102. Teupser, D., Thiery, J., and Seidel, D. (1999) Alpha-Tocopherol Down-Regulates Scavenger Receptor Activity in Macrophages, *Atherosclerosis 144*, 109–115.

103. Wu, D., Hayek, M.G., and Meydani, S. (2001) Vitamin E and Macrophage Cyclooxygenase Regulation in the Aged, *J. Nutr. 131*, 382S–388S.

104. Kumar, R.K., Edwards, K.N., and Bury, G. (2000) Haemolytic Anaemia Secondary to Vitamin E Deficiency in Premature Infants, *Indian J. Pediatr. 67*, 537–538.

105. Feranchak, A.P., Sontag, M.K., Wagener, J.S., Hammond, K.B., Accurso, F.J., and Sokol, R.J. (1999) Prospective, Long-Term Study of Fat-Soluble Vitamin Status in Children with Cystic Fibrosis Identified by Newborn Screen, *J. Pediatr. 135*, 601–610.

106. Kayden, H. (2001) The Genetic Basis of Vitamin E Deficiency in Humans, *Nutrition 17*, 797–798.

107. Cavalier, L., Ouahchi, K., Kayden, H., DiDonato, S., Reutenauer, L., Mandel, J.-L., and Koenig, M. (1998) Ataxia with Isolated Vitamin E Deficiency: Heterogeneity of Mutations and Phenotypic Variability in a Large Number of Families, *Am. J. Hum. Gen. 62*, 301–310.

108. Gordon, N. (2001) Hereditary Vitamin-E Deficiency, *Dev. Med. Child Neurol. 43*, 133–135.

109. Sokol, R. (1990) Vitamin E and Neurologic Deficits, *Adv. Pediatr. 37*, 119–148.

110. Gabsi, S., Gouider-Khouja, N., Belal, S., Fki, M., Kefi, M., Turki, I., Ben Hamida, M., Kayden, H., Mebazaa, R., and Hentati, F. (2001) Effect of Vitamin E Supplementation in Patients with Ataxia with Vitamin E Deficiency, *Eur. J. Neurol. 8*, 477–481.

111. Bendich, A. (1992) Vitamin E Status of US Children, *J. Am. Coll. Nutr. 11*, 441–444.

112. Christen, W.G., Gaziano, J.M., and Hennekens, C.H. (2000) Design of Physicians' Health Study II—A Randomized Trial of Beta-Carotene, Vitamins E and C, and Multivitamins, in Prevention of Cancer, Cardiovascular Disease, and Eye Disease, and Review of Results of Completed Trials, *Ann. Epidemiol. 10*, 125–134.

Chapter 10

Bioavailability and Biopotency of Vitamin E in Humans: An Ongoing Controversy

Peter P. Hoppe and Klaus Kraemer

BASF Aktiengesellschaft, Ludwigshafen 67056, Germany

Introduction

Vitamin E (α-tocopherol) occurs in nature as a single compound in which carbon atoms 2, 4' and 8' are in the R position and the phytyl side chain is straight (*RRR*, Fig. 10.1). Synthetic vitamin E (*all-rac*-α-tocopherol) is a mixture of eight stereoisomers including *RRR*. Half of the material is in the $2R$-conformation, half in the $2S$. It is believed that biopotency depends largely on the $2R$-position; this is evident from the relative potencies in Fig. 10.1 (in brackets) as determined in the rat gestation-resorption test (1).

The bioavailability and biopotency of natural (*RRR*) and synthetic (*all-rac*) α-tocopherol have been discussed controversially for decades. Studies in humans that tried to assess potency of vitamin E are rare and the methods used do not appear to be sensitive enough to allow measurement of graded dose responses. Therefore bioavailability studies have been used in lieu of potency studies. The terms bioavailability and biopotency have often been used interchangeably, resulting in potentially misleading and invalid conclusions. The confusion in terminology is widespread in the literature and has hampered progress in determining potency. The aim of this presentation is a critical review of the studies on potency and bioavailability.

Definitions of Bioavailability and Potency

Bioavailability is defined as the rate and extent of a drug's appearance in blood. The purpose of conducting bioavailability studies comparing two drugs is to determine whether they are bioequivalent. Bioequivalence has been defined as follows: "Two medicinal products are bioequivalent if they are *pharmaceutical equivalents* (containing the same amount of the same active substance in the same form) or *pharmaceutical alternatives* (containing the same active moiety but in a different chemical form, e.g., salt or ester) and if their bioavailabilities following the same molar dose are similar to such a degree that their effects, with respect to efficacy and safety, will be essentially the same" (2). *Essentially the same* has been defined as follows: "The 90% confidence interval for the ratio of means [e.g., C_{max}, area under the curve (AUC) or steady-state concentration] should lie within an acceptance interval of 0.80 to 1.25"

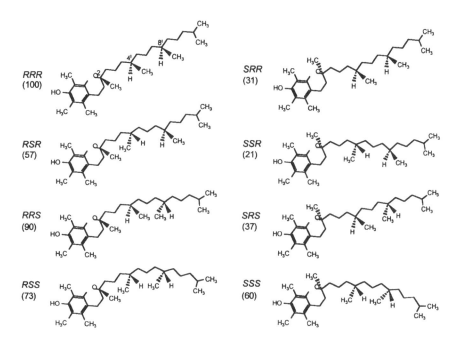

Fig. 10.1. The eight stereoisomers of *all-rac*-α-tocopherol. Given in brackets are the potencies relative to *RRR* α-tocopherol as determined by the rat-gestation test. *Source*: Ref. 1.

(2). The acceptance range was established to allow for the variation in response between subjects and because the clinical effects from dosages within the range cannot be distinguished. Consequently, comparative bioavailability studies are accepted as a surrogate method for potency studies, provided that the compounds to be compared are pharmaceutical equivalents or alternatives. It is obvious, that this proviso does not apply to the comparison of *RRR* and *all-rac*, because *RRR* as a single compound and *all-rac* as a mixture of eight compounds are neither pharmaceutical equivalents nor alternatives.

Potency is a measure of the effects of a compound. Typical vitamin E effects are the prevention of fetal resorption (death) in the rat, the prevention or cure of myopathy in laboratory animals, and the prevention of erythrocyte hemolysis. The relative potencies of *RRR* vs. *all-rac* that are presently accepted for humans were derived from the gestation-resorption test in rats. They are as follows: (i) 1 mg *all-rac*-α-tocopheryl acetate = 1 USP unit; and (ii) 1 mg *RRR*-α-tocopheryl acetate = 1.36 USP units (3).

Novel Definition of Vitamin E Activity by Food and Nutrition Board

Recently, in establishing Dietary Reference Intakes, the Food and Nutrition Board (FNB) suggested a redefinition of vitamin E activity (4). It was suggested that the vitamin E activity of α-tocopherol is limited to the natural _RRR_-form and the three other 2_R_-stereoisomeric forms in synthetic α-tocopherol, excluding the four 2_S_-stereoisomers. This would result in a theoretical potency ratio (_RRR_ vs. _all-rac_) of 2:1. The natural (_RRR_) β-, γ- and δ- homologues and the respective tocotrienols were not considered to have vitamin E activity in humans. This contrasts with activities of 25–40, 1–11, 1, 28 and 5% reported for β-, γ- and δ- tocopherols, and α- and β-tocotrienols, respectively, relative to α-tocopherol determined in the rat gestation-resorption test (5). The redefinition is a marked change from the present definition of vitamin E that includes all tocopherols and tocotrienols that exhibit qualitatively the biological activity of α-tocopherol (5). The FNB argued that hepatic α-tocopherol transfer protein (α-TTP) has a preference for _RRR_ compared with the other forms, resulting in preferential secretion into and "maintenance in blood," whereas the stereoisomers with 2_S_-configuration and the non-α-tocopherol homologues are "not maintained" in plasma and tissues. Note that the term "maintenance" is not defined unlike appropriate bioavailability terminology such as mean residence time or half-life. It is a vague term because γ-tocopherol and _SRR_-α-tocopherol (one of the stereoisomers that is considered by FNB as not having vitamin E activity) are also absorbed into blood (6) and are present in blood, albeit with an apparent shorter half-life than _RRR_ (7). γ-Tocopherol constitutes as much as 30–50% of total vitamin E in human skin, muscle, vein, and adipose tissue (8). As recently reviewed, it is a more effective trap for lipophilic electrophiles than α-tocopherol, and as its urinary metabolite, γ-carboxy-ethyl-hydroxy-chroman (γ-CEHC) also has natriuretic and anti-inflammatory activities. Thus, it may be more important to human health than previously appreciated (9).

The Challenge to Determine Vitamin E Potency in Humans

Designing a true potency study for humans is a challenge for several reasons. Overt vitamin E deficiency is virtually nonexistent, and healthy subjects with normal plasma α-tocopherol (~25 μmol/L) cannot be used. Vitamin E depletion would require >1 y and would be unethical. Myopathy that can be monitored by plasma creatine kinase does not normally occur in human vitamin E deficiency. Finally, vitamin E deficiency presents as subtle neurological symptoms (peripheral neuropathy) that are not useful for establishing a graded dose-response relationship, unlike myopathy in animals. At present there are no clinical end points that appear suitable to assay potency in humans.

Biochemical end points reflecting antioxidant potency such as breath pentane (10) and peroxide-induced hemolysis _ex vivo_ in vitamin E–depleted subjects have been used (11). Data obtained by the latter method were taken by FNB to draw a line between vitamin E deficiency and adequacy at 12 μmol/L to determine the requirement (4). α-Tocopherol in plasma low density lipoprotein (LDL; a bioavailability measure) and the rate and lag time of conjugated diene formation, thiobarbituric acid-reactive substances, and macrophage degradation of LDL (measures of potency) after copper-oxidation _in vitro_ were measured in a study comparing _RRR_ and _all-rac_ at supplement doses of 1600 mg/d for 8 wk (12). α-Tocopherol in LDL increased and the susceptibility to oxidation decreased at similar rates for both groups, indicating that either form of α-tocopherol provided equal antioxidant protection at this high dose. However, the sensitivity of this method to detect minor differences in potency is not sufficient for a precise assessment of the ratio.

Bioavailability Studies

Comparing RRR and all-rac in Parallel Groups

A total of nine studies were published that compared *RRR* and *all-rac* in parallel groups of subjects, three using a single dose and six using repetitive dosing ranging from 10 to 56 d [for a review, see (13)]. The doses ranged from 100 to 1600 mg/d. Plasma concentration at 24 h postdosing, C_{max}, the AUC for plasma or red blood cells, the plasma steady-state concentration, and the concentration in LDL were variously used as parameters of bioavailability. An overview of the ratios found for *RRR:all-rac* in these studies is shown in Figure 10.2. The majority of the ratios are clustered around the line indicating the presently accepted potency ratio, 1.36, and within the range of acceptance. A single value outside the range (2.62) was reported by Horwitt (14) based on the percentage increase at 24 h, a one-time point measurement. If the AUC_{6-48h} is calculated from the author's data, a ratio of 1.56 is obtained. Figure 10.2 is in striking contrast to the assumption that *all-rac* should be allotted half the potency of *RRR* (14). Despite the fact that, as outlined above, the bioavailability factor does not reflect the biopotency factor, it is interesting to note that the ratio is consistently below 2:1. This indicates that, contrary to the opinion of the FNB (4), more than one half of *all-rac* may be retained in plasma. In tissues, the ratio tends to be even lower than in plasma (8).

Intraindividual Biokinetic Comparisons of d3-RRR and d6-all-rac

In these studies, differentially deuterated forms (d_3-*RRR* and d_6-*all-rac*) were dosed simultaneously at a 1:1 ratio by weight (15). The technique was developed to obtain a deeper insight into the *kinetics* of absorption, plasma transport, and tissue distribution (15) and to elucidate the sites of biodiscrimination between vitamin E homologues and stereoisomers. Using this approach, intraindividual comparisons were made of plasma and tissue kinetics (8,16) including plasma kinetics in smokers and nonsmokers (17), hepatic VLDL secretion (6), metabolic breakdown to α-CEHC (18), maternal-fetal transfer (19), and of biodiscrimination between vitamin E homologues and stereoisomers (20). Studies measuring the plasma response to differentially labeled *RRR*-α-tocopherol, *SRR*-α-tocopherol (one of the eight stereoisomers of *all-rac*), and γ-tocopherol have led to the hypothesis that absorption takes place with roughly equal efficiencies indicating little or no discrimination among isomers in the digestive tract (6). In contrast, biodiscrimination occurs in the liver, resulting in preferential resecretion of *RRR via* nascent very low density lipoproteins (VLDL) into blood (6), and this

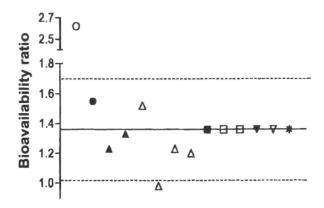

Fig. 10.2. Ratios of *RRR:all-rac*-α-tocopherol in blood and plasma low density lipopro-teins reported from parallel-group studies with unlabeled materials. Points, from left to right, are from *single-dose studies* based on plasma concentration at 24 h (○) reported by Horwitt (14) and on the area under the curve $(AUC)_{6-48h}$ (●) calculated from Horwitt (14); on plasma C_{max} and AUC (▲), Yoshikawa *et al.* (25); on AUC and C_{max} in plasma, and on AUC and C_{max} in red blood cells (△), Ferslew *et al.* (26); and from *repetitive-dose studies* based on plasma steady state (■), Baker *et al.* (27); plasma steady state and protection from LDL oxidation (□), Devaraj et al. (28) ; α-tocopherol in low density lipoprotein (LDL) (▼), Reaven and Witztum (12); plasma steady state (▽),Winklhofer-Roob *et al.* (29); and plasma steady state (*) Chopra and Bhagavan (30). The solid and hatched lines denote the presently accepted biopotency factor of 1.36 and the upper and lower limits of acceptance, respectively.

has been attributed to a higher affinity of α-TTP for *RRR* (21). It is not known whether discrimination by α-TTP also occurs after high (supplemental) doses. The main sites and potential events leading to discrimination are given in Table 10.1.

From these studies, the ratios (d_3-*RRR*:d_6-*all-rac*) of concentrations in plasma and various tissues ("bioavailability ratios") were also reported. The ratios obtained in all published studies are shown in Figure 10.3, as reviewed recently (13). Included in Figure 10.3 are data reported by the authors and data calculated from the authors' data. Not included is the ratio of 3.42 found for umbilical cord plasma (19). Figure 10.3 indicates that roughly one half of the data fall into the range of acceptance as defined above, whereas the remainder lies between the upper limit of the range and the value of 2. This is at variance with the statements of the authors who reported ratios of ~2 (17,19,21). Traber *et al.* (17) claimed that the ratio was roughly 2, while specifying in the same sentence, that the average value for all subjects and time points was 1.7 ± 0.2. It appears that reiteration of this claim that was originally raised by Horwitt (14), one of the leading vitamin E researchers of his time, has superceded subsequent attempts at estimating the ratio precisely.

In total, the ratios obtained with the competitive uptake method are somewhat higher than those derived from the parallel group studies with unlabeled material. This

TABLE 10.1

Sites and Potential Events of Discrimination Between *RRR-* and *all-rac-*α-Tocopherol[a]

Site	Event	Discrimination	Reference
Pancreas	Carboxylester hydrolase	?	
Small intestinal chyme	Release from matrix	No	
	Tocopherylester hydrolysis	No	
	Entry into mixed micelles	No	
	Micelle diffusion	No	
Enterocyte	Intracellular transport	No	(32)
	Incorporation into chylomicrons	No	
	Secretion of chylomicrons	No	
Plasma chylomicrons	Endothelial lipoprotein lipase	No	
	Formation of chylomicron remnants	No	
	Kinetics	No	(6)
Plasma	Biokinetics	Yes	(6,8,16, 20)
Liver	Binding to remnant receptor	No	
	Binding to α-TTP	Dietary doses: Yes	(21)
		Supplements (≥ 50mg): ?	(33)
	VLDL secretion	Yes	(6)
	Biliary secretion	?	
	α-CEHC formation	?	
Kidneys	α-CEHC excretion	Yes	(18)
Other tissues	Rates of uptake, release and metabolism	Yes (e.g., CNS)	(8)
		No (e.g., liver, muscle)	(8)
Placenta	Maternal/fetal transfer	Yes	(19)

[a]Abbreviations: α-TTP, α-tocopherol transfer protein; VLDL, very low density lipoprotein; α-CEHC, α-carboxy-ethyl-hydroxychroman; CNS, central nervous system.

may accrue from experimental bias. Cohn (22) pointed out that, because the total dose contains 50% d_3-2R-stereoisomers, 25% d_6-2R-stereoisomers, and 25% d_6-2S-stereoisomers, 75% 2R-forms compete with 25% 2S-forms. Because of the threefold higher abundance of the 2-R-form and because of the preference of α-TTP for 2R-stereoisomers, an overestimation of *RRR* and discrimination against *all-rac* result.

How to explain the differences between the ratios shown in Figure 10.3 and the purported ratio of 2? One explanation is that some authors reported *maximum* ratios (8,18,23). These were found at time points during the elimination phase rather than at established time points such as C_{max} or during steady state. Because elimination rates of *RRR* and *all-rac* are different, the ratio of concentrations is not stable but becomes wider with increasing distance from the last dose (13). This results in variable ratios depending on the time point. Furthermore, as shown by kinetic modeling (13), the ratio also depends on the dose, with supplement doses resulting in lower discrimination than dietary doses. Thus, because the bioavailability ratio is affected by the time point and the dose, it is arbitrary and not a true reflection of the ratio of potency.

There is one example of a comparative bioavailability study using differentially labeled vitamin E forms that truly reflects their relative potencies (24). In that study, the free phenolic form and the acetate ester of natural-source vitamin E were given

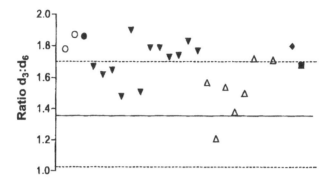

Fig. 10.3. Ratios of d$_3$-*RRR*:d$_6$-*all-rac*-α-tocopherol in human blood and tissues following simultaneous intraindividual dosing. Points from left to right are ratios reported for the following: plasma and red blood cells (○), Acuff *et al.* (31); maternal plasma (●), Acuff *et al.* (19); presumed plasma C_{max} and steady state (▼) and tissues (△) upon various dosages, Burton *et al.* (8); plasma C_{max} (◆), Traber *et al.* (18); and plasma (■), Traber *et al.* (17). The solid and hatched lines denote the presently accepted biopotency factor of 1.36 and the upper and lower limits of acceptance, respectively.

jointly as d$_3$-α-*RRR*-tocopherol and d$_6$-*RRR*-α-tocopheryl acetate, respectively, at the same molar dose. The mean ratio d$_3$/d$_6$ for plasma and the AUC was 1.0 and the ratio was constant at all time points. Because in this case the compounds were pharmacologic alternatives containing the same active moiety, *RRR*-α-tocopherol, the bioavailability ratio truly reflects the potency ratio.

Conclusions

Rat studies have shown that *RRR* has 1.36-fold the potency of *all-rac*. In humans, a lack of sensitive clinical end points or biomarkers has precluded biopotency comparisons. Bioavailability studies were conducted in lieu of potency studies and these have resulted in estimates of the ratio of bioavailability of up to 2. This has resulted in the inference, literally or indirectly, that this ratio reflects the ratio of potency. This inference is challenged on the basis of the differences between *RRR* and *all-rac* in stereochemistry and hence, kinetics. Because of these differences, potency studies that measure functional end points must be conducted. At present there is clearly a lack of adequate methods to conduct such studies in humans. Until sensitive and specific clinical end points or biomarkers have been developed, the discussion about the potency of *all-rac* relative to *RRR* will be ongoing.

References

1. Weiser, H., and Vecchi, M. (1982) Stereoisomers of α-Tocopheryl Acetate. II. Biopotencies of All Eight Stereoisomers, Individually or in Mixtures, as Determined by Rat Resorption-Gestation Tests, *Int. J. Vitam. Nutr. Res. 52*, 351–370.

2. European Agency for Evaluation of Medicinal Products (2001) Note for Guidance on the Investigation of Bioavailability and Bioequivalence, London.

3. Anonymous (2002) *The United States Pharmacopeia*, 25th rev., pp. 1804–1806, Pharmacopoeial Convention, Rockville, MD.

4. Food and Nutrition Board, Institute of Medicine (2000) *Dietary Reference Intakes for Vitamin C, Vitamin E, Selenium and Carotenoids*, National Academy Press, Washington.

5. Combs, G.F. (1998) *The Vitamins. Fundamental Aspects in Nutrition and Health*, 2nd edn., Academic Press, New York.

6. Traber M.G., Rudel, L.L., Burton, G.W., Hughes, L., Ingold, K.U., and Kayden, H.J. (1990) Nascent VLDL from Liver Perfusions of Cynomolgus Monkeys Are Preferentially Enriched in *RRR*-Compared with SRR-α-Tocopherol; Studies Using Deuterated Tocopherols, *J. Lipid Res. 31*, 687–694.

7. Galli, F., Lee, R., Dunster, C., and Kelly, F.J. (2002) Gas Chromatography Mass Spectrometry Analysis of Carboxymethyl-Hydroxychroman Metabolites of α- and γ-Tocopherol in Human Plasma, *Free Radic. Biol. Med. 32*, 333–340.

8. Burton, G.W., Traber, M.G., Acuff, R.V., Walters, D.N., Kayden, H., Hughes, L., and Ingold, K.U. (1998) Human Plasma and Tissue Alpha-Tocopherol Concentrations in Response to Supplementation with Deuterated Natural and Synthetic Vitamin E, *Am. J. Clin. Nutr. 67*, 669–684.

9. Jiang, Q., Christen, S., Shigenaga, M.K., and Ames, B.N. (2001) γ-Tocopherol, the Major Form of Vitamin E in the US Diet, Deserves More Attention, *Am. J. Clin. Nutr. 74*, 714–22.

10. Lemoyne, M., van Gossum, A., Kurian, R., Ostro, M., Axler, J., and Jeejeebhoy, K.N. (1987) Breath Pentane Analysis as an Index of Lipoperoxidation: A Functional Test of Vitamin E Status, *Am. J. Clin. Nutr. 46*, 267–272.

11. Horwitt, M. (1962) Interrelations Between Vitamin E and Polyunsaturated Fatty Acids in Adult Men, *Vitam. Horm. 20*, 541–558.

12. Reaven, P.D., and Witztum, J.L. (1993) Comparison of Supplementation of *RRR*-Alpha-Tocopherol and Racemic-Alpha-Tocopherol in Humans: Effects on Lipid Levels and Lipoprotein Susceptibility to Oxidation, *Arterioscler. Thromb. 13*, 601–608.

13. Hoppe, P.P., and Krennrich, G. (2000) Bioavailability and Potency of Natural-Source and All-Racemic α-Tocopherol in the Human: A Dispute, *Eur. J. Nutr. 39*, 183–193.

14. Horwitt, M.K. (1980). Relative Biological Values of *d*-α-Tocopheryl Acetate and *all-rac*-α-Tocopheryl Acetate in Man, *Am. J. Clin. Nutr. 33*, 1856–1860.

15. Burton, G.W., and Ingold, K.U. (1993) Biokinetics of Vitamin E Using Deuterated Tocopherols, in *Vitamin E in Health and Disease* (Packer, L., and Fuchs, J., eds.) pp. 329–344, Marcel Dekker, New York.

16. Ingold, K.U., Burton, G.W., Foster, D.O., Hughes, L., Lindsay, D.A., and Webb, A. (1987) Biokinetics of and Discrimination Between Dietary *RRR*- and *SRR*-α-Tocopherols in the Male Rat, *Lipids 22*, 163–172.

17. Traber, M., Winklhofer-Roob, B.M., Roob, J.M., Khoschsorur, G., Aigner, R., Cross, C., Ramakrishnan, R., and Brigelius-Flohe, R. (2001) Vitamin E Kinetics in Smokers and Nonsmokers, *Free Radic. Biol. Med. 31*, 1368–1374.

18. Traber, M.G., Elsner, A., and Brigelius-Flohe, R. (1998) Synthetic as Compared with Natural Vitamin E Is Preferentially Excreted as α-CEHC in Human Urine: Studies Using Deuterated α-Tocopheryl Acetates, *FEBS Lett. 437*, 145–148.

19. Acuff, R.V., Dunworth, R.G., Webb, L.W., and Lane, J.R. (1998) Transport of Deuterium Labeled Tocopherols During Pregnancy, *Am. J. Clin. Nutr. 67*, 459–464.

20. Burton, G.W., Ingold, K.U., Cheeseman, K.H., and Slater, T.F. (1990) Application of Deuterated α-Tocopherols to the Biokinetics and Bioavailability of Vitamin E, *Free Radic. Res. Commun. 11*, 99–107.

21. Hosomi, A., Arita, M., Sato, Y., Kiyose, C., Ueda, T., Igarashi, O., Arai, H., and Inoue, K. (1997) Affinity for Alpha-Tocopherol Transfer Protein as a Determinant of the Biological Activities of Vitamin E Analogs, *FEBS Lett. 40*, 105–108.

22. Cohn, W. (1999) Evaluation of Vitamin E Potency [Letter], *Am. J. Clin. Nutr. 69*, 157.

23. Horwitt, M.K., Elliott, W.H., Kanjanggulpan, P., and Fitch, C.D. (1984) Serum Concentrations of α-Tocopherol After Ingestion of Various Vitamin E Preparations, *Am. J. Clin. Nutr. 40*, 240–245.

24. Cheeseman, K.H., Holley, A.E., Kelly, F.J., Wasil, M., Hughes, L., and Burton, G. (1995) Biokinetics in Humans of *RRR*-α-Tocopherol: the Free Phenol, Acetate Ester, and Succinate Ester Forms of Vitamin E, *Free Radic. Biol. Med. 19*, 591–598.

25. Yoshikawa, T., Miyagawa, H., Takemura, T. (1988) Absorption of *d*- and *dl*-α-tocopherol in Healthy Human Volunteers, in *Proc. 4th Biennial Meeting, Society of Free Radical Research* (Hayaishi,O., Niki, E., Kondo, M., and Yoshikawa, T., eds.) pp. 295–298, 9–13 April, Kyoto, Japan.

26. Ferslew, K.E., Acuff, R.V., Daigneault, E.A., Woolley, T.W., and Stanton, P.E., Jr. (1993) Pharmacokinetics and Bioavailability of the *RRR* and All Racemic Stereoisomers of α-Tocopherol in Humans After Single Oral Administration, *J. Clin. Pharmacol. 33*, 84–88.

27. Baker, H., Handelman, G.J., Short, S., Machlin, L.J., Bhagavan, H.N., Dratz, E.A., and Frank, O. (1986) Comparison of Plasma α- and γ-Tocopherol Levels Following Chronic Oral Administration of Either *all-rac*-α-Tocopheryl or *RRR*-α-Tocopheryl Acetate in Normal Adult Male Subjects, *Am. J. Clin. Nutr. 43*, 382–387.

28. Devaraj, S., Adams-Huet, B., Fuller, C.J., and Jialal, I. (1997) Dose-Response Comparison of *RRR*-α-Tocopherol and all-racemic-α-Tocopherol on LDL Oxidation, *Arterioscler. Thromb. Vasc. Biol. 17*, 2273–2279.

29. Winklhofer-Roob, B.M., van't Hof, M.A., and Shmerling, D.H. (1996) Long-Term Oral Vitamin E Supplementation in Cystic Fibrosis Patients: *RRR*-α-Tocopheryl Acetate Preparations, *Am. J. Clin. Nutr. 63*, 722–728.

30. Chopra, R.K., and Bhagavan, H.N. (1999) Relative Bioavailabilities of Natural and Synthetic Vitamin E Formulations Containing Mixed Tocopherols in Human Subjects, *Int. J. Vitam. Nutr. Res. 69*, 92–95.

31. Acuff, R.V., Thedford, S.S., Hidiroglou, N.N., Papas, A.M., and Odom, T.A., Jr. (1994) Relative Bioavailability of *RRR*- and *all-rac*-α-Tocopheryl Acetate in Humans: Studies Using Deuterated Compounds, *Am. J. Clin. Nutr. 60*, 397–402.

32. Catignani, D.L., and Bieri, J.G. (1977) Rat Liver α-Tocopherol Binding Protein, *Biochim. Biophys. Acta 497*, 349–357

33. Traber, M.G., Rader, D., Acuff, R.V. et al (1998) Vitamin E Dose-Response Studies in Humans with Use of Deuterated *RRR*-α-Tocopherol, *Am. J. Clin. Nutr. 68*, 847–853.

Chapter 11

Vitamin E: Evidence for the 2:1 Preference for *RRR*-Compared with *all-rac*-α-Tocopherols

Maret G. Traber and David Blatt

Linus Pauling Institute, Oregon State University, Corvallis, OR 97331

Vitamin E Structures vs. Activities

Almost since the discovery of vitamin E in 1922 (1), there have been disagreements concerning the biologic activities of the various tocopherols and tocotrienols. α-Tocopherol is found in highest concentrations in animal plasma and tissues. Currently, only α-tocopherol has been demonstrated to reverse human vitamin E deficiency symptoms; it is the only form of vitamin E that meets the year 2000 vitamin E recommended dietary allowance (RDA) (2). However, in the plant kingdom, a variety of compounds that have vitamin E antioxidant activity have been described. The latest of these was described in plankton and cold water fishes that consume plankton (3).

Vitamin E was originally defined as a required nutrient because it is essential for maintaining the fetus during pregnancy (1). Thus, tests to evaluate biologic activities of the naturally occurring vitamin E forms were based on amounts required to protect the fetus (4). Unfortunately, most of the studies (5,6) were conducted long before modern methods of chromatography along with sensitive detection methods were available. It is therefore impossible to assess how much contaminating α-tocopherol might have been in the preparations used to test the biologic activity of other vitamin E forms. The presence of contaminating α-tocopherol is important because the α-tocopherol transfer protein (TTP) is expressed during pregnancy in the uterus and placenta (7). This protein selectively transfers α-tocopherol compared with other tocopherols and tocotrienols (8) and may be critical for maintaining the maternal/fetal unit α-tocopherol concentrations.

The rationale for possible differences in biologic activities between various naturally occurring vitamin E forms has been based on slight differences in antioxidant activities. However, antioxidant activities cannot be the explanation for differences in the biologic activities between naturally occurring and synthetic α-tocopherol. When α-tocopherol is chemically synthesized, eight stereoisomers are formed arising from the three chiral centers in the tail, which can be *R* or *S*. The naturally occurring α-tocopherol is in the *RRR*-form, whereas the synthetic contains equimolar concentrations of *RRR*, *RSR*, *RRS*, *RSS*, *SRR*, *SSR*, *SRS*, and *SSS*. These forms have identical chromanol rings and thus identical antioxidant activities.

Lack of Bioequivalence of *RRR*- and all-*rac*-α-Tocopherols

Bioequivalence implies comparable bioavailability and biopotency (9). Different formulations are *bioequivalent* if the rate and extent of absorption of the active ingredients are not significantly different when administered under similar conditions, or result in the same concentration vs. time curves, or if all of their effects are identical (10). *RRR*- and all-*rac*-α-tocopherols have not been proven to meet any of these criteria for bioequivalence; thus, they must be presumed to be different compounds, not different formulations of the same compound. Because the active moieties (stereoisomers) differ in *RRR*- and all-*rac*-α-tocopherols, and have different or unknown potencies for each physiologic effect, bioavailability and biopotency ratios cannot be assumed to be equal.

Determination of Biological Activities Using Deuterium-Labeled Vitamin E

Bioavailability of α-Tocopherol

Plasma concentrations of α-tocopherols reflect the sum of the four primary pharmacokinetic processes: absorption, distribution, metabolism, and excretion. An individual molecule goes through these processes sequentially, but all four processes occur simultaneously. Clearance from plasma occurs by both distribution and elimination. Distribution and elimination (metabolism and excretion) begin as soon as the first molecule is absorbed. Slower absorption allows more distribution and elimination to occur before absorption is complete. Thus, the area under the curve (AUC) of plasma concentrations vs. time is related to the rate and extent of absorption and distribution and to the rate of elimination (9).

Bioavailability is a measure of an individual's total exposure to a substance. The ratio of steady-state plasma concentrations of a substance after oral compared with intravenous administration is proportional to its fractional absorption. *Bioavailability* is the rate and extent to which the active moiety is absorbed, enters the systemic circulation, and is available at the site of action (9,10).

Calculation of deuterated α-tocopherol (AUC) may be the best method for determining vitamin bioavailability E. When healthy adults ingested 15, 75, or 150 mg deuterated *RRR*-α-tocopheryl acetate (11), the AUC calculated from plasma d$_3$-*RRR*-α-tocopherol concentrations increased linearly with dose (i.e., a 10-fold increase in dose resulted in a 10-fold increase in AUC). Linear increases in the AUC with dose reflect linear increases in the sum of absorption, distribution, metabolism, and excretion; this suggests that these processes were not saturated.

However, there was little change in plasma total (labeled plus unlabeled) α-tocopherol concentrations, which were 12 μmol/L at baseline and averaged 13.3, 15.4, and 16.7 over 96 h after the 15, 75, and 150 mg doses, respectively. Thus, the newly absorbed labeled vitamin E was preferentially used to replace the "old" circulating vitamin E. This observation also makes it clear that interpretation of vitamin E bioavailability cannot be carried out using unlabeled vitamin E. Thus, studies comparing the efficacy of supplements containing either natural or synthetic vitamin E are invalid because the amount of newly absorbed vitamin E from the supplement cannot be assessed.

Mechanisms of Transport and Discrimination in Humans

The use of stable isotope–labeled vitamin E has been instrumental in describing how differences in the absorption and plasma transport of vitamin E forms lead to differences in their biological activities. Because these topics have been reviewed extensively elsewhere (12,13), they will be discussed only briefly here. Studies using *RRR*- and *SRR*-α-tocopherols and *RRR*-γ-tocopherol labeled with different amounts of deuterium demonstrated that all of these forms were equally absorbed and secreted in chylomicrons into the plasma (14,15). Subsequently, there was a preferential secretion of *RRR*-α-tocopherol from the liver into the plasma in very low density lipoproteins (VLDL). This preference for α-tocopherol is dependent upon the function of TTP (16) because patients with a genetic defect in TTP become spontaneously vitamin E deficient when consuming diets adequate in vitamin E (17).

Studies in Mice Lacking the α-Tocopherol Transfer Protein

Although studies in humans with defects in the TTP gene have suggested that TTP was required for maintenance of plasma α-tocopherol concentrations (17), only limited information is available on these vitamin E–deficient subjects' tissue α-tocopherol concentrations (18). Mice in which the gene for TTP was deleted have extraordinarily low vitamin E plasma and tissue concentrations (19–21) and express vitamin E deficiency symptoms (7,19). To determine whether TTP knockout mice were unable to discriminate between natural and synthetic vitamin E (21), adult TTP knockout (*Ttpa$^{-/-}$*, n = 5), heterozygous (*Ttpa$^{+/-}$*, n = 7), and control (*Ttpa$^{+/+}$*, n = 3) mice consumed equimolar d$_6$ *RRR*- and d$_3$ *all-rac*-α-tocopheryl acetates (30 mg each/kg diet) for 3 mo; labeled and unlabeled α-tocopherols in plasma and 17 tissues were then measured by liquid chromatography/mass spectrometry (LC/MS) (21). Although deuterium-labeled α-tocopherols represented >85% of the plasma α tocopherol in all groups, *Ttpa$^{-/-}$* mice had plasma total α-tocopherol concentrations that were only 5.4% of *Ttpa$^{+/+}$* and 7.7% of *Ttpa$^{+/-}$* mice. *Ttpa$^{-/-}$* tissue (except liver) total α-tocopherol concentrations were 2–20% of those in *Ttpa$^{+/+}$* mice. These data are consistent with the concept that vitamin E is absorbed and transported to the liver, and only if TTP is expressed is α-tocopherol exported from the liver into the plasma for tissue delivery. Clearly, TTP is required not only to maintain plasma α-tocopherol, but also tissue α-tocopherol.

The other proposed function of TTP is the preferential secretion of *RRR*-α-tocopherol from the liver into the plasma. In mice fed 1:1 d$_6$ *RRR*- and d$_3$ *all-rac*-α-tocopheryl acetates for 3 mo, the d$_6$:d$_3$ ratios in plasma and 16 tissues from

Ttpa[+/+] and *Ttpa*[+/-] mice were double those of *Ttpa*[-/-] mice (Fig. 11.1) (21). In TTP-expressing mice, tissue enrichment of natural over synthetic α-tocopherol appears to be due to nonspecific uptake of α-tocopherol from the plasma, which contained 2:1 d_6:d_3 α-tocopherols. *Ttpa*[+/-] mice that expressed half the amount of hepatic TTP (20) also had d_6:d_3 ratios of nearly 2. These data suggest that plasma α-tocopherol concentrations are highly dependent upon the function of TTP and that this protein preferentially selects only the 2*R*-α-tocopherol forms from *all-rac*-α-tocopherol for secretion into plasma. Importantly, patients who were α-TTP heterozygotes (expressing two different mutations) were also able to discriminate between *RRR*- and *SRR*-α-tocopherols (17,22). These 2:1 ratios are also consistent with findings in normal humans administered labeled *RRR*- and *all-rac*-α-tocopheryl acetates (23–25).

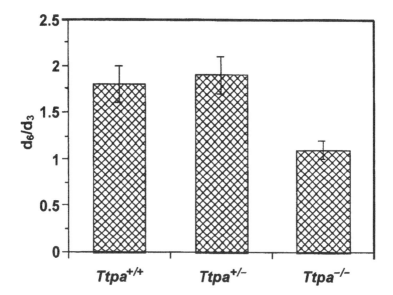

Fig. 11.1. α-Tocopherol transfer protein (TTP)-knockout mice do not discriminate between natural and synthetic vitamin E. Shown are the d_6:d_3 α-tocopherol ratios in adult α-TTP knockout (*Ttpa*[-/-], n = 5), heterozygous (*Ttpa*[+/-], n = 7), and wild-type (*Ttpa*[+/+], n = 3) mice that consumed a 1:1 d_6-*RRR*- and d_3-all-*rac*-α-tocopheryl acetate–containing diet (30 mg each/kg diet) for 3 mo. Labeled and unlabeled α-tocopherols in plasma and 17 tissues were measured. The d_6:d_3 α-tocopherol ratios in plasma and 16 tissues (excluding liver) from *Ttpa*[+/+], *Ttpa*[+/-] and *Ttpa*[-/-] mice were 1.8 ± 0.2, 1.9 ± 0.2, and 1.1 ± 0.1, respectively (*P* < 0.0001, *Ttpa*[-/-] vs. *Ttpa*[+/+] or *Ttpa*[+/-]). *Source:* Adapted from (21).

Vitamin E Kinetics in Pigs

Vitamin E supplements are often provided to farm animals; therefore, the relative bioavailabilities of *RRR-* and *all-rac-*α-tocopherols were evaluated in swine (26). Deuterium-labeled vitamin E (150 mg each of d_3-*RRR-* and d_6-*all rac-*α-tocopheryl acetates) was administered orally to female pigs (n = 3) with the morning feed. Blood samples were obtained at 0, 3, 6, 9, 12, 24, 36, 48, and 72 h. The maximum observed plasma concentration of d_3-*RRR-*α-tocopherol was achieved at 12 h (1.12 μmol/L), whereas d_6-α-tocopherol peaked earlier (at 9 h) at a lower concentration (0.66 μmol/L, $P < 0.05$, Fig. 11.2). The d_3/d_6 ratios were 1.35 ± 0.73 at 3 h and increased to 2.00 ± 0.14 at 72 h ($P < 0.03$). The d_3-α- and d_6-α-tocopherol disappearance rates were similar for both tocopherols and estimated to be 0.029 μmol/L per hour. The AUC for d_3-α- and d_6-α-tocopherol were 38.2 and 18.5, respectively. These studies suggest that pigs rapidly discriminate between forms of vitamin E and selectively retain 2*R*-forms in the plasma.

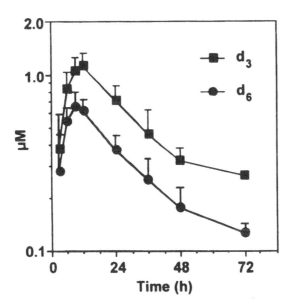

Fig. 11.2. Discrimination between natural and synthetic vitamin E in swine. The plasma d_3- and d_6-α-tocopherol concentrations after deuterium-labeled vitamin E (150 mg each d_3-*RRR-* and d_6-all-*rac-*α-tocopheryl acetates) was administered orally to adult female pigs (n = 3) with the morning feed are shown. The ratio of d_3-/d_6-α-tocopherols increased from 1.35 ± 0.73 at h 3 to 2.0 ± 0.14 at h 72 ($P < 0.03$). *Source:* Adapted from (26).

Vitamin E Kinetics During Pregnancy and Lactation in Sows

The biological activities of RRR- and all-rac-α-tocopherols were also evaluated during pregnancy and lactation, using pigs as a model system (27). Vitamin E delivery to fetuses and to suckling piglets was monitored by feeding 150 mg each of d_3-RRR-α- and d_6-all-rac-α-tocopheryl acetates daily to 3 pregnant sows from 7 d before to 7 d after birth. Blood from fasting sows was obtained in the morning of d −7, −4, 1, and 4 (during supplementation, d 1 was delivery day) and on d 7, 14, 21. Sow's milk, as well as piglet (n = 9) plasma and tissues were obtained at birth, 7 and 21 d. Labeled and unlabeled vitamin E concentrations were measured by LC/MS.

At birth, despite markedly elevated sow plasma deuterated α-tocopherol concentrations, no labeled α-tocopherol was detected in piglet plasma or tissues. After initiation of suckling by the piglets, a dramatic increase in plasma and tissue α-tocopherol concentrations was observed. These data emphasize the limited placental vitamin E transfer and the importance of milk as a delivery system for enhancing the vitamin E status of the newborn. At d 7 compared with birth, total α-tocopherol concentrations in most piglet tissues had increased 10-fold. The highest deuterated vitamin E concentrations in piglet tissues were found in liver, followed by lung, heart, kidney, muscle, intestine, and brain. Plasma d_3-α-tocopherol concentrations were approximately double those of d_6-α-tocopherol in both sows and piglets.

Biologic Activity of the Tocotrienols: Importance of Metabolism

Because tocotrienols have both antioxidant activities and reportedly decrease serum cholesterol synthesis, the serum cholesterol response and LDL oxidation after α-, γ-, or δ-tocotrienyl acetate supplementation was studied in hypercholesterolemic subjects (28). The subjects consumed a low-fat diet for 4 wk, then were randomly assigned to placebo (n = 13), α- (n = 13), γ- (n = 12), or δ- (n = 13) tocotrienyl acetate supplements (250 mg/d). Supplements were eaten with dinner for 8 wk while subjects continued to consume the low-fat diet.

No differences were observed in plasma cholesterol levels, LDL cholesterol levels, or LDL apolipoprotein B concentrations, suggesting that the tocotrienols were ineffective in modulating cholesterol synthesis. Importantly, tocotrienyl supplements did not affect plasma α- or γ-tocopherol concentrations. Twelve hours after the last dose of supplemental tocotrienyl acetates, fasting plasma concentrations of α-, γ-, and δ-tocotrienols were ~1, 0.5, and 0.1 μmol/L, respectively. Apparently, tocotrienyl acetate supplements are hydrolyzed and absorbed, but plasma tocotrienol concentrations remain barely detectable. The reason for these relatively low concentrations compared with the ~20 μmol/L α-tocopherol concentration is the very fast clearance of the plasma tocotrienols, as demonstrated by Yap et al. (29).

The mode of tocotrienol elimination from the body is not known. Hypothetically, the rapid disappearance of the tocotrienols could be a result of metabolism.

Carboxyethyl-hydroxychromans (α- and γ-CEHC) are human urinary metabolites of α- and γ-tocopherols, respectively (30). Lodge *et al.* (31) demonstrated that α- or γ-tocotrienols are metabolized to the same carboxyethyl-hydroxychromans (α- and γ-CEHC) as are α- and γ-tocopherols, respectively, and that these CEHC are detectable in urine. The subjects (n = 5 men, 1 woman) were healthy, nonsmokers, who did not take antioxidant supplements. A single supplement was consumed with breakfast each week for 4 wk in the following order: 125 mg γ-tocotrienyl acetate, 500 mg γ-tocotrienyl acetate, 125 mg α-tocotrienyl acetate, then 500 mg α-tocotrienyl acetate. Complete urine collections were obtained 24 h before, as well as on the day of supplementation, and on the subsequent 2 d. The fraction of the dose excreted as CEHC, as shown in Figure 11.3, was ~1 and 5% for α- and γ-tocotrienols, respectively. These relatively low percentages suggest that tocotrienols (and perhaps other vitamin E forms) that are rapidly cleared from plasma are eliminated apparently by a mechanism other than metabolism, or the metabolites are excreted by other routes.

Conclusions

The use of stable isotope–labeled vitamin E has allowed the estimation of vitamin E pharmacokinetic parameters. However, unlike drugs that are generally absorbed,

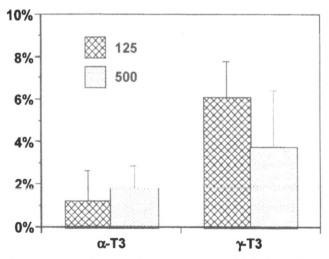

Fig. 11.3. Humans metabolize tocotrienols to carboxyethyl-hydroxychromans (CEHC) and excrete them in the urine. Complete urine collections (24 h) were obtained for 1 d before (baseline), the day of, and 2 d after human subjects (n = 6) ingested tocotrienol supplements. The subjects consumed 125 mg γ-tocotrienyl acetate in wk 1 and 500 mg in wk 2; then 125 mg α-tocotrienyl acetate was administered in wk 3 and 500 mg in wk 4. The percentage of the administered dose recovered as the respective urinary CEHC metabolites is shown for each dose. *Source*: Adapted from (31).

transported, and excreted by nonspecific mechanisms, the use of labeled vitamin E forms has demonstrated how differences in the plasma transport of vitamin E forms lead to differences in their biologic activities. It is quite apparent that TTP activity determines that only 2R-α-tocopherols are specifically maintained in the plasma. This leads to the question concerning what function α-tocopherol plays specifically and why α-tocopherol if other tocopherols and tocotrienols have potent antioxidant activities.

References

1. Evans, H.M., and Bishop, K.S. (1922) On the Existence of a Hitherto Unrecognized Dietary Factor Essential for Reproduction, *Science 56*, 650–651.
2. Food and Nutrition Board, and Institute of Medicine (2000) *Dietary Reference Intakes for Vitamin C, Vitamin E, Selenium, and Carotenoids*, p. 506, National Academy Press, Washington.
3. Yamamoto, Y., Fujisawa, A., Hara, A., and Dunlap, W.C. (2001) An Unusual Vitamin E Constituent (Alpha-Tocomonoenol) Provides Enhanced Antioxidant Protection in Marine Organisms Adapted to Cold-Water Environments, *Proc. Natl. Acad. Sci. USA 98*, 13144–13148.
4. Evans, H.M., Emerson, O.H., and Emerson, G.A. (1936) The Isolation from Wheat Germ Oil of an Alcohol, Alpha-Tocopherol, Having the Properties of Vitamin E, *J. Biol. Chem. 113*, 319–332.
5. Bunyan, J., McHale, D., Green, J., and Marcinkiewicz, S. (1961) Biological Potencies of ε- and ξ$_1$-Tocopherol and 5-Methyltocol, *Br. J. Nutr. 15*, 253–257.
6. Weiser, H., and Vecchi, M. (1982) Stereoisomers of α-Tocopheryl Acetate. II. Biopotencies of All Eight Stereoisomers, Individually or in Mixtures, as Determined by Rat Resorption-Gestation Tests, *Int. J. Vitam. Nutr. Res. 52*, 351–370.
7. Jishage, K., Arita, M., Igarashi, K., Iwata, T., Watanabe, M., Ogawa, M., Ueda, O., Kamada, N., Inoue, K., Arai, H., and Suzuki, H. (2001) Alpha-Tocopherol Transfer Protein Is Important for the Normal Development of Placental Labyrinthine Trophoblasts in Mice, *J. Biol. Chem. 273*, 1669–1672.
8. Hosomi, A., Arita, M., Sato, Y., Kiyose, C., Ueda, T., Igarashi, O., Arai, H., and Inoue, K. (1997) Affinity for Alpha-Tocopherol Transfer Protein as a Determinant of the Biological Activities of Vitamin E Analogs, *FEBS Lett. 409*, 105–108.
9. Shargel, L., and Yu, A.B.C. (1999) *Applied Biopharmaceutics and Pharmacokinetics*, 4th edn., Appleton-Lange, Stamford, CT.
10. Rescigno, A. (1997) Fundamental Concepts in Pharmacokinetics, *Pharm. Res. 35*, 363–390.
11. Traber, M.G., Rader, D., Acuff, R., Ramakrishnan, R., Brewer, H.B., and Kayden, H.J. (1998) Vitamin E Dose-Response Studies in Humans with Use of Deuterated *RRR*-α-Tocopherol, *Am. J. Clin. Nutr. 68*, 847–853.
12. Traber, M.G. (1999) in *Modern Nutrition in Health and Disease*, 9th edn. (Shils, M.E., Olsen, J.A. Shike, M., and Ross, A.C., eds.) pp. 347–362, Williams & Wilkins, Baltimore.
13. Traber, M.G., and Arai, H. (1999) Molecular Mechanisms of Vitamin E Transport, *Annu. Rev. Nutr. 19*, 343–355.
14. Traber, M.G., Burton, G.W., Hughes, L., Ingold, K.U., Hidaka, H., Malloy, M., Kane, J., Hyams, J., and Kayden, H.J. (1992) Discrimination Between Forms of Vitamin E by

Humans with and Without Genetic Abnormalities of Lipoprotein Metabolism, *J. Lipid Res. 33*, 1171–1182.

15. Traber, M.G., Burton, G.W., Ingold, K.U., and Kayden, H.J. (1990) *RRR-* and *SRR-*Alpha-Tocopherols Are Secreted Without Discrimination in Human Chylomicrons, but *RRR-*Alpha-Tocopherol Is Preferentially Secreted in Very Low Density Lipoproteins, *J. Lipid Res. 31*, 675–685.

16. Traber, M.G., Sokol, R.J., Burton, G.W., Ingold, K.U., Papas, A.M., Huffaker, J.E., and Kayden, H.J. (1990) Impaired Ability of Patients with Familial Isolated Vitamin E Deficiency to Incorporate Alpha-Tocopherol into Lipoproteins Secreted by the Liver, *J. Clin. Investig. 85*, 397–407.

17. Cavalier, L., Ouahchi, K., Kayden, H.J., Di Donato, S., Reutenauer, L., Mandel, J.L., and Koenig, M. (1998) Ataxia with Isolated Vitamin E Deficiency: Heterogeneity of Mutations and Phenotypic Variability in a Large Number of Families, *Am. J. Hum. Genet. 62*, 301–310.

18. Traber, M.G., Sokol, R.J., Ringel, S.P., Neville, H.E., Thellman, C.A., and Kayden, H.J. (1987) Lack of Tocopherol in Peripheral Nerves of Vitamin E-Deficient Patients with Peripheral Neuropathy, *N. Engl. J. Med. 317*, 262–265.

19. Yokota, T., Igarashi, K., Uchihara, T., Jishage, K., Tomita, H., Inaba, A., Li, Y., Arita, M., Suzuki, H., Mizusawa, H., and Arai, H. (2001) Delayed-Onset Ataxia in Mice Lacking Alpha-Tocopherol Transfer Protein: Model for Neuronal Degeneration Caused by Chronic Oxidative Stress, *Proc. Natl. Acad. Sci. USA 98*, 15185–15190.

20. Terasawa, Y., Ladha, Z., Leonard, S.W., Morrow, J.D., Newland, D., Sanan, D., Packer, L., Traber, M.G., and Farese, R.V.J. (2000) Increased Atherosclerosis in Hyperlipidemic Mice Deficient in Alpha-Tocopherol Transfer Protein and Vitamin E, *Proc. Natl. Acad. Sci. USA 97*, 13830–13834.

21. Leonard, S.W., Terasawa, Y., Farese, R.V., and Traber, M.G. (2002) Incorporation of Deuterium Labeled *RRR-* and all *rac-*α-Tocopherols into Plasma and Tissues of α-Tocopherol Transfer Protein Deficient Mice, *Am. J. Clin. Nutr. 75*, 555–560.

22. Traber, M.G. (1996) in *Antioxidants in Disease Mechanisms and Therapeutic Strategies* (Sies, H., ed.) pp. 49–63, Academic Press, San Diego.

23. Traber, M.G., Rader, D., Acuff, R., Brewer, H.B., and Kayden, H.J. (1994) Discrimination Between *RRR-* and all *rac-*α-Tocopherols Labeled with Deuterium by Patients with Abetalipoproteinemia, *Atherosclerosis 108*, 27–37.

24. Burton, G.W., Traber, M.G., Acuff, R.V., Walters, D.N., Kayden, H., Hughes, L., and Ingold, K.U. (1998) Human Plasma and Tissue Alpha-Tocopherol Concentrations in Response to Supplementation with Deuterated Natural and Synthetic Vitamin E, *Am. J. Clin. Nutr. 67*, 669–684.

25. Acuff, R.V., Dunworth, R.G., Webb, L.W., and Lane, J.R. (1998) Transport of Deuterium-Labeled Tocopherols During Pregnancy, *Am. J. Clin. Nutr. 67*, 459–464.

26. Lauridsen, C., Engel, H., Craig, A.M., and Traber, M.G. (2002) Relative Bioavailability of Dietary *RRR-* and all-*rac-*α-Tocopheryl Acetates in Swine Assessed Using Deuterium-Labeled Vitamin E, *J. Anim. Sci. 80*, 702–707.

27. Lauridsen, C., Engel, H., Jensen, S.K., Craig, A.M., and Traber, M.G. (2002) Discrimination Between *RRR-* and all *rac-*α-Tocopherol by Lactating Sows and Their Progeny Using Deuterated Vitamin E, *J. Nutr. 132*, 1258–1264.

28. O'Byrne, D., Grundy, S., Packer, L., Devaraj, S., Baldenius, K., Hoppe, P.P., Kraemer, K., Jialal, I., Traber, M.G., and Wilkinson, G.R. (2000) Studies of LDL Oxidation

Following α-, γ-, or δ-Tocotrienyl Acetate Supplementation of Hypercholesterolemic Humans, *Free Radic. Biol. Med. 29*, 834–845.

29. Yap, S.P., Yuen, K.H., and Wong, J.W. (2001) Pharmacokinetics and Bioavailability of α-, γ-, or δ-Tocotrienols Under Different Food Status, *J. Pharm. Pharmacol. 53*, 67–71.
30. Brigelius-Flohe, R., and Traber, M.G. (1999) Vitamin E: Function and Metabolism, *FASEB J. 13*, 1145–1155.
31. Lodge, J.K., Riddlington, J., Vaule, H., Leonard, S.W., and Traber, M.G. (2001) α- and γ-Tocotrienols are Metabolized to Carboxyethyl-Hydroxychroman (CEHC) Derivatives and Excreted in Human Urine, *Lipids 36*, 43–48.

Chapter 12

Mechanisms of Vitamin E Metabolism

Regina Brigelius-Flohé, Dirk Kluth, Nico Landes, Paul Pfluger, and Marc Birringer

German Institute of Human Nutrition, Potsdam-Rehbrücke, Germany

Introduction

Vitamin E is a term that includes 4 tocopherols (α, β, γ, δ) and 4 tocotrienols (α, β, γ, δ) (1). Naturally occurring tocopherols have an *RRR* stereochemistry, whereas the synthetic all-*rac*-tocopherols comprise all 8 possible stereoisomers. Tocotrienols have an unsaturated side chain and thus contain only one chiral center, which in the natural form has the *R* configuration. *RRR*-α-tocopherol is preferentially retained in the human body indicating that it may have functions that cannot be exerted by other forms of vitamin E. The special aspects of handling α-tocopherol include sorting, degradation, and possibly also absorption.

Absorption

The human diet contains the range of vitamin E types, with γ-tocopherol as the most abundant form in the typical American diet (2). Despite the higher γ-tocopherol intake, its concentration in human plasma is 5–10 times less than that of α-tocopherol. Administration of 250 mg α-, γ-, or δ-tocotrienol/d for 8 wk did not lead to plasma levels >1 μmol/L (3). It is generally believed that all forms of vitamin E are equally absorbed in the intestine [for review see (4)] and that the preferential treatment of α-tocopherol occurs in the liver by means of the α-tocopherol transfer protein (α-TTP; see below). Studies with children with cholestatic liver disease (5) and with patients suffering from cystic fibrosis (6) demonstrated that uptake of vitamin E into intestinal mucosa cells requires bile acids and pancreatic enzymes. Absorption is believed to be a passive process facilitated by fat intake. Release of vitamin E into the circulation occurs *via* chylomicrons. The mechanism of chylomicron formation and lipid loading is relatively well known (7), whereas it remains unclear how chylomicrons are loaded with different forms of vitamin E. In cultured hepatocytes, fibroblasts, and macrophages, an α-TTP–dependent but lipoprotein assembly–independent process was reported (8). Similar to the cholesterol reverse transport, it appears to be mediated by the ABCA1 transporter (9). Whether such a pathway is also functioning in the intestine requires investigation.

Sorting

In the liver, α-TTP selects the α-tocopherols with *R*-configuration at C-atom 2 and delivers them to very low density lipoproteins by a mechanism that is not yet known. The affinities of α-TTP for β-, γ-, δ-tocopherols and α-tocotrienol are 38, 9, 2, and 11% of that for α-tocopherol, respectively (10), explaining the high levels of α-tocopherol in plasma. The vital function of α-TTP is evident from patients with mutations in the gene for α-TTP. These patients have extremely low α-tocopherol plasma levels and suffer from neurological symptoms typical of vitamin E deficiency, e.g., cerebellar ataxia (11,12). α-TTP is expressed in Bergmann glia cells surrounding and alimenting Purkinje cells (13), which are involved in the coordination of intentional movements. α-TTP is obviously needed in the Purkinje layer to regulate the transfer of α-tocopherol, a role that other tocopherols cannot perform.

In 1922, Evans and Bishop (14) reported on the indispensability of vitamin E for the reproduction of female rats. The detection of α-TTP in the mouse uterus links α-TTP to α-tocopherol–dependent reproduction (15). In α-TTP knockout mice, the embryos die between d 9.5 and 10.5 of gestation and are resorbed (15). Circumstantial evidence suggests that α-TTP is also operative in the human placenta. Pregnant women were given equal amounts of d_3-*RRR*-α- and d_6-all-*rac*-α-tocopheryl acetate 5 d before delivery (16). The ratio of d_3/d_6 was found to be 1.86 in the mothers' plasma and 3.42 in the cord blood. This means that α-tocopherol has been sorted twice, first in the mothers' liver and obviously again at the maternal-fetal interface.

Degradation

Tocopherols and tocotrienols are metabolized by side-chain degradation. The final products, carboxyethyl hydroxychromans (CEHC), are conjugated with glucuronic acid or sulfate and excreted in the urine (17–21). Irrespective of the dosage, only 1–3% of the ingested *RRR*-α-tocopherol appears in the urine as α-CEHC (22). In contrast, at least 50% of γ-tocopherol is degraded and eliminated this way (23). Up to 2 and 6% of α- and γ-tocotrienol administered orally have been found as urinary α- and γ-CEHC, respectively (21). Thus, different forms of vitamin E are metabolized at different metabolic rates. Pertinent degradation pathways were therefore studied in cultured HepG2 cells.

β-*Oxidation*

The structures of the final degradation products, CEHC, suggest that side-chain degradation must have occurred *via* a β-oxidation pathway (Scheme 12.1). The identification of immediate precursors of α- and γ-CEHC, α- and γ-carboxymethylbutyl hydroxychroman (CMBHC) in human urine (22,24–26) and of α-carboxymethylhexyl hydroxychroman (α-CMHHC) in HepG2 cells (27) established this mechanism. For γ-tocopherol, all theoretical intermediates of the β-oxidation pathway were confirmed recently (28).

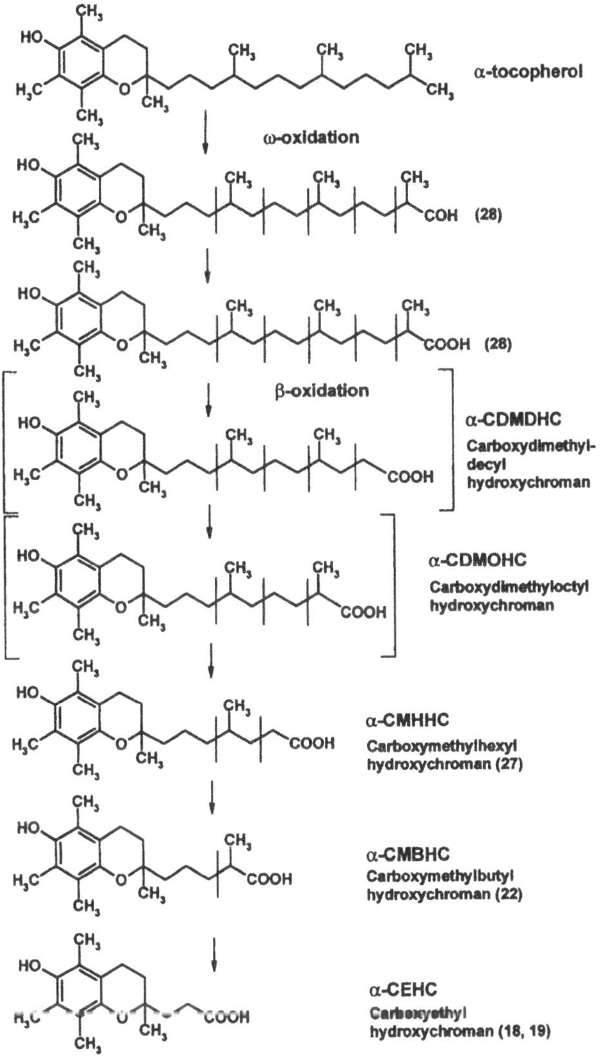

SCHEME 12.1. Mechanism of α-tocopherol side-chain degradation. Shown are the metabolites identified to date with respective references. For references of identified metabolites from other tocopherols, see text. Putative intermediates, currently identified only for γ-tocopherol (28), are set in brackets.

Although the β-oxidation scheme is straightforward for the tocopherols, auxiliary enzymes are required for the degradation of tocotrienols because of their unsaturated side chain (Scheme 12.2). Formation of the α,β-unsaturated fatty acid in the second and fourth round in the β-oxidation of tocotrienols leads to compounds with two conjugated

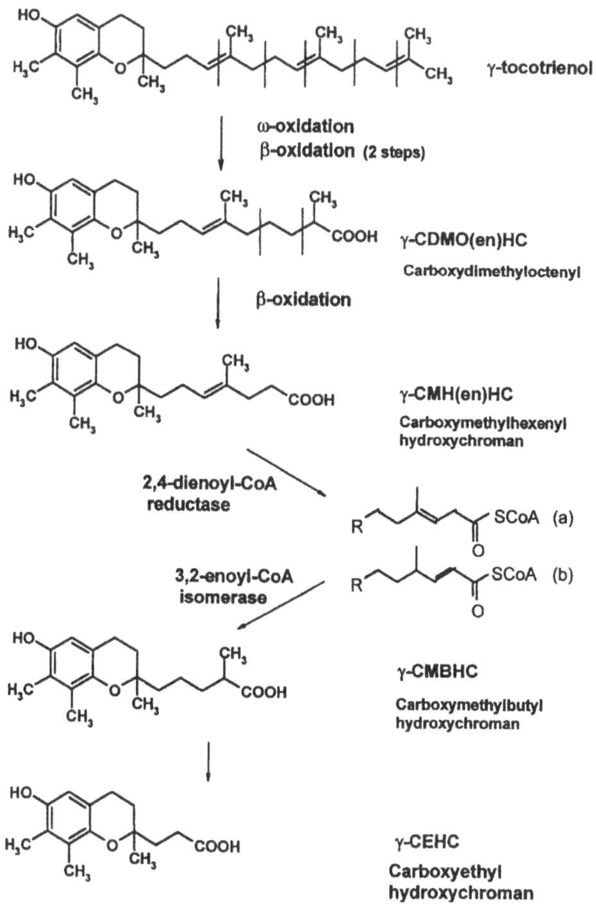

SCHEME 12.2. Mechanism of γ-tocotrienol side-chain degradation. β-Oxidation intermediates of γ-tocotrienol, γ-carboxymethylbutyl hydroxychroman (γ-CMBHC), the γ-car-
boxymethylhexyl hydroxychroman (CMHHC) precursor γ-carboxymethylhexenyl
hydroxychroman [γ-CMH(en)HC], and γ-carboxydimethyloctenyl [γ-CDMO(en)HC] as
well as the final product γ-carboxyethyl hydroxychroman (γ-CEHC) were identified in the
culture medium of HepG2 cells. The cells were incubated with 50 μmol/L γ-tocotrienol
for 2 d. Then, medium was extracted, metabolites separated by high-performance liquid
chromatography (HPLC) with electrochemical detection and identified by gas chromatography (GC)/mass spectrometry (MS) (27). All MS spectra showed an *m/z* typical of
the fragmented chroman structure, which is 223 in case of γ-substitution, and an *m/z* of
73 characteristic for the cleaved trimethylsilyl group. Molecular ion *m/z* was 408, 450,
476, and 518 indicating γ-CEHC, γ-CMBHC, γ-CMH(en)HC, and γ-CDMO(en)HC.
Compounds (a) and (b) are hypothetical intermediates produced by 2,4-dienoyl-CoA
reductase and 3,2-enoyl-CoA isomerase, respectively.

double bonds that are not accepted by enoyl-CoA hydratase that usually forms the β-hydroxyacyl-CoA. In linoleic acid metabolism, this impasse is resolved by the action of two auxiliary enzymes, 2,4-dienoyl-CoA reductase and 3,2-enoyl-CoA isomerase (see biochemistry texts). The identification of γ-carboxydimethyloctenyl hydroxychroman [γ-CDMO(en)HC] and γ-CMBHC lacking the double bond originally present in the tocotrienol (Scheme 12.2) confirms that the unsaturated side chain of tocotrienols is indeed metabolized like unsaturated fatty acids. Production of the CMBHC precursor, γ-carboxymethylhexenyl hydroxychroman [γ-CMH(en)HC], does not require auxiliary steps and has been found to be released as such from HepG2 cells incubated with α- or γ-tocotrienol. In principle, the enzymatic steps implicated are possible in peroxisomes and in mitochondria. At present, however, it is not clear where the side-chain degradation of tocopherols and tocotrienols is performed. The release of CEHC precursors by HepG2 cells indicates that the formation of the final product CEHC might involve rate-limiting steps. Whether precursors accumulate to relevant concentrations also *in vivo* and to which compartment they are released remains to be investigated.

ω-Oxidation

β-Oxidation of tocopherol and tocotrienol depends on initial ω-hydroxylation and subsequent oxidation of the hydroxyl group to an aldehyde and carboxylic acid function. The involvement of cytochrome P450 (CYP) enzymes appears plausible and has been deduced from inhibition and induction studies. Ketoconazole and sesamin, then believed to affect CYP3A-type enzymes specifically, have been shown to inhibit the release of γ- and δ-CEHC and γ- and δ-CMBHC from HepG2 cells (29). Rifampicin, known as a CYP3A inducer, stimulated the release of α-CEHC from HepG2 cells when treated with all-*rac*-α-tocopherol but not when treated with *RRR*-α-tocopherol (27). Recently, however, microsomes from insect cells transfected with human CYP4F2 have been shown to be most active in oxidizing the ω-methyl group of *RRR*-α- and γ-tocopherol, whereas a variety of other cytochromes including CYP3A4 were inactive (28). In this system, CYP4F2 preferentially degraded γ-tocopherol, whereas α-tocopherol was bound slightly better but metabolized more slowly. This was interpreted to explain the comparatively slow turnover of *RRR*-α-tocopherol in general. Difficulties in detecting α-CEHC release from HepG2 cells might, however, equally result from other reasons. We have shown that *RRR*-α-tocopherol degrades only after a long incubation time or in cells already adapted to all-*rac*-α-tocopherol (27). The inability of rifampicin to up-regulate *RRR*-α-tocopherol degradation could point to the involvement of a different CYP in the metabolism of this stereoisomer. In addition, the experimental results shown in Figure 12.1 may be interpreted along these lines. The yield of degradation products from HepG2 cells was substantially higher with γ-tocopherol and could not be further increased by rifampicin treatment, whereas rifampicin stimulated all-*rac*-α-tocopherol–derived α-CEHC and α-CMBHC release. If not due to different CYP, the differential rates and inducibilities of the tocopherol metabolism may result from the following: (i) highly different affinities or metabolizing rates of the hydroxylation system for α- and γ-tocopherol as suggested (28); (ii)

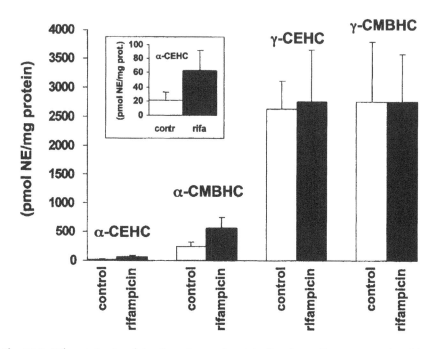

Fig. 12.1. Rifampicin stimulates the release of metabolites from all-*rac*-α-tocopherol but not from γ-tocopherol from HepG2 cells into the culture medium. Tocopherol-adapted HepG2 cells (27) were deprived of α-tocopherol by incubation in tocopherol-free medium for 4 d (wash-out). After that, no metabolites were detectable in the cell culture medium. During the last 2 d of the wash-out period, 50 μmol/L rifampicin was added to the culture medium. Control cells did not receive rifampicin. After the complete wash-out phase, 100 μmol/L all-*rac*-α- or γ-tocopherol was added. Afer 72 h, the medium was collected and 1-naphtol added as internal standard (final concentration 0.033 μmol/L). Thereafter, the medium was acidified to pH 4.5 and extracted 3 times with 10 mL *t*-butylmethylether. Solvent was removed by evaporation; the resulting residue was dissolved in high-performance liquid chromatography (HPLC) solvent and analyzed by HPLC-electrochemical detection as described (27). To obtain faster elution, acetonitrile was mixed with the solvent from min 32 to reach 40% acetonitrile after 52 min. Because an authentic standard was not available for γ-carboxymethylbutyl hydroxychroman (γ-CMBHC), data are expressed as 1-naphtol equivalents (NE). Response factors were 1.46 and 1.48 for α- and γ-carboxyethyl hydroxychromans (CEHC), respectively, and 2.44 for α-CMBHC. Actual concentrations (pmol/mg protein) were: 29.6 ± 16.2 for α-CEHC (control); 91.3 ± 42.5 for α-CEHC (rifampicin); 602 ± 176 for α-CMBHC (control); 1471.5 ± 397.5 for α-CMBHC (rifampicin). The concentrations of γ-CEHC (nmol/mg protein) were: 3.89 ± 0.71 (control) and 4.08 ± 1.32 (rifampicin). Thus, concentrations of α-CMBHC are somewhat underestimated in the figure and probably also those of γ-CMBHC. Values are means ± SD (n = 3). For better clarity, the α-CEHC is shown again in the insert on a different scale.

effective binding of α-tocopherol to α-TTP or to one of the tocopherol-associated proteins (TAP) (30,31), thereby preventing degradation; or (iii) upregulation of the metabolizing system by γ-tocopherol itself. Which of the possibilities finally turns out to be true remains to be determined.

Relevance of Side-Chain Degradation In Vivo

Although only 1–3% of the administered dose of *RRR*-α-tocopherol is found as α-CEHC in human urine, the percentage of all-*rac*-α-tocopherol–derived CEHC is 3–4 times higher (32). In rats administered a single dose of equal amounts of radioactively labeled *SRR*- and *RRR*-α-[5-methyl-^{14}C] tocopherol, 87.6 and 83% of the respective radioactivity was recovered in feces after 96 h (33). Of the administered dose, 7.8% was observed in the urine as α-CEHC from *SRR*- and 1.3% from *RRR*-α-tocopherol. This means that <10% of *RRR*-α-tocopherol that was not excreted *via* the feces was eliminated in the urine, whereas this fraction was up to 60% for *SRR*-α-tocopherol. This corresponds to the calculation made by Swanson *et al.* (23) who reported that 50% of the γ-tocopherol consumed may be converted to CEHC. Both studies demonstrated that urinary excretion of CEHC appears to be more relevant for tocopherols distinct from *RRR*-α-tocopherol. A detailed study of tocotrienol degradation has not yet been undertaken. First results show a higher CEHC excretion from γ- than from α-tocotrienol (21). However, accumulated precursors as observed in HepG2 cells and probably excreted via the bile *in vivo* should be considered for balancing. In addition, fecal elimination may be important.

Taken together, our knowledge of the mechanism of vitamin E metabolism has increased tremendously during the last few years. These studies further demonstrate the unique and preferential role played by α-tocopherol. Different metabolic rates certainly will influence biopotency and bioequivalence of individual forms of vitamin E. The possibility that CYP, i.e., drug or xenobiotic metabolizing enzymes, are involved in the degradation of vitamin E opens the possibility of an interference of vitamin E with the drug metabolism that should be investigated thoroughly in view of the large dosages of vitamin E taken for self-supplementation.

Acknowledgments

We thank Elvira Krohn and Stefanie Deubel for skillful technical assistance, D.P.R. Muller, London, for α-CMBHC, Bill Wechter for LLU-α (γ-CEHC), and Peter Hoppe, BASF, for α- and γ-tocotrienol. This work was supported by the Deutsche Forschungsgemeinschaft, DFG, Br778/6–1.

References

1. IUPAC-IUB Joint Commission on Biochemical Nomenclature (JCBN) (1982) Nomenclature of Tocopherols and Related Compounds. Recommendations 1981, *Eur. J. Biochem. 123*, 473–475.
2. Bieri, J.G., and Evarts, R.P. (1973) Tocopherols and Fatty Acids in American Diets. The Recommended Allowance for Vitamin E, *J. Am. Diet. Assoc. 62*, 147–151.

3. O'Byrne, D., Grundy, S., Packer, L., Devaraj, S., Baldenius, K., Hoppe, P.P., Kraemer, K., Jialal, I., and Traber, M.G. (2000) Studies of LDL Oxidation Following Alpha-, Gamma-, or Delta-Tocotrienyl Acetate Supplementation of Hypercholesterolemic Humans, *Free Radic. Biol. Med. 29*, 834–845.

4. Kayden, H.J., and Traber, M.G. (1993) Absorption, Lipoprotein Transport, and Regulation of Plasma Concentrations of Vitamin E in Humans, *J. Lipid Res. 34*, 343–358.

5. Sokol, R.J., Heubi, J.E., Iannaccone, S., Bove, K.E., and Balistreri, W.F. (1983) Mechanism Causing Vitamin E Deficiency During Chronic Childhood Cholestasis, *Gastroenterology 85*, 1172–1182.

6. Elias, E., Muller, D.P.R., and Scott, J. (1981) Association of Spinocerebellar Disorders with Cystic Fibrosis or Chronic Childhood Cholestasis and Very Low Serum Vitamin E, *Lancet 2*, 1319–1321.

7. Hussain, M.M. (2000) A Proposed Model for the Assembly of Chylomicrons, *Atherosclerosis 148*, 1–15.

8. Arita, M., Nomura, K., Arai, H., and Inoue, K. (1997) α-Tocopherol Transfer Protein Stimulates the Secretion of α-Tocopherol from a Cultured Liver Cell Line Through a Brefeldin A-Insensitive Pathway, *Proc. Natl. Acad. Sci. USA 94*, 12437–12441.

9. Oram, J.F., Vaughan, A.M., and Stocker, R. (2001) ATP-Binding Cassette Transporter A1 Mediates Cellular Secretion of Alpha-Tocopherol, *J. Biol. Chem. 276*, 39898–39902.

10. Hosomi, A., Arita, M., Sato, Y., Kiyose, C., Ueda, T., Igarashi, O., Arai, H., and Inoue, K. (1997) Affinity for Alpha-Tocopherol Transfer Protein as a Determinant of the Biological Activities of Vitamin E Analogs, *FEBS Lett. 409*, 105–108.

11. Ouahchi, K., Arita, M., Kayden, H., Hentati, F., Ben Hamida, M., Sokol, R., Arai, H., Inoue, K., Mandel, J.L., and Koenig, M. (1995) Ataxia with Isolated Vitamin E Deficiency Is Caused by Mutations in the Alpha-Tocopherol Transfer Protein, *Nat. Genet. 9*, 141–145.

12. Gotoda, T., Arita, M., Arai, H., Inoue, K., Yokota, T., Fukuo, Y., Yazaki, Y., and Yamada, N. (1995) Adult-Onset Spinocerebellar Dysfunction Caused by a Mutation in the Gene for the α-Tocopherol-Transfer Protein, *N. Engl. J. Med. 333*, 1313–1352.

13. Hosomi, A., Goto, K., Kondo, H., Iwatsubo, T., Yokota, T., Ogawa, M., Arita, M., Aoki, J., Arai, H., and Inoue, K. (1998) Localization of Alpha-Tocopherol Transfer Protein in Rat Brain, *Neurosci. Lett. 256*, 159–162.

14. Evans, H.M., and Bishop, K.S. (1922) On the Existence of a Hithero Unrecognized Dietary Factor Essential for Reproduction, *Science 56*, 650–651.

15. Jishage, K., Arita, M., Igarashi, K., Iwata, T., Watanabe, M., Ogawa, M., Ueda, O., Kamada, N., Inoue, K., Arai, H., and Suzuki, H. (2001) Alpha-Tocopherol Transfer Protein Is Important for the Normal Development of Placental Labyrinthine Trophoblasts in Mice, *J. Biol. Chem. 276*, 1669–1672.

16. Acuff, R.V., Dunworth, R.G., Webb, L.W., and Lane, J.R. (1998) Transport of Deuterium-Labeled Tocopherols During Pregnancy, *Am. J. Clin. Nutr. 67*, 459–464.

17. Chiku, S., Hamamura, K., and Nakamura, T. (1984) Novel Urinary Metabolite of Delta-Tocopherol in Rats, *J. Lipid Res. 25*, 40–48.

18. Schönfeld, A., Schultz, M., Petrzika, M., and Gassmann, B. (1993) A Novel Metabolite of RRR-Alpha-Tocopherol in Human Urine, *Die Nahrung. 37*, 498–500.

19. Schultz, M., Leist, M., Petrzika, M., Gassmann, B., and Brigelius-Flohé, R. (1995) Novel Urinary Metabolite of α-Tocopherol, 2,5,7,8-Tetramethyl-2(2'-Carboxyethyl)-6-Hydroxychroman, as an Indicator of an Adequate Vitamin E Supply? *Am. J. Clin. Nutr. 62*, 1527S–1534S.

20. Wechter, W.J., Kantoci, D., Murray, E.D., Jr., D'Amico, D.C., Jung, M.E., and Wang, W.H. (1996) A New Endogenous Natriuretic Factor: LLU-α, *Proc. Natl. Acad. Sci. USA 93*, 6002–6007.
21. Lodge, J.K., Ridlington, J., Leonard, S., Vaule, H., and Traber, M.G. (2001) α- and γ-Tocotrienols Are Metabolized to Carboxyethyl-Hydroxychroman Derivatives and Excreted in Human Urine, *Lipids 36*, 43–48.
22. Schuelke, M., Elsner, A., Finckh, B., Kohlschütter, A., Hübner, C., and Brigelius-Flohé, R. (2000) Urinary Alpha-Tocopherol Metabolites in Alpha-Tocopherol Transfer Protein-Deficient Patients, *J. Lipid Res. 41*, 1543–1551.
23. Swanson, J.E., Ben, R.N., Burton, G.W., and Parker, R.S. (1999) Urinary Excretion of 2,7, 8-Trimethyl-2-(beta-carboxyethyl)-6-hydroxychroman Is a Major Route of Elimination of Gamma-Tocopherol in Humans, *J. Lipid Res. 40*, 665–671.
24. Parker, R.S., and Swanson, J.E. (2000) A Novel 5′-Carboxychroman Metabolite of Gamma-Tocopherol Secreted by HepG2 Cells and Excreted in Human Urine, *Biochem. Biophys. Res. Commun. 269*, 580–583.
25. Pope, S.A., Clayton, P.T., and Muller, D.P.R. (2000) A New Method for the Analysis of Urinary Vitamin E Metabolites and the Tentative Identification of a Novel Group of Compounds, *Arch. Biochem. Biophys. 381*, 8–15.
26. Pope, S.A., Burtin, G.E., Clayton, P.T., Madge, D.J., and Muller, D.P.R. (2001) New Synthesis of (+/−)-α-CMBHC and Its Confirmation as a Metabolite of Alpha-Tocopherol (Vitamin E), *Bioorg. Med. Chem. 9*, 1337–1343.
27. Birringer, M., Drogan, D., and Brigelius-Flohé, R. (2001) Tocopherols Are Metabolized in HepG2 Cells by Side Chain ω-Oxidation and Consecutive β-Oxidation, *Free Radic. Biol. Med. 31*, 226–232.
28. Sontag, T.J., and Parker, R.S. (2002) Cytochrome P450 ω-Hydroxylase Pathway of Tocopherol Catabolism: Novel Mechanism of Regulation of Vitamin E Status, *J. Biol. Chem., 277*, 25290–25246.
29. Parker, R.S., Sontag, T.J., and Swanson, J.E. (2000) Cytochrome P4503A-Dependent Metabolism of Tocopherols and Inhibition by Sesamin, *Biochem. Biophys. Res. Commun. 277*, 531–534.
30. Stocker, A., Zimmer, S., Spycher, S.E., and Azzi, A. (1999) Identification of a Novel Cytosolic Tocopherol-Binding Protein: Structure, Specificity, and Tissue Distribution, *IUBMB Life 48*, 49–55.
31. Zimmer, S., Stocker, A., Sarbolouki, M.N., Spycher, S.E., Sassoon, J., and Azzi, A. (2000) A Novel Human Tocopherol-Associated Protein: Cloning, in Vitro Expression, and Characterization, *J. Biol. Chem, 275*, 25672–25680.
32. Traber, M.G., Elsner, A., and Brigelius-Flohé, R. (1998) Synthetic as Compared with Natural Vitamin E Is Preferentially Excreted as Alpha-CEHC in Human Urine: Studies Using Deuterated Alpha-Tocopheryl Acetates, *FEBS Lett. 437*, 145–148.
33. Kaneko, K., Kiyose, C., Ueda, T., Ichikawa, H., and Igarashi, O. (2000) Studies of the Metabolism of α-Tocopherol Stereoisomers in Rats Using [5-methyl-(14)C]*SRR*- and *RRR*-α-tocopherol, *J. Lipid Res. 41*, 357–367.

Chapter 13

γ-Tocopherol Metabolism and Its Relationship with α-Tocopherol in Humans

Rosalind Lee, Francesco Galli, and Frank J. Kelly

School of Health & Life Sciences, Kings College London, London, UK

Introduction

Vitamin E is widely associated with lipids in nature, especially monounsaturated and polyunsaturated fatty acids. In this context, it plays an important role in minimizing and, in some cases, preventing the oxidation of susceptible lipid molecules. Although this antioxidant function has been considered to be the primary role of vitamin E for several decades, it is becoming increasingly clear that this antioxidant vitamin also has a range of other important functions in cell biology. This multiplicity of function is due, in part, to the range of molecules that comprise the vitamin E family, namely, α, β, δ, and γ-tocopherols and α, β, δ, and γ-tocotrienols. These eight isoforms and some of their respective metabolites perform a wide range of roles within the body.

Humans have lost the ability to make vitamin E; as a consequence, they must acquire it from their diet. In contrast to other antioxidant vitamins such as vitamin C and β-carotene that are abundant in fruit and vegetables, members of the vitamin E family are more prevalent in plant seeds and oils. Even though α-tocopherol is by far the most prevalent form of vitamin E in humans, this is not the case in the plant kingdom. In many instances such as with soybean, corn, and sesame oil, γ-tocopherol is the predominant form of vitamin E (1). Indeed, γ-tocopherol predominates so much in our natural food chain that it represents ~70% of all vitamin E consumed in the typical American diet. Equivalent information is not available for Europe but there also, γ-tocopherol is likely to dominate vitamin E intake in most countries.

Given the large intakes of γ-tocopherol by humans, it is somewhat surprising that blood contains a relatively small proportion in comparison to α-tocopherol. Typically, the ratio is ~10:1 in favor of α-tocopherol (2). This intriguing observation was eventually explained by a series of interesting studies by Traber and colleagues (3–5). They showed that intestinal cells take up all forms of vitamin E equally and then release them into the circulation with chylomicrons. The isoforms then reach the liver in association with chylomicron remnants, although there is some opportunity for direct transfer to tissues during this transfer process. This may explain why certain tissues such as muscle and adipose tissue have higher than expected concentrations of γ-tocopherol. On reaching the liver, the chylomicron remnants are taken up and a specific protein, the α-tocopherol transfer protein (α-TTP), preferentially selects α-tocopherol from other incoming tocopherols for incorporation into very low density lipoproteins (6). Non-α-tocopherol forms are much less well retained; as a consequence, they are metabolized and either excreted *via* the bile or transported *via* the circulation to the kidney for subsequent elimination in urine (7,8).

γ-Tocopherol and Disease

A low γ-tocopherol concentration and a high α- to γ-tocopherol ratio have been reported in patients with coronary heart disease (CHD) compared with controls (9,10) and in a population with a high incidence of CHD (11). These observations suggest that a serum tocopherol profile with low γ-tocopherol levels and a high α- to γ-tocopherol ratio may be an indicator of an increased risk for CHD. In support of such a conclusion, Kushi *et al*. (12) reported that intake of dietary-derived vitamin E (i.e., mainly γ-tocopherol), but not supplemental vitamin E (α-tocopherol) was beneficial in reducing cardiovascular disease. Moreover, the same group recently arrived at a similar conclusion for postmenopausal women and death from stroke (13).

In response to these and other positive observations about γ-tocopherol, there is renewed interest in approaches to improve γ-tocopherol status by dietary change. For example, it has been suggested that rapeseed oil–rich diets help to increase the levels of γ-tocopherol in the body. In the Lyons Diet Heart Study it was found that a Mediterranean diet rich in α-linolenic acid (high in rapeseed oil) was beneficial in secondary prevention of CHD (14). Similarly, Lemcke-Norojarvi *et al*. (15) reported that substitution of corn and sesame oils in the diet (which contain substantial quantities of γ-tocopherol) leads to an increase in serum γ-tocopherol in healthy women without affecting the serum α-tocopherol concentrations. Similar conclusions were reached by Clooney and colleagues (16) who demonstrated that the source of dietary γ-tocopherol is an important determinant of its subsequent plasma concentrations. They found that the consumption of γ-tocopherol in the form of sesame seeds is the most efficient way of increasing blood γ-tocopherol concentrations. For example, consumption or 5 mg of γ-tocopherol/d for 3 d in the form of sesame seeds, but not walnut or soybean oil significantly elevated serum γ-tocopherol levels (19%) while decreasing β-tocopherol concentrations (34%)

Specific Beneficial Effects of γ-Tocopherol

Recently, Saldeen and colleagues (17) examined the differential effects of α- and γ-tocopherol on thrombogenesis. In a rat model, they found that γ-tocopherol was significantly more potent than α-tocopherol in inhibiting platelet aggregation and thrombogenesis. Moreover, they also report that γ-tocopherol reduced superoxide anion generation, lipid peroxidation, and low density lipoprotein oxidation and increased superoxide dismutase activity to a much greater extent than α-tocopherol.

γ-Tocopherol Metabolism

A number of studies have demonstrated that α-2,7,8-trimethyl-2-(β-carboxyethyl)-6-hydroxychroman (α-CEHC) is the main metabolite of the α-tocopherol in both plasma (18,19) and urine (20,21). Swanson *et al.* (22) suggested that the formation of γ-CEHC and its consequent excretion in urine represent a primary route of elimination of γ-tocopherol in humans. They also identified a putative metabolite analog to γ-CEHC, namely, γ-CMBHC (γ-carboxymethylbutyl hydroxychroman), which is excreted in urine as a minor form (1–4% of the concentration of γ-CEHC) and secreted *in vitro* by hepatoma cells (23). In the same study, the authors described in one subject, the time course of γ-CEHC and γ-CMBHC excretion after *RRR*-γ-tocopherol supplementation (150 mg/d for 2 d). The increase in the urinary concentration of γ-CEHC and γ-CMBHC 14 h after supplementation was approximately eight- and threefold, respectively. In general, studies undertaken in North America record high baseline levels of urinary γ-CEHC excretion (2.5–31.5 μmol/L), reflecting the higher intake of γ-tocopherol in the U.S. diet compared with that of Europe.

Interestingly, γ-CEHC was first identified by Wechter and colleagues in 1996 (24) when they were searching for molecules that elicit natriuretic activity. γ-CEHC, or LLU-α as it was named by Wechter, was shown to inhibit the 7O pS ATP-sensitive K^+ channel in the thick ascending limb cells of the rat kidney (24). In this respect, γ-CEHC activity may be unique in its ability to regulate extracellular fluid volume. Recently, however, Hatton and colleagues (25), reported that administration of either γ-tocopherol or γ-tocotrienol to male rats gave rise to increased plasma concentrations of γ-CEHC. Moreover, γ-tocotrienol supplementation in humans has been reported to increase urinary γ-CEHC concentrations (26). These findings weaken the case for a single class of nutrients being the immediate precursor of an important regulatory factor. Tocotrienols have a structure similar to that of tocopherol except that they have an unsaturated side chain. Previously, tocotrienols have been reported to lead to cholesterol suppression (27), tumor suppression (28), and a blockage of increased blood pressure (29). But studies using purified tocotrienols in hypercholesterolemic subjects did not confirm these findings (30).

Use of Deuterated γ-Tocopherol to Examine Its Biokinetics in Normal Subjects

Given the potentially unique qualities of γ-tocopherol and its metabolites and the fact that supplementation with α-tocopherol has a profound influence on body γ-tocopherol levels (31), we thought it worthwhile to examine further the bioavailability and metabolism in human subjects.

The objective of the study was to establish the biokinetics of γ-tocopherol within the body. We approached this problem by supplementing healthy subjects with deuterium-labeled γ-tocopherol and followed its appearance in the plasma pool and subsequent urinary excretion (32). Twenty-one subjects were recruited, (11 men/10 women) between 22 and 47 y of age (mean 32.6 ± 8.3 y). Subject characteristics are

TABLE 13.1
Subject Characteristics

Subject	Sex	Age (y)	Body mass index (kg/m^2)
1	F	47	20.2
2	M	28	19.6
3	M	45	23.9
4	F	27	22.9
7	F	26	22.7
8	F	25	19.1
9	M	26	23.1
10	M	50	35.7
11	F	22	26.0
13	F	34	22.8
14	F	30	19.9
15	M	35	22.3
16	F	27	21.2
17	F	34	21.6
18	F	43	23.7
19	M	44	24.0
20	M	31	21.8
21	M	29	22.4
22	M	29	24.5
23	M	29	20.7
24	M	24	21.8
Mean		32.6	22.9
SD		8.3	3.4

provided in Table 13.1. All were healthy nonsmoking individuals who were not taking any medication or vitamin supplements for the duration of this study or 4 wk before the start. The study was approved by the St. Thomas' Hospital Research Ethics Committee.

Protocol

Subjects were asked to make five visits to the laboratory when early morning urine and blood samples were collected. On the first visit when baseline samples were collected, subjects were given a capsule containing 100 mg of *RRR* d^2-γ-tocopherol acetate. This was taken with a supplied breakfast (a cake or chocolate bar containing a similar amount of fat). Subsequent visits were made on d 1, 3, 7, and 10 when early morning urine and blood samples were also collected. In addition to this, two subjects provided samples 6, 9 and 12 h postsupplementation. Plasma was obtained by centrifugation (13,000 × *g*, 15 min at 4°C) and immediately stored at −80°C until processing. Urine samples were stored at −20°C.

HPLC Analysis of Plasma α- and γ-Tocopherol

The plasma concentration of α- and γ-tocopherol were measured by high-performance liquid chromatography (HPLC) as described previously (33). Briefly, an internal standard of α-tocopherol acetate (Sigma, Dorset, UK) was added to 200 μL of plasma and vortexed. This was then mixed with cold hexane (500 μL) and vortexed again. The hexane layer was removed and evaporated to dryness under nitrogen. The dry extract was resuspended in methanol (400 μL) for analysis on a Gilson HPLC system (Anachem, Beds, UK) using reverse-phase HPLC. The extracted sample (100 μL) was then injected onto an Apex II Octadecyl 5-μm 10-cm column preceded by an Apex Bio300 Guard column 4.6-mm i.d. cartridge (Jones Chromatography, Glamorgan, UK). The separation was achieved using a mobile phase of 98% methanol/2% water set at a flow rate of 1 mL/min. The α-tocopherol, γ-tocopherol, and tocopherol acetate were measured using ultraviolet light detection at 292 nm. External standards of α-tocopherol (Sigma) and γ-tocopherol (Roche, Herts, UK) were also prepared. The final α- and γ-tocopherol plasma concentrations were corrected for cholesterol. Plasma cholesterol levels were determined by enzymatic colorimetric test using CHOD/PAP methods and Unimate 5 Chol kits, (Roche Diagnostic). Analysis was carried out by the Chemical Pathology Department, St. Thomas' Hospital).

Gas Chromatography-Mass Spectrometry (GC-MS) Analysis of Urinary Metabolites QL and CEHC

Internal standards d^3 quinone lactone (QL), d^2 CEHC and d^9 CEHC (4 nmol in 25 μL of acetonitrile or ethanol) were added to an aliquot of urine (4 mL). Enzymatic hydrolysis of the glucuronide conjugate of the metabolites was carried out using *Escherichia coli* β-glucuronidase (Sigma, EC: 3.2.1.31) 650 U in 200 μL of 0.25 mol/L sodium acetate buffer (pH 6.2). The samples were flushed with nitrogen and incubated at 30°C in the dark for 14 h. The resulting solution was acidified to pH 1.5 and extracted with hex-ane/*tert*-butyl methyl ether (10 mL, 99:1). After centrifugation (13,000 × g at 4°C for 15 min), the organic layer was removed and evaporated under nitrogen. The residue was dissolved in anhydrous pyridine (100 μL) and silylated at 65°C for 1 h with 50 μL N,O-bis (trimethylsilyl) trifluoroacetamide containing 1% trimethyl-chlorosilane (Pierce Chemical, Rockford, IL), after flushing the tube head space with nitrogen. The solvents were evaporated under nitrogen and the residue dissolved in hexane for GC-MS analysis.

The disilyl ester of CEHC metabolites and unsilylated QL metabolite were quantified by GC-MS using a Hewlett-Packard 6990 GC (Agilent Technologies, Stockport, UK) coupled to a Hewlett-Packard 5973 Mass Selective Detector. The carrier gas was helium. The separation of the compounds was achieved using a HP-1 cross-linked methylsioxane column (25 m × 0.2-mm i.d. and film thickness 0.33 μm). The oven temperature was set to 50°C for 2 min followed by ramp of 50°C/min to 240°C; this was held for 4 min before a further increase of 25°C/min to 285°C, which was maintained for 6 min. The injection volume was 1 μL injected in splitless mode. The transfer line temperature was 290°C and the MS source temperature was 230°C.

In selected ion mode, the following ions corresponding to the molecular ion of the metabolites were monitored: m/z 276 (d^0 QL), 279 (d^3 QL), 285 (d^9 QL), 408 (d^0 γ-CEHC), 410 (d^2 γ-CEHC), 422 (d^0 α-CEHC), 425 (d^3 α-CEHC), and 431 (d^9 α-CEHC). Concentrations of d^0 QL, γ- and α-CEHC were calculated from the peak area relative to the corresponding internal standard (d^3 QL, d^2 CEHC and d^9 CEHC, respectively). The concentration of the d^0 γ-CEHC was calculated using a response factor and the d^9 α-CEHC as internal standard (Fig. 13.1).

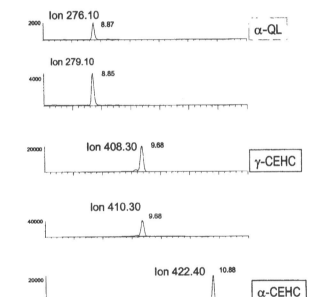

Fig. 13.1. Gas chromatography-mass spectrometry trace of a typical urine sample. Ion 276 represents d^0 α-QL and the corresponding internal standard ion 279 d^3 α-QL; ion 408 represents d^0 γ-2,7,8-trimethyl-2-(β-carboxyethyl)-6-hydroxychroman (γ-CEHC) and the corresponding internal standard ion 410 d^2 γ-CEHC; and ion 422 represents d^0 α-CEHC and the corresponding internal standard ion 431 d^9 α-CEHC.

The final urinary CEHC and QL concentrations were adjusted for creatinine concentrations. The concentration of creatinine was determined spectrophotometrically using a Sigma Diagnostics creatinine kit and following manufacturer's instructions (555A, Sigma Chemical, St. Louis, MO). Regression analysis was carried out with robust standard errors. The nonparametric data were log transformed to allow for parametric analysis.

Results

Plasma responses. Baseline α-tocopherol, γ-tocopherol, cholesterol, and cholesterol-corrected tocopherol data are shown in Table 13.2. As expected, γ-tocopherol at baseline was <10% of the α-tocopherol concentration. Moreover, considerable interindividual differences in both α- and γ-tocopherol existed among the subjects (Fig. 13.2; d 0 data). Ingestion of 100 mg of *RRR* γ-tocopherol acetate increased plasma γ-tocopherol concentrations 2.6-fold on d 1 compared with d 0 (0.37 ± 0.15 to 0.98 ± 0.42 μmol/mmol cholesterol, $P < 0.0001$). However, this response disappeared by d 3

TABLE 13.2
Plasma Tocopherol (Toc) Concentrations, Cholesterol (Chol), and
Tocopherol/Cholesterol Concentrations at Baseline

Subject	α-Toc (μmol)	γ-Toc (μmol)	Chol (mmol)	α-Toc/Chol (μmol/mmol)	γ-Toc/Chol (μmol/mmol)
1	28.38	2.54	6.71	4.23	0.38
2	29.89	2.46	4.25	7.03	0.58
3	26.25	1.06	4.36	6.02	0.24
4	28.80	2.08	4.87	5.91	0.43
7	35.65	1.28	5.51	6.47	0.23
8	24.79	1.98	4.86	5.10	0.41
9	19.84	1.44	4.19	4.74	0.34
10	39.91	3.33	5.22	7.65	0.64
11	19.82	1.50	3.68	5.39	0.41
13	33.12	2.16	5.82	5.69	0.37
14	25.58	1.48	3.95	6.48	0.38
15	33.83	1.83	6.72	5.03	0.27
16	22.97	0.97	4.88	4.71	0.20
17	22.99	1.27	4.84	4.75	0.26
18	42.59	2.42	6.53	6.52	0.37
19	34.16	3.59	5.29	6.46	0.68
20	20.29	1.53	3.83	5.30	0.40
21	25.87	1.27	4.92	5.26	0.26
22	29.90	1.60	4.89	6.12	0.33
23	29.65	1.10	5.04	5.88	0.22
24	25.99	2.86	3.98	6.53	0.72
Mean	28.6	1.89	4.97	5.77	0.39
SD	6.27	0.74	0.90	0.87	0.15

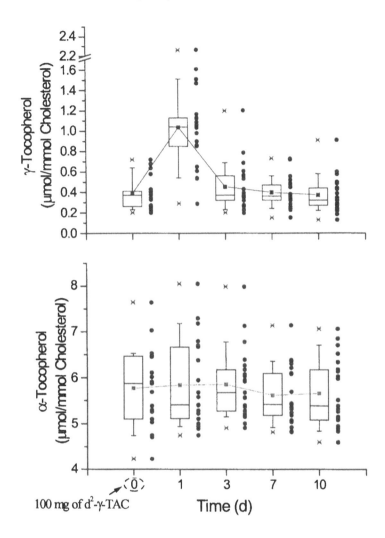

Fig. 13.2. Plasma α- and γ tocopherol concentrations after ingestion of 100 mg d²-γ-tocopherol at d 0. Data are provided both as medians and interquartile ranges as individual data points.

when plasma γ-tocopherol concentrations had returned to baseline (Fig. 13.2). Ingestion of 100 mg of *RRR* γ-tocopherol acetate did not alter the plasma α-tocopherol concentration (Fig. 13.2).

Urinary metabolite responses. After γ-tocopherol supplementation, urinary d² γ-CEHC increased markedly, rising to 2.24 ± 1.96 μmol/g creatinine on d 1 (Fig. 13.3). As seen with the plasma γ-tocopherol response, d² γ-CEHC concentrations returned to

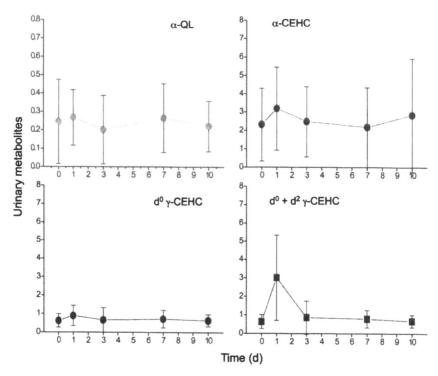

Fig. 13.3. Mean urinary metabolite concentrations given as μmol/g creatinine (the error bars show SD) after a 100-mg dose of d^2-γ-tocopherol at d 0. α-QL excretion rate did not change significantly from baseline. In contrast, α-2,7,8-trimethyl-2-(β-carboxyethyl)-6-hydroxychroman (α-CEHC) increased significantly ($P < 0.05$) at d 1 compared with baseline but then returned to baseline concentrations by d 3. The same pattern of increase was seen in the d^0-γ-CEHC metabolite ($P < 0.05$) at d 1. The d^2-γ-CEHC metabolite increased significantly at d 1 ($P < 0.0001$) and remained significantly increased at d 3 ($P < 0.005$) but then returned to baseline concentrations by the next sampling point at d 5.

baseline by d 3. In contrast, the urinary d^0 γ-CEHC concentrations changed only marginally, increasing from 0.80 ± 0.55 to 1.07 ± 0.68 μmol/g creatinine on d 1. Similarly, the α-tocopherol metabolites in urine showed very little response to γ-tocopherol supplementation (Fig. 13.3). There was a small increase in α-CEHC on d 1 (2.21 ± 2.12 to 3.19 ± 2.78 μmol/g creatinine), whereas α-QL was unaffected or slightly decreased after supplementation with γ-tocopherol.

Interindividual responses. A wide range in response to γ-tocopherol supplementation was seen in these healthy individuals (Fig. 13.4). Plasma responses over the 24-h postsupplementation period ranged from no change to a threefold increase. Such differences were even more marked in the urinary excretion of the d^2 γ-CEHC metabo-

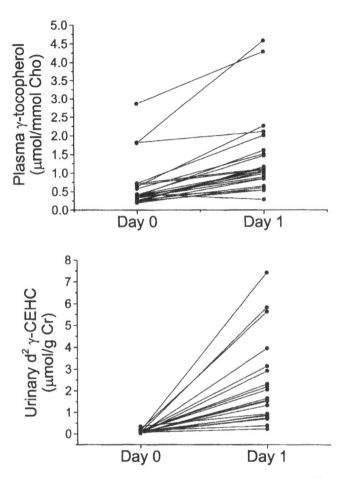

Fig. 13.4. Interindividual responses to supplementation with 100 mg of d^2-γ-tocopherol acetate. The plasma concentrations of γ-tocopherol and its urinary metabolite, γ-2,7,8-trimethyl-2-(β-carboxyethyl)-6-hydroxychroman (γ-CEHC) are shown for 21 healthy volunteers before and 24 h postsupplementation with 100 mg of d^2-γ-tocopherol acetate.

lite with 40-fold differences in response apparent (Fig. 13.4). To examine these early responses in more detail, two subjects were asked to provide blood and urine samples 6, 9, 12, 24, and 48 h postsupplementation with 100 mg d^2-γ-tocopherol.

Early responses to γ-tocopherol supplementation. In the two subjects examined, the major plasma responses to γ-tocopherol supplementation occurred within the first 24 h (Fig. 13.5). In both, peak plasma responses occurred at or before 6 h postsupplementation; by 24 h, these levels had fallen to approximately the same as the unlabeled, endogenous γ-tocopherol concentrations. Examination of the appearance of γ-CEHC in plasma revealed that this also occurs, rapidly building to a peak

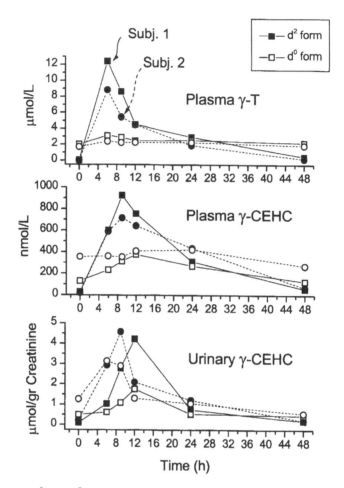

Fig. 13.5. The d^0 and d^2 plasma γ-tocopherol, γ-2,7,8-trimethyl-2-(β-carboxyethyl)-6-hydroxychroman (γ-CEHC) and urinary γ-CEHC concentrations for 2 subjects during the first 48 h postsupplementation with 100 mg d^2-γ-tocopherol are shown.

plasma level by 8 h postsupplementation. A similar pattern of γ-CEHC appearance occurred in urine (Fig. 13.5).

Discussion

Supplementation with 100 mg of γ-tocopherol resulted in a transient increase in plasma γ-tocopherol and urinary γ-CEHC concentrations. In most subjects, the plasma response was rapid and complete within 72 h. The majority of the changes in the plasma pool likely occurred within the first 24 hours. At the peak level, plasma γ-tocopherol concentration increased approximately threefold, whereas urinary γ-CEHC concentrations increased >20-fold. Importantly, however, ingestion of this large bolus of γ-tocopherol had no apparent effect on the plasma α-tocopherol concentration and produced only a slight increase in urinary α-CEHC. Thus, supplementation with γ-tocopherol, even at this large dose, does not displace α-tocopherol from the circulation. This is most likely due to the high specificity that α-TTP has for binding α-tocopherol over any other vitamin E homolog (34).

The considerably greater increase in urinary γ-metabolite excretion compared with plasma γ-tocopherol increase suggests that the metabolic clearance of γ-tocopherol increased considerably after ingestion of the bolus dose. From the combined excretion curves, it was possible to calculate that >90% of the γ-CEHC was eliminated in the first 72 h after d^2-γ-tocopherol administration. From a more detailed examination of the early response in two subjects, it is likely that the majority of this occurred in the first 24 h.

Considerable interindividual variability was observed in the response to γ-tocopherol loading (Fig. 13.4). For example, there was little difference in plasma γ-tocopherol values between 0 and 24 h in some subjects, whereas others showed marked increase. This was not simply due to a difference in clearance rates among subjects because those subjects with the greatest urinary excretion rates were not those with the lowest plasma values at 24 h. Similarly, these sizable differences in γ-tocopherol bioavailability were not due to dietary differences at the time of the study because the fat content of the food taken with the supplement was the same for each subject. Similar differences have been reported previously for α-tocopherol bioavailability (35) and the response appears to be specific to the individual.

In the United States, the average daily intake of γ-tocopherol has been estimated to be 18 mg (22). Based on this and a measure of γ-CEHC metabolite excretion, Swanson and colleagues could account for ~50% of the dietary intake. In the present study, ~38% was accounted for as this metabolite.

In conclusion, this stable isotope study illustrates the rapid metabolism and loss of ingested γ-tocopherol in humans. Urinary γ-CEHC appears to represent a main, but not exclusive route for excretion of γ-tocopherol metabolites. Furthermore, it is of interest to note that the concentration and metabolism of α-tocopherol in plasma are not influenced by γ-tocopherol supplementation.

Acknowledgments

We are grateful to Dr. Joy Swanson for supplying the deuterium-labeled metabolite standards and Dr. Jeffery Atkinson for supplying the deuterium-labeled γ-tocopherol acetate used in this study. We also acknowledge the Food Standards Agency (UK) for funding this work under the Antioxidant Program.

References

1. McLaughlin, P.J., and Weihrauch, J.L. (1979) Vitamin E Contents of Foods, *J. Am. Diet Assoc. 75*, 647–665.

2. Behrens, W.A, and Madere, R. (1986) Alpha and Gamma Tocopherol Concentrations in Human Serum, *J. Am. Coll. Nutr. 5*, 91–96.
3. Traber, M.G., Kayden, H.J., Balmer-Green, J., and Green, M.H. (1986) Absorption of Water-Miscible Forms of Vitamin E in a Patient with Cholestasis and in Thoracic Duct-Cannulated Rats, *Am. J. Clin. Nutr. 44*, 914–923.
4. Traber, M.G., and Kayden, H.J., (1989) Preferential Incorporation of α-Tocopherol vs. γ-Tocopherol in Human Lipoproteins, *Am. J. Clin. Nutr. 49*, 517–526.
5. Traber, M.G., Burton, G.W., Hughes, L., Ingold, K.U., Hidaka, H., Malloy, M., Kane, J., Hyams, J., and Kayden, H.J., (1992) Discrimination Between Forms of Vitamin E by Humans with and Without Genetic Abnormalities of Lipoprotein Metabolism, *J. Lipid Res. 33*, 1171–1182.
6. Kuklenkamp, J., Ronk, M., Yusin, M., Stolz, A., and Kaplowitz, N. (1993) Identification and Purification of a Human Liver Cytosolic Tocopherol Binding Protein, *Protein Expr. Purif. 4*, 382–389.
7. Simon, E.J., Eisengart, A., Sundheim, L., and Milhorat, A.T. (1956) The Metabolism of Vitamin E Purification and Characterisation of Urinary Metabolites of Alpha-Tocopherol, *J. Biol. Chem. 221*, 807–817.
8. Schultz, M., Leist, M., Petrzika, M., Gassmann, B., and Brigelius-Flohé, R. (1995) Novel Urinary Metabolite of Apha-Tocopherol 2,5,7,8-Tetramethyl-2(2'-carboxyethyl)-6-hydroxychroman as An Indicator of an Adequate Vitamin E Supply? *Am. J. Clin. Nutr: 62*,1527S–1534S.
9. Öhrval, L.M., Sundlöf, G., and Vessby, B. (1996) γ- but Not α-Tocopherol Levels in Serum Are Reduced in Coronary Heart Disease Patients, *J. Intern. Med. 239*, 111–117.
10. Kontush, A., Spranger, T., Reich, A., Baum, K., and Beisiegel, U. (1999). Lipophilic Antioxidants in Blood Plasma as Markers of Atherosclerosis: The Role of α-Carotene and γ-Tocopherol, *Atherosclerosis 144*, 117–122.
11. Kristenson, M., Ziedén, B., Kucinskienë, Z., Schäfer-Elinder, L., Bergdahl, B., Elwing, B., Abaravicius, A., Razinkovienë, L., Calkauskas, H., and Olsson, A.G. (1997) Antioxidant State and Mortality from Coronary Heart Disease in Lithuanian and Swedish Men: Concomitant Cross Sectional Study of Men Aged 50, *Br. Med. J. 314*, 629–633.
12. Kushi, L.H., Folsom, A.R., Prineas, R.J., Mink, P.J., Wu, Y., and Bostick, R.M. (1996) Dietary Antioxidant Vitamins and Death from Coronary Heart Disease in Postmenopausal Women, *N. Engl. J. Med. 334*, 1156–1162.
13. Yochum, L.A., Folsom, A.R., and Kushi, L.H. (2000) Intake of Antioxidant Vitamins and Risk of Death from Stroke in Postmenopausal Women, *Am. J. Clin. Nutr. 72*, 476–483.
14. De Lorgeril, M., Salen, P., Martin, J.L., Monjaud, I., Delaye, J., and Mamelle, N. (1999) Mediterranean Diet, Traditional Risk Factors and the Rate of Cardiovascular Complications After Myocardial Infarction: Final Report of the Lyons Diet Heart Study, *Circulation 99*, 779–785.
15. Lemcke-Norojarvi, M., Kamal-Eldin, A., Appelqvist, L.A., Dimberg, L.H., Ohrvall, M., and Vessby, B. (2001) Corn and Sesame Oils Increase Serum α-Tocopherol Concentrations in Healthy Swedish Women, *J. Nutr. 131*, 1195–1201.
16. Cooney, R.V., Custer, L.J., Okinaka, L., and Franke, A.A. (2001) Effects of Dietary Sesame Seeds on Plasma Tocopherol Levels, *Nutr. Cancer 39*, 66–71.
17. Saldeen,T., Li, D., and Mehta, J.L. (1999) Differential Effects of α- and γ-Tocopherol on Low-Density Lipoprotein Oxidation, Superoxide Activity, Platelet Aggregation and Arterial Thrombogenesis, *J. Am. Coll. Cardiol. 34*, 1208–1215.

18. Stahl, W., Graf, P., Brigelius-Flohé, R., Wechter, W., and Sies, H. (1999) Quantification of the Alpha- and Gamma-Tocopherol Metabolites 2,5,7,8-Tetramethyl-2-(2'-carboxyethyl)-6-hydroxychroman and 2,7,8-Trimethyl-2-(2'-carboxyethyl)-6-hydroxychroman in Human Serum, *Anal. Biochem. 275*, 254–259.

19. Galli, F., Lee, R., Dunster, C., and Kelly, F.J. (2002) Gas Chromatography Mass Spectrometry Analysis of Cardoxyethyl-Hydroxychroman Metabolites of α- and γ-Tocopherol in Human Plasma, *Free Radic. Biol. Med. 32*, 333–340.

20. Lodge, J.K, Traber, M.G., Elsner, A., and Brigelius-Flohé, R. (2000) A Rapid Method for the Extraction and Determination of Vitamin E Metabolites in Human Urine, *J. Lipid Res. 41*, 148–154.

21. Pope, S.A.S, Clayton, P.T, and Muller, D.P.R. (2000) A New Method for the Analysis of Urinary Vitamin E Metabolites and Tentative Identification of a Novel Group of Compounds, *Arch. Biochem. Biophys. 381*, 8–15.

22. Swanson, J.E., Ben, R.N, Burton, G.W., and Parker, R.S. (1999) Urinary Excretion of 2,7, 8-Trimethyl-2-(beta-carboxyethyl)-6-Hydroxychroman Is a Major Route of Elimination of Gamma-Tocopherol in Humans, *J. Lipid Res. 40*, 665–671.

23. Parker, R.S., and Swanson, J.E. (2000) A Novel 5'-Carboxychroman Metabolite of Gamma-Tocopherol Secreted by HepG2 Cells and Excreted in Human Urine, *Biochem. Biophys. Res. Commun. 269*, 580–583.

24. Wechter, W.J., Kantoci, D., Murray, E.D., D'Amico, D.C., Jung, M.E., and Wang, W.-H. (1996) A New Endogenous Natriuretic Factor: LLU-α, *Proc. Natl. Acad. Sci. USA 93*, 6002–6007.

25. Hattori, A., Fukushima, T., Yoshimura, H., Abe, K., and Ima, K. (2000) Production of LLU-α Following an Oral Administration of γ-Tocotrienol or γ-Tocopherol to Rats, *Biol. Pharm. Bull. 23*, 1395–1397.

26. Lodge, J.K., Ridlington, J., Leonard, S., Vaule, H., and Traber, M.G. (2001) α- and γ-Tocotrienols Are Metabolized to Carboxyethyl-Hydrochroman Derivatives and Excreted in Human Urine, *Lipids 36*, 43–48.

27. Quereshi, A.A., Bradlow, B.A., Brace, L., Mananello, J., Peterson, D.M., Pearce, B.C., Wright, J.J.K., Gapon, A., and Elson, C.E. (1995) Response of Hypercholesterolemic Subjects to Administration of Tocotrienols, *Lipids 30*, 1171–1177.

28. Nesaretnam, K., Stephen, R., Dils, R., and Darbre, P. (1998) Tocotrienols Inhibit the Growth of Human Breast Cancer Cells Irrespective of Estrogen Receptor Status, *Lipids 33*, 401–409.

29. Newaz, M.A., and Nawal, N.N. (1999) Effects of Gamma-Tocotrienol on Blood Pressure, Lipid Peroxidation and Total Antioxidant Status in Spontaneously Hypertensive Rats (SHR), *Clin. Exp. Hypertens. 2*, 1297–1313.

30. O'Byrne, D., Grundy, S., Packer, L., Devaraj, S., Baldenius, K., Hoppe, P.P., Kraemer, K., Jialal, I., and Traber M.G. (2000) Studies of LDL Oxidation Following α-, γ-, or δ-Tocotrienyl Acetate Supplementation of Hypocholesterolemic Humans, *Free Radic. Biol. Med. 29*, 834–845.

31. Handelman, G.J., Machlin, L.J., Fitch, K., Weiter, J.J., and Dratz, E.A. (1985) Oral α-Tocopherol Supplements Decrease Plasma γ-Tocopherol Levels in Humans, *J. Nutr. 15*, 807–813.

32. Galli, F., Lee, R., Dunster, C., Atkinson, J., Floridi, A., Kelly, F.J. (2001) γ-Tocopherol Metabolism and Its Relationship with α-Tocopherol in Humans: A Stable Isotope Supplementation Study, *Biofactors 15*, 65–69.

33. Kelly, F.J., Rodgers, W., Handle, J., Smith, S., and Hall, M.A. (1990) Time Course of Vitamin E Repletion in the Premature Infant, *Br. J. Nutr. 6*, 631–638.
34. Hosomi, A., Arita, M., Sato, Y., Kiyose, C., Ueda, T., Igarashi, O., Arai, H., and Inoue, K. (1997) Affinity for α-Tocopherol Transfer Protein as a Determinant of the Biological Activities of Vitamin E Analogs, *FEBS Lett. 409*, 105–108.
35. Roxborough, H.E., Burton, G.W., and Kelly, F.J. (2000) Inter- and Intra-Individual Variation in Plasma and Red Blood Cell Vitamin E After Supplementation, *Free Radic. Res. 33*, 437–445.

Chapter 14

Vitamin E in Cell Signaling

Angelo Azzi

Institute of Biochemistry and Molecular Biology, University of Bern, 3012 Bern, Switzerland

Protection by Vitamin E Against Disease

Mutations of the α-tocopherol transfer protein (α-TTP) gene lead to reduced α-toco-pherol concentrations in plasma and tissues that ultimately lead to a severe syndrome called ataxia with vitamin E deficiency (AVED) (1). Vitamin E therapy prevents progression of the syndrome and reverses some of the neurological symptoms (2,3). Vitamin E supplementation is also useful in a number of disorders, especially atherosclerosis, ischemic heart disease, and some cancers (4–6). The simple antioxidant function of vitamin E is not sufficient to explain all of the effects shown, and different biological roles have to be considered.

Protection against human atherosclerosis has been observed in subjects taking larger vitamin E quantities with the diet (7). A large prospective cohort study (Nurses' Health Study) revealed that those who obtained vitamin E from supplements had a relative risk of nonfatal myocardial infarction or death from coronary disease of 0.54 (8).

Arterial Imaging Studies

Ultrasound measurements of the arterial intima-media wall thickness (IMT) allow early evaluation of subclinical stages of atherosclerosis (9,10). The Etude sur le Vieillisement Arteriel (EVA) trial showed that higher red blood cell vitamin E was correlated with less thickening of the arterial wall (11). A very significant inverse correlation between the progression of carotid artery narrowing and vitamin E plasma levels was found also in the Kuopio Ischemic Heart Disease Study (12). The Antioxidant Supplementation in Atherosclerosis Prevention study (ASAP) (13) showed that over 3 y, progression of carotid atherosclerosis measured by IMT was reduced by 74% in men receiving vitamins E and C. No effect on the arterial wall thickness was found in women.

Using data from the Cholesterol Lowering Atherosclerosis Study (CLAS) (14), less carotid IMT progression was found for high supplementary vitamin E users compared with low vitamin E users (15). However, in the Study to Evaluate Carotid Ultrasound changes in patients treated with Ramipril and vitamin E (SECURE), no vitamin E–induced differences in atherosclerosis progression (16) were observed. The majority of these studies indicate that vitamin E protects against carotid thickening.

Controlled Intervention Trials

The α-Tocopherol β Carotene (ATBC) trial, the Linxian China trial, the Cambridge Heart Antioxidant Study (CHAOS), and Gruppo Italiano per lo Studio della Sopravvivenza nell' Infarto Miocardico (GISSI) have provided different results. The CHAOS study (a secondary prevention trial on subjects with established heart disease) (17) showed a 77% reduction in the risk for nonfatal myocardial infarction by vitamin E. However, the GISSI trial, which was among patients who had survived a myocardial infarction, did not show a significant difference for the group given vitamin E (18,19). A reevaluation of the data (20–22) suggested a more favorable interpretation if partial end points were taken. The Linxian China Study (23) showed small but significant reductions in total and cancer mortality in subjects receiving β-carotene, vitamin E, and selenium.

The ATBC trial (24) indicated that in heavy smokers, vitamin E produced a 32% lower risk of prostate cancer and a 41% lower mortality from prostate cancer (25). SPACE, Secondary Prevention with Antioxidants of Cardiovascular Disease in Endstage Renal Disease (26), showed a 46% reduction in myocardial infarction, ischemic stroke, peripheral vascular disease, and unstable angina. The HPS [Medical Research Council (MRC)/British Heart Foundation (BHF) Heart Protection Study] demonstrated that statins can reduce the risk of heart attack or stroke by up to one third, but vitamin C and/or Vitamin E were without benefit (Congress of the American Heart Association Scientific Sessions, 2001, November 11–14, Anaheim, California).

Of the most important intervention studies, CHAOS and SPACE are consistent with each other and a careful analysis of the GISSI study reveals that α-tocopherol supplementation resulted in significant effects. However the HOPE and the HPS yielded decisively negative outcomes. It is likely that the type of population, the amount of tocopherol, the ability to be absorbed, and the genotypic and nutritional aspects of the population studied may be important in the understanding of these discrepancies.

Tocopherol Binding Proteins in Tissues

α-TTP is expressed in the liver, in some parts of the brain, in the retina lymphocytes and fibroblasts, as well as in the labyrinthine trophoblast region of the placenta (27–30). It is still unclear how many other α-tocopherol binding proteins exist, and which mechanism regulates α-tocopherol transfer and its concentration in peripheral cells or subcellular compartments. Recently, a novel tocopherol binding protein was identified, the 46-kDa tocopherol-associated protein (TAP) (31,32). Ubiquitously expressed, TAP may be specifically involved in the intracellular transport of tocopherol, for example, between membrane compartments and the plasma membrane, similarly to the yeast secretory protein (SEC14p).

Furthermore, cloning and expression of two novel genes that are highly similar to the human tocopherol-associated protein (hTAP), were obtained recently in our laboratory. Immunoprecipitation of the three hTAP and extraction of bound ligands indicates that they bind not only tocopherol, but also phosphatidylinositol, phosphatidylcholine, and phosphatidylglycerol. Ligand association analyses indicate competition among ligands for hTAP binding. Recombinant hTAP have low GTPase. We propose that hTAP represent a novel type of ligand-dependent regulatory protein, which is modulated by tocopherol. These proteins may regulate a number of phosphatidylinositol- and phosphatidylcholine-requiring enzymes in a tocopherol-dependent fashion as well as a number of cell-signaling pathways.

Molecular Properties of α-Tocopherol: Antioxidant and Nonantioxidant Functions

It is commonly believed that phenolic compounds such as vitamin E play only a protective role against free radical damage and that vitamin E is the major hydrophobic chain-breaking antioxidant that prevents the propagation of free radical reactions in membranes and lipoproteins. The antioxidant properties of vitamin E are well known (33), especially in connection with the prevention of low density lipoprotein (LDL) oxidation (34), although the correlation between LDL oxidation and atherosclerosis is not always evident (35,36). Alternative studies have suggested that α-tocopherol protection against LDL oxidation may be secondary to the inhibition of protein kinase C (PKC). This enzyme is responsible for triggering the release of reactive oxygen species with consequent lipid oxidation (37,38) (Table 14.1).

The nonantioxidant properties of tocopherol have been indicated by several experiments in which the four tocopherol analogs had effects that could not be correlated with their antioxidant capacity. Furthermore, the selective uptake and transport

TABLE 14.1

Effects of α-Tocopherol and Their Supposed Molecular Mechanisms[a]

Reaction	Proposed Mechanism	Reference
Inhibition of protein kinase C	NA/A	(39,40,47,48,77,84,87–95)
Inhibition of cell proliferation	NA	(76,77)
Inhibition of platelet adhesion and aggregation	NA/ND/A	(69,70,78,79)
Inhibition of monocyte-endothelial adhesion	NA/ND/A	(67,80–82)
Inhibition of ROS in monocytes and neutrophils	A/NA	(48,83–86)
Inhibition of α-tropomyosin expression	NA	(56)
Inhibition of liver collagen α1 (I) expression	ND	(96)
Inhibition of collagenase expression	NA	(58)
Modulation of α-TTP expression	ND	(59)
Inhibition of scavenger receptors class A (SR-A)	NA	(97)
Inhibition of scavenger receptors CD36	NA	(98)
Inhibition of ICAM-1 and VCAM-1 expression	ND	(67)

[a]Abbreviations: NA, non-antioxidant; A, antioxidant; ND, not discussed; ROS, reactive oxygen species; α-TTP, α-tocopherol transfer protein; SR, scavenger receptor; ICAM, intercellular adhesion molecule; VCAM, vascular cell adhesion molecule.

of α-tocopherol appears to represent the evolutionary selection of a molecule with unique functions not shared by other antioxidants.

Effects of α-Tocopherol at the Cellular Level

PKC inhibition was found to be at the basis of the vascular smooth muscle cell growth arrest induced by α-tocopherol (39–41) (Fig. 14.1). It occurs at α-tocopherol concentrations close to those measured in healthy adults (42). β-Tocopherol, *per se* ineffective, prevents the inhibitory effect of α-tocopherol. The mechanism involved is not related to the radical scavenging properties of these two molecules, which are essentially equal (43). This phenomenon is present in a number of different cell types, including monocytes, macrophages, neutrophils, fibroblasts, and mesangial cells (44–51) (Table 14.2). α-Tocopherol, but not β-tocopherol, inhibits thrombin-induced PKC activation and endothelin secretion in endothelial cells (52). It also inhibits PKC-dependent phosphorylation and translocation of the cytosolic factor p47(phox) in monocytes, with consequent impairment of the NADPH-oxidase assembly and of superoxide production (38).

In vitro inhibition of recombinant PKC by α-tocopherol is not caused by a tocopherol-protein interaction nor does α-tocopherol inhibit PKC expression. Inhibition of PKC activity by α-tocopherol occurs at a cellular level by causing dephosphorylation of the enzyme, whereby α-tocopherol is much less potent (53). Dephosphorylation of

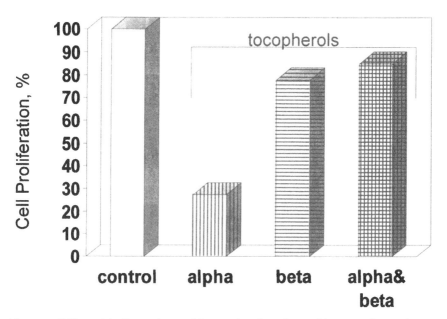

Fig. 14.1. Differential effects of α- and β-tocopherol on the proliferation of smooth muscle cells. For details see Ref. 53.

TABLE 14.2
α-Tocopherol Sensitivity of Different Cell Lines

Tissue of Origin	Sensitive	Insensitive
Rat aorta smooth muscle	A10/A7r5	
Primary rat aorta smooth muscle	hAI/hAIII	
Human Tenon's fibroblasts	hTF	
Human skin fibroblasts	CCD-SK	
Mouse neuroblastoma	NB2A	
Human pigmented primary		
Retinal epithelial cells	hPRE	
Human leukemia	U-937	
Human prostate tumor	DU-145	
Mouse fibroblast	Balb/c-3T3	
Glioma	C6	
		HeLa
Chinese hamster lung		LR73
Chinese hamster ovary		CHO
Human osteosarcoma		Saos-2
Mouse macrophage		P388 Dl

PKC occurs *via* the protein phosphatase (PP)$_2$A, which was found to be activated by treatment with α-tocopherol (53–55) (Fig. 14.2).

Transcriptional regulation by α-tocopherol. Upregulation of α-tropomyosin expression by α-tocopherol, and not by β-tocopherol occurs *via* a nonantioxidant mechanism (56,57). In human skin fibroblasts, age-dependent increases in collagenase expression can be reduced by α-tocopherol (58). Liver α-TTP and its mRNA are modulated by dietary vitamin E (59). Scavenger receptors are particularly important in the formation of atherosclerotic foam cells (60) and disruption of CD36 protects against atherosclerotic lesion. In smooth muscle cells and monocytes/macrophages, the oxidized LDL scavenger receptors (SR)-A and CD36 are down-regulated at the transcriptional level by α-tocopherol (Fig. 14.3) but not by β-tocopherol (61–63).

Inhibition of monocyte-endothelial adhesion. α-Tocopherol enrichment of monocytes, as well as neutrophils, decreases adhesion to human endothelial cells both *in vivo* and *in vitro* (64,65) and depends on the expression of adhesion molecules (66–68).

Inhibition of platelet adhesion and aggregation. α-Tocopherol inhibits aggregation of human platelets by a PKC-dependent mechanism both *in vitro* and *in vivo* (47,69) and delays intra-arterial thrombus formation (70). The studies reported above are consistent with the conclusions of Iuliano *et al.* (71) that circulating LDL accumulates in human atherosclerotic plaques and that such accumulation by macrophages is prevented by α-tocopherol *in vivo*. The protection offered by α-tocopherol may not be due only to the prevention of LDL oxidation, but also to the down-regulation of the scavenger receptor CD36 and to the inhibition of PKC activity (Fig. 14.4).

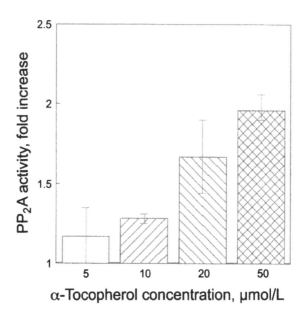

Fig. 14.2. Effect of α-tocopherol on protein phosphatase (PP)₂A. For details see Ref. 53.

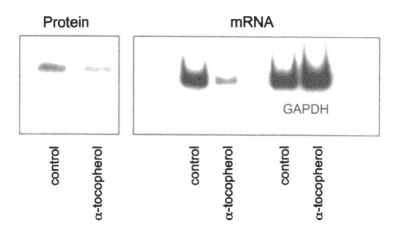

Fig. 14.3. Effect of α-tocopherol on the expression of CD36. For details see Ref. 98.

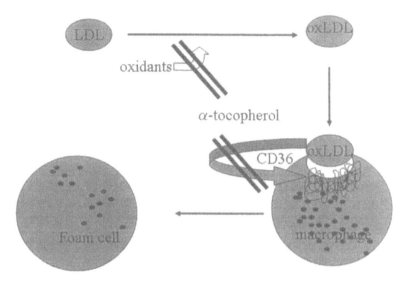

Fig. 14.4. Role of α-tocopherol in prevention of foam cell formation. Oxidation of low density lipoproteins (LDL) is inhibited by α-tocopherol through the inhibition (protein kinase C–sensitive) of superoxide production as well as by the scavenging of radicals. Further inhibition of CD36 expression prevents oxLDL uptake and foam cell formation.

Despite general agreement on the α-tocopherol inhibitory action at the PKC level, the expression of several genes, such as CD36 (62), SR class A (72), collagenase (58), and intercellular adhesion molecule (ICAM)-1 (66), appears to be regulated by α-tocopherol in a PKC-independent way. Furthermore, a number of observations, such as PP$_2$A (53) and diacylglycerol kinase (73) activation, 5-lipoxygenase (74) and cyclooxygenase (75) inhibition, continue to lack a mechanistic explanation.

Conclusions

The studies reported above lead to the following conclusions: The effects of α-tocopherol, described at a molecular and animal level, are relevant to the protection against atherosclerosis and some tumors. The basis for the unclear results achieved by similar clinical trials on the effect of vitamin E in humans remains obscure. The trial conditions should be possibly modified to select particular populations who have low plasma vitamin E, moderate progression of atherosclerosis, and the ability to readily absorb vitamin E; in addition, they should be followed for a longer period of time.

Acknowledgments

The studies reported here were supported by the Swiss National Science Foundation, by the Foundation for Nutrition in Switzerland, and Bayer Vital.

References

1. Ben Hamida, C., Doerflinger, N., Belal, S., Linder, C., Reutenauer, L., Dib, C., Gyapay, G., Vignal, A., Le Paslier, D., and Cohen, D. (1993) Localization of Friedreich Ataxia Phenotype with Selective Vitamin E Deficiency to Chromosome 8q by Homozygosity Mapping [see comments], *Nat. Genet. 5*, 195–200.

2. Labauge, P., Cavalier, L., Ichalalene, L., and Castelnovo, G. (1998) [Friedreich's ataxia and hereditary vitamin E deficiency. Case study], *Rev. Neurol. (Paris) 154*, 339–341.

3. Benomar, A., Yahyaoui, M., Marzouki, N., Birouk, N., Bouslam, N., Belaidi, H., Amarti, A., Ouazzani, R., and Chkili, T. (1999) Vitamin E Deficiency Ataxia Associated with Adenoma, *J. Neurol. Sci. 162*, 97–101.

4. Sigounas, G., Anagnostou, A., and Steiner, M. (1997) *dl*-α-Tocopherol Induces Apoptosis in Erythroleukemia, Prostate, and Breast Cancer Cells, *Nutr. Cancer 28*, 30–35.

5. Pratico, D., Tangirala, R.K., Rader, D.J., Rokach, J., and FitzGerald, G.A. (1998) Vitamin E Suppresses Isoprostane Generation In Vivo and Reduces Atherosclerosis in ApoE-Deficient Mice, *Nat. Med. 4*, 1189–1192.

6. Keaney, J.F., Jr., Simon, D.I., and Freedman, J.E. (1999) Vitamin E and Vascular Homeostasis: Implications for Atherosclerosis, *FASEB J. 13*, 965–975.

7. Gey, K.F., Moser, U.K., Jordan, P., Stahelin, H.B., Eichholzer, M., and Ludin, E. (1993) Increased Risk of Cardiovascular Disease at Suboptimal Plasma Concentrations of Essential Antioxidants: An Epidemiological Update with Special Attention to Carotene and Vitamin C, *Am. J. Clin. Nutr. 57*, 787S–797S.

8. Stampfer, M.J., Hennekens, C.H., Manson, J.E., Colditz, G.A., Rosner, B., and Willett, W.C. (1993) Vitamin E Consumption and the Risk of Coronary Disease in Women, *N. Engl. J. Med. 328*, 1444–1449.

9. Hodis, H.N., Mack, W.J., and Barth, J. (1996) Carotid Intima-Media Thickness as a Surrogate End Point for Coronary Artery Disease, *Circulation 94*, 2311–2312.

10. Selzer, R.H., Hodis, H.N., Kwong-Fu, H., Mack, W.J., Lee, P.L., Liu, C.R., and Liu, C.H. (1994) Evaluation of Computerized Edge Tracking for Quantifying Intima-Media Thickness of the Common Carotid Artery from B-Mode Ultrasound Images, *Atherosclerosis 111*, 1–11.

11. Bonithon-Kopp, C., Coudray, C., Berr, C., Touboul, P.J., Feve, J.M., Favier, A., and Ducimetiere, P. (1997) Combined Effects of Lipid Peroxidation and Antioxidant Status on Carotid Atherosclerosis in a Population Aged 59–71 y: The EVA Study. Etude sur le Vieillisement Arteriel, *Am. J. Clin. Nutr. 65*, 121–127.

12. Salonen, J.T., and Salonen, R. (1993) Ultrasound B-Mode Imaging in Observational Studies of Atherosclerotic Progression, *Circulation 87*, II56–65.

13. Salonen, J.T., Nyyssonen, K., Salonen, R., Lakka, H.M., Kaikkonen, J., Porkkala-Sarataho, E., Voutilainen, S., Lakka, T.A., Rissanen, T., Leskinen, L.,Tuomainen, T.-P., Valkonen, V.-P., Ristonmaa, U., and Poulsen, H.E.(2000) Antioxidant Supplementation in Atherosclerosis Prevention (ASAP) Study: A Randomized Trial of the Effect of Vitamins E and C on 3-Year Progression of Carotid Atherosclerosis, *J. Intern. Med. 248*, 377–386.

14. Blankenhorn, D.H., Johnson, R.L., Nessim, S.A., Azen, S.P., Sanmarco, M.E., and Selzer, R.H. (1987) The Cholesterol Lowering Atherosclerosis Study (CLAS): Design, Methods, and Baseline Results, *Control. Clin. Trials 8*, 356–387.

15. Azen, S.P., Qian, D., Mack, W.J., Sevanian, A., Selzer, R.H., Liu, C.R., Liu, C.H., and Hodis, H.N. (1996) Effect of Supplementary Antioxidant Vitamin Intake on Carotid

Arterial Wall Intima-Media Thickness in a Controlled Clinical Trial of Cholesterol Lowering, *Circulation 94*, 2369–2372.

16. Lonn, E., Yusuf, S., Dzavik, V., Doris, C., Yi, Q., Smith, S., Moore-Cox, A., Bosch, J., Riley, W., and Teo, K. (2001) Effects of Ramipril and Vitamin E on Atherosclerosis: The Study to Evaluate Carotid Ultrasound Changes in Patients Treated with Ramipril and Vitamin E (SECURE), *Circulation 103*, 919–925.

17. Stephens, N.G., Parsons, A., Schofield, P.M., Kelly, F., Cheeseman, K., and Mitchinson, M.J. (1996) Randomised Controlled Trial of Vitamin E in Patients with Coronary Disease: Cambridge Heart Antioxidant Study (CHAOS), *Lancet 347*, 781–786.

18. Brown, M. (1999) Do Vitamin E and Fish Oil Protect Against Ischaemic Heart Disease? *Lancet 354*, 441–442.

19. Marchioli, R. (1999) Antioxidant Vitamins and Prevention of Cardiovascular Disease: Laboratory, Epidemiological and Clinical Trial Data, *Pharmacol. Res. 40*, 227–238.

20. Jialal, I., Devaraj, S., Huet, B.A., and Traber, M. (1999) GISSI-Prevenzione Trial, *Lancet 354*, 1554; Discussion 1556–1557.

21. Salen, P., and de Lorgeril, M. (1999) GISSI-Prevenzione Trial, *Lancet 354*, 1555; Discussion 1556–1557.

22. Ng, W., Tse, H.F., and Lau, C.P. (1999) GISSI-Prevenzione Trial, *Lancet 354*, 1555–1556; Discussion 1556–1557.

23. Blot, W.J., Li, J.Y., Taylor, P.R., Guo, W., Dawsey, S., Wang, G.Q., Yang, C.S., Zheng, S.F., Gail, M., and Li, G.Y. (1993) Nutrition Intervention Trials in Linxian, China: Supplementation with Specific Vitamin/Mineral Combinations, Cancer Incidence, and Disease-Specific Mortality in the General Population, *J. Natl. Cancer Inst. 85*, 1483–1492.

24. Albanes, D., Heinonen, O.P., Taylor, P.R., Virtamo, J., Edwards, B.K., Rautalahti, M., Hartman, A.M., Palmgren, J., Freedman, L.S., Haapakoski, J.,Barrett, M.J., Pietinen, P., Malila, N., Tala, E., Liippo, K., Salomaa, E.R., Tangrea, J.A., Teppo, L., Askin, F.B., Taskinen, E., Erozan, Y., Greenwald, P., and Huttunen, J.K. (1996) α-Tocopherol and β-Carotene Supplements and Lung Cancer Incidence in the α-Tocopherol, β-Carotene Cancer Prevention Study: Effects of Base-Line Characteristics and Study Compliance, *J. Natl. Cancer Inst. 88*, 1560–1570.

25. Heinonen, O.P., Albanes, D., Virtamo, J., Taylor, P.R., Huttunen, J.K., Hartman, A.M., Haapakoski, J., Malila, N., Rautalahti, M., Riatti, S., Maenpaa, H., Teerenhovi, L., Koss, L., Virolainen, M., and Edwards, B.K. (1998) Prostate Cancer and Supplementation with α-Tocopherol and β-Carotene: Incidence and Mortality in a Controlled Trial, *J. Natl. Cancer Inst. 90*, 440–446.

26. Boaz, M., Smetana, S., Weinstein, T., Matas, Z., Gafter, U., Iaina, A., Knecht, A., Weissgarten, Y., Brunner, D., Fainaru, M., and Green, M.S. (2000) Secondary Prevention with Antioxidants of Cardiovascular Disease in Endstage Renal Disease (SPACE): Randomised Placebo-Controlled Trial, *Lancet 356*, 1213–1218.

27. Copp, R.P., Wisniewski, T., Hentati, F., Larnaout, A., Ben Hamida, M., and Kayden, H.J. (1999) Localization of α-Tocopherol Transfer Protein in the Brains of Patients with Ataxia with Vitamin E Deficiency and Other Oxidative Stress Related Neurodegenerative Disorders, *Brain Res. 822*, 80–87.

28. Yokota, T., Shiojiri, T., Gotoda, T., Arita, M., Arai, H., Ohga, T., Kanda, T., Suzuki, J., Imai, T., Matsumoto, H., Harino, S., Kiyosawa, M., Mizusawa, H., and Inoue, K. (1997) Friedreich-Like Ataxia with Retinitis Pigmentosa Caused by the His101Gln Mutation of the α-Tocopherol Transfer Protein Gene, *Ann. Neurol. 41*, 826–832.

29. Tamaru, Y., Hirano, M., Kusaka, H., Ito, H., Imai, T., and Ueno, S. (1997) α-Tocopherol Transfer Protein Gene: Exon Skipping of All Transcripts Causes Ataxia, *Neurology 49*, 584–588.

30. Jishage, K., Arita, M., Igarashi, K., Iwata, T., Watanabe, M., Ogawa, M., Ueda, O., Kamada, N., Inoue, K., Arai, H., and Suzuki, H. (2001) α-Tocopherol Transfer Protein Is Important for the Normal Development of Placental Labyrinthine Trophoblasts in Mice, *J. Biol. Chem. 276*, 1669–1672.

31. Stocker, A., Zimmer, S., Spycher, S.E., and Azzi, A. (1999) Identification of a Novel Cytosolic Tocopherol-Binding Protein: Structure, Specificity, and Tissue Distribution, *IUBMB Life 48*, 49–55.

32. Zimmer, S., Stocker, A., Sarbolouki, M.N., Spycher, S.E., Sassoon, J., and Azzi, A. (2000) A Novel Human Tocopherol-Associated Protein: Cloning, In Vitro Expression, and Characterization, *J. Biol. Chem. 275*, 25672–25680.

33. Packer, L., Weber, S.U., and Rimbach, G. (2001) Molecular Aspects of α-Tocotrienol Antioxidant Action and Cell Signaling, *J. Nutr. 131*, 369S–373S.

34. Esterbauer, H., Schmidt, R., and Hayn, M. (1997) Relationships Among Oxidation of Low-Density Lipoprotein, Antioxidant Protection, and Atherosclerosis, *Adv. Pharmacol. 38*, 425–456.

35. Perugini, C., Bagnati, M., Cau, C., Bordone, R., Paffoni, P., Re, R., Zoppis, E., Albano, E., and Bellomo, G. (2000) Distribution of Lipid-Soluble Antioxidants in Lipoproteins from Healthy Subjects. II. Effects of In Vivo Supplementation with α-Tocopherol, *Pharmacol. Res. 41*, 67–74.

36. Perugini, C., Bagnati, M., Cau, C., Bordone, R., Zoppis, E., Paffoni, P., Re, R., Albano, E., and Bellomo, G. (2000) Distribution of Lipid-Soluble Antioxidants in Lipoproteins from Healthy Subjects. I. Correlation with Plasma Antioxidant Levels and Composition of Lipoproteins, *Pharmacol. Res. 41*, 55–65.

37. Devaraj, S., Li, D., and Jialal, I. (1996) The Effects of Alpha Tocopherol Supplementation on Monocyte Function. Decreased Lipid Oxidation, Interleukin 1 Beta Secretion, and Monocyte Adhesion to Endothelium, *J. Clin. Investig. 98*, 756–763.

38. Cachia, O., Benna, J.E., Pedruzzi, E., Descomps, B., Gougerot-Pocidalo, M.A., and Leger, C.L. (1998) α-Tocopherol Inhibits the Respiratory Burst in Human Monocytes. Attenuation of p47(Phox) Membrane Translocation and Phosphorylation, *J. Biol. Chem. 273*, 32801–32805.

39. Boscoboinik, D., Szewczyk, A., and Azzi, A. (1991) α-Tocopherol (Vitamin E) Regulates Vascular Smooth Muscle Cell Proliferation and Protein Kinase C Activity, *Arch. Biochem. Biophys. 286*, 264–269.

40. Boscoboinik, D., Szewczyk, A., Hensey, C., and Azzi, A. (1991) Inhibition of Cell Proliferation by α-Tocopherol. Role of Protein Kinase C, *J. Biol. Chem. 266*, 6188–6194.

41. Tasinato, A., Boscoboinik, D., Bartoli, G.M., Maroni, P., and Azzi, A. (1995) *d*-α-Tocopherol Inhibition of Vascular Smooth Muscle Cell Proliferation Occurs at Physiological Concentrations, Correlates with Protein Kinase C Inhibition, and Is Independent of Its Antioxidant Properties, *Proc. Natl. Acad. Sci. USA 92*, 12190–12194.

42. Eichholzer, M., Stahelin, H.B., Gey, K.F., Ludin, E., and Bernasconi, F. (1996) Prediction of Male Cancer Mortality by Plasma Levels of Interacting Vitamins: 17-Year Follow-Up of the Prospective Basel Study, *Int. J. Cancer 66*, 145–150.

43. Pryor, A.W., Cornicelli, J.A., Devall, L.J., Tait, B., Trivedi, B.K., Witiak, D.T., and Wu, M. (1993) A Rapid Screening Test to Determine the Antioxidant Potencies of Natural and Synthetic Antioxidants, *J. Org. Chem. 58*, 3521–3532.

44. Devaraj, S., and Jialal, I. (1996) Oxidized Low-Density Lipoprotein and Atherosclerosis, *Int. J. Clin. Lab. Res. 26*, 178–184.

45. Devaraj, S., Adams-Huet, B., Fuller, C.J., and Jialal, I. (1997) Dose-Response Comparison of *RRR*-α-Tocopherol and all-Racemic α-Tocopherol on LDL Oxidation, *Arterioscler. Thromb. Vasc. Biol. 17*, 2273–2279.

46. Devaraj, S., and Jialal, I. (1998) The Effects of α-Tocopherol on Critical Cells in Atherogenesis, *Curr. Opin. Lipidol. 9*, 11–15.

47. Freedman, J.E., Farhat, J.H., Loscalzo, J., and Keaney, J.F., Jr. (1996) α-Tocopherol Inhibits Aggregation of Human Platelets by a Protein Kinase C-Dependent Mechanism, *Circulation 94*, 2434–2440.

48. Kanno, T., Utsumi, T., Kobuchi, H., Takehara, Y., Akiyama, J., Yoshioka, T., Horton, A.A., and Utsumi, K. (1995) Inhibition of Stimulus-Specific Neutrophil Superoxide Generation by α-Tocopherol, *Free Radic. Res. 22*, 431–440.

49. Koya, D., Lee, I.K., Ishii, H., Kanoh, H., and King, G.L. (1997) Prevention of Glomerular Dysfunction in Diabetic Rats by Treatment with *d*-α-Tocopherol, *J. Am. Soc. Nephrol. 8*, 426–435.

50. Studer, R.K., Craven, P.A., and DeRubertis, F.R. (1997) Antioxidant Inhibition of Protein Kinase C-Signaled Increases in Transforming Growth Factor-β in Mesangial Cells, *Metabolism 46*, 918–925.

51. Tada, H., Ishii, H., and Isogai, S. (1997) Protective Effect of D-α-Tocopherol on the Function of Human Mesangial Cells Exposed to High Glucose Concentrations, *Metabolism 46*, 779–784.

52. Martin-Nizard, F., Boullier, A., Fruchart, J.C., and Duriez, P. (1998) α-Tocopherol but Not Beta-Tocopherol Inhibits Thrombin-Induced PKC Activation and Endothelin Secretion in Endothelial Cells, *J. Cardiovasc. Risk 5*, 339–345.

53. Ricciarelli, R., Tasinato, A., Clement, S., Ozer, N.K., Boscoboinik, D., and Azzi, A. (1998) α-Tocopherol Specifically Inactivates Cellular Protein Kinase C α by Changing Its Phosphorylation State, *Biochem. J. 334*, 243–249.

54. Clement, S., Tasinato, A., Boscoboinik, D., and Azzi, A. (1997) The Effect of α-Tocopherol on the Synthesis, Phosphorylation and Activity of Protein Kinase C in Smooth Muscle Cells After Phorbol 12-Myristate 13-Acetate Down-Regulation, *Eur. J. Biochem. 246*, 745–749.

55. Neuzil, J., Weber, C., and Kontush, A. (2001) The Role of Vitamin E in Atherogenesis: Linking the Chemical, Biological and Clinical Aspects of the Disease, *Atherosclerosis 157*, 257–283.

56. Aratri, E., Spycher, S.E., Breyer, I., and Azzi, A. (1999) Modulation of α-Tropomyosin Expression by α-Tocopherol in Rat Vascular Smooth Muscle Cells, *FEBS Lett. 447*, 91–94.

57. Azzi, A., Boscoboinik, D., Fazzio, A., Marilley, D., Maroni, P., Ozer, N.K., Spycher, S., and Tasinato, A. (1998) *RRR*-α-Tocopherol Regulation of Gene Transcription in Response to the Cell Oxidant Status, *Z. Ernaehrwiss. 37*, 21–28.

58. Ricciarelli, R., Maroni, P., Ozer, N., Zingg, J.M., and Azzi, A. (1999) Age-Dependent Increase of Collagenase Expression Can Be Reduced by α-Tocopherol Via Protein Kinase C Inhibition, *Free Radic. Biol. Med. 27*, 729–737.

59. Shaw, H.M., and Huang, C. (1998) Liver α-Tocopherol Transfer Protein and Its mRNA Are Differentially Altered by Dietary Vitamin E Deficiency and Protein Insufficiency in Rats, *J. Nutr. 128*, 2348–2354.

60. Febbraio, M., Podrez, E., Smith, J., Hajjar, D., Hazen, S., Hoff, H., Sharma, K., and Silverstein, R. (2000) Targeted Disruption of the Class B Scavenger Receptor CD36 Protects Against Atherosclerotic Lesion Development in Mice, *J. Clin. Investig. 105*, 1049–1056.

61. Teupser, D., Thiery, J., and Seidel, D. (1999) α-Tocopherol Down-Regulates Scavenger Receptor Activity in Macrophages, *Atherosclerosis 144*, 109–115.

62. Ricciarelli, R., Zingg, J.M., and Azzi, A. (2000) Vitamin E Reduces the Uptake of Oxidized LDL by Inhibiting CD36 Scavenger Receptor Expression in Cultured Aortic Smooth Muscle Cells, *Circulation 102*, 82–87.

63. Devaraj, S., Hugou, I., and Jialal, I. (2001) α-Tocopherol Decreases CD36 Expression in Human Monocyte-Derived Macrophages, *J. Lipid Res. 42*, 521–527.

64. Islam, K.N., Devaraj, S., and Jialal, I. (1998) α-Tocopherol Enrichment of Monocytes Decreases Agonist-Induced Adhesion to Human Endothelial Cells, *Circulation 98*, 2255–2261.

65. Martin, A., Foxall, T., Blumberg, J.B., and Meydani, M. (1997) Vitamin E Inhibits Low-Density Lipoprotein-Induced Adhesion of Monocytes to Human Aortic Endothelial Cells In Vitro, *Arterioscler. Thromb. Vasc. Biol. 17*, 429–436.

66. Wu, D., Koga, T., Martin, K.R., and Meydani, M. (1999) Effect of Vitamin E on Human Aortic Endothelial Cell Production of Chemokines and Adhesion to Monocytes, *Atherosclerosis 147*, 297–307.

67. Yoshikawa, T., Yoshida, N., Manabe, H., Terasawa, Y., Takemura, T., and Kondo, M. (1998) α-Tocopherol Protects Against Expression of Adhesion Molecules on Neutrophils and Endothelial Cells, *Biofactors 7*, 15–19.

68. Steiner, M., Li, W., Ciaramella, J.M., Anagnostou, A., and Sigounas, G. (1997) *dl*-α-Tocopherol, a Potent Inhibitor of Phorbol Ester Induced Shape Change of Erythro- and Megakaryoblastic Leukemia Cells, *J. Cell. Physiol. 172*, 351–360.

69. Williams, J.C., Forster, L.A., Tull, S.P., Wong, M., Bevan, R.J., and Ferns, G.A. (1997) Dietary Vitamin E Supplementation Inhibits Thrombin-Induced Platelet Aggregation, but Not Monocyte Adhesiveness, in Patients with Hypercholesterolaemia, *Int. J. Exp. Pathol. 78*, 259–266.

70. Saldeen, T., Li, D., and Mehta, J.L. (1999) Differential Effects of α- and γ-Tocopherol on Low-Density Lipoprotein Oxidation, Superoxide Activity, Platelet Aggregation and Arterial Thrombogenesis, *J. Am. Coll. Cardiol. 34*, 1208–1215.

71. Iuliano, L., Mauriello, A., Sbarigia, E., Spagnoli, L.G., and Violi, F. (2000) Radiolabeled Native Low-Density Lipoprotein Injected into Patients with Carotid Stenosis Accumulates in Macrophages of Atherosclerotic Plaque: Effect of Vitamin E Supplementation, *Circulation 101*, 1249–1254.

72. Teupser, D., Stein, O., Burkhardt, R., Nebendahl, K., Stein, Y., and Thiery, J. (1999) Scavenger Receptor Activity Is Increased in Macrophages from Rabbits with Low Atherosclerotic Response: Studies in Normocholesterolemic High and Low Atherosclerotic Response Rabbits, *Arterioscler. Thromb. Vasc. Biol. 19*, 1299–1305.

73. Lee, I.K., Koya, D., Ishi, H., Kanoh, H., and King, G.L. (1999) d-α-Tocopherol Prevents the Hyperglycemia Induced Activation of Diacylglycerol (DAG)-Protein Kinase C (PKC) Pathway in Vascular Smooth Muscle Cell by an Increase of DAG Kinase Activity, *Diabetes Res. Clin. Pract. 45*, 183–190.

74. Jialal, I., Devaraj, S., and Kaul, N. (2001) The Effect of α-Tocopherol on Monocyte Proatherogenic Activity, *J. Nutr. 131*, 389S–394S.

75. Wu, D., Hayek, M.G., and Meydani, S. (2001) Vitamin E and Macrophage Cyclooxygenase Regulation in the Aged, *J. Nutr. 131*, 382S–388S.

76. Azzi, A., Boscoboinik, D., Marilley, D., Özer, N.K., Stäuble, B., and Tasinato, A. (1995) Vitamin E: A Sensor and an Information Transducer of the Cell Oxidation State, *Am. J. Clin. Nutr. 62 (Suppl.)*, 1337S–1346S.

77. Chatelain, E., Boscoboinik, D.O., Bartoli, G.M., Kagan, V.E., Gey, F.K., Packer, L., and Azzi, A. (1993) Inhibition of Smooth Muscle Cell Proliferation and Protein Kinase C Activity by Tocopherols and Tocotrienols, *Biochim. Biophys. Acta 1176*, 83–89.

78. Freedman, J.E., Farhat, J.H., Loscalzo, J., and Keaney, J.F., Jr. (1996) α-Tocopherol Inhibits Aggregation of Human Platelets by a Protein Kinase C-Dependent Mechanism, *Circulation 94*, 2434–2440.

79. Mabile, L., Bruckdorfer, K.R., and Rice-Evans, C. (1999) Moderate Supplementation with Natural α-Tocopherol Decreases Platelet Aggregation and Low-Density Lipoprotein Oxidation, *Atherosclerosis 147*, 177–185.

80. Islam, K.N., O'Byrne, D., Devaraj, S., Palmer, B., Grundy, S.M., and Jialal, I. (2000) α-Tocopherol Supplementation Decreases the Oxidative Susceptibility of LDL in Renal Failure Patients on Dialysis Therapy, *Atherosclerosis 150*, 217–224.

81. Martin, A., Janigian, D., Shukitt-Hale, B., Prior, R.L., and Joseph, J.A. (1999) Effect of Vitamin E Intake on Levels of Vitamins E and C in the Central Nervous System and Peripheral Tissues: Implications for Health Recommendations, *Brain Res. 845*, 50–59.

82. Wu, C.G., Hoek, F.J., Groenink, M., Reitsma, P.H., Van Deventer, S.J.H., and Chamuleau, R.A.F.M. (1997) Correlation of Repressed Transcription of α-Tocopherol Transfer Protein with Serum α-Tocopherol During Hepatocarcinogenesis, *Int. J. Cancer 71*, 686–690.

83. Brillant, L., Léger, C.L., and Descomps, B. (1999) Vitamin E Inhibition of $O_2^{\cdot-}$ Production in the Promonocyte Cell Line THP-1 Is Essentially Due to *RRR*-δ-Tocopherol, *Lipids 34 (Suppl.)*, S293.

84. Devaraj, S., Li, D., and Jialal, I. (1996) The Effects of Alpha Tocopherol Supplementation on Monocyte Function. Decreased Lipid Oxidation, Interleukin 1β Secretion, and Monocyte Adhesion to Endothelium, *J. Clin. Investig. 98*, 756–763.

85. Cachia, O., El Benna, J., Pedruzzi, E., Descomps, B., Gougerot-Pocidalo, M.A., and Leger, C.L. (1998) α-Tocopherol Inhibits the Respiratory Burst in Human Monocytes—Attenuation of p47(Phox) Membrane Translocation and Phosphorylation, *J. Biol. Chem. 273*, 32801–32805.

86. Wu, Z.G., Wu, J., Jacinto, E., and Karin, M. (1997) Molecular Cloning and Characterization of Human JNKK2, a Novel Jun NH 2-Terminal Kinase-Specific Kinase, *Mol. Cell. Biol. 17*, 7407–7416.

87. Azzi, A., Boscoboinik, D., and Hensey, C. (1992) The Protein Kinase C Family, *Eur. J. Biochem. 208*, 547–557.

88. Azzi, A., Boscoboinik, D., Chatelain, E., Özer, N.K., and Stäuble, B. (1993) *d*-α-Tocopherol Control of Cell Proliferation, *Mol. Aspects Med. 14*, 265–271.

89. Boscoboinik, D., Chatelain, E., Bartoli, G.M., and Azzi, A. (1992) in *Free Radicals and Aging* (Emerit, I., and Chance, B., eds.) pp. 164–177, Birkhäuser Verlag, Basel.

90. Tasinato, A., Boscoboinik, D., Bartoli, G.M., Maroni, P., and Azzi, A. (1995) *d*-α-Tocopherol Inhibition of Vascular Smooth Muscle Cell Proliferation Occurs at Physiological Concentrations, Correlates with Protein Kinase C Inhibition, and Is Independent of Its Antioxidant Properties, *Proc. Natl. Acad. Sci. USA 92*, 12190–12194.

91. Devaraj, S., Adams-Huet, B., Fuller, C.J., and Jialal, I. (1997) Dose-Response Comparison

of *RRR*-α-Tocopherol and all-Racemic α-Tocopherol on LDL Oxidation, *Arterioscler. Thromb. Vasc. Biol. 17*, 2273–2279.

92. Devaraj, S., and Jialal, I. (1998) The Effects of α-Tocopherol on Critical Cells in Atherogenesis, *Curr. Opin. Lipidol. 9*, 11–15.

93. Koya, D., Lee, I.K., Ishii, H., Kanoh, H., and King, G.L. (1997) Prevention of Glomerular Dysfunction in Diabetic Rats by Treatment with *d*-α-Tocopherol, *J. Am. Soc. Nephrol. 8*, 426–435.

94. Studer, R.K., Craven, P.A., and DeRubertis, F.R. (1997) Antioxidant Inhibition of Protein Kinase C-Signaled Increases in Transforming Growth Factor-Beta in Mesangial Cells, *Metabolism 46*, 918–925.

95. Tada, H., Ishii, H., and Isogai, S. (1997) Protective Effect of D-α-Tocopherol on the Function of Human Mesangial Cells Exposed to High Glucose Concentrations, *Metabolism 46*, 779–784.

96. Chojkier, M., Houglum, K., Lee, K.S., and Buck, M. (1998) Long- and Short-Term D-α-Tocopherol Supplementation Inhibits Liver Collagen Alpha 1(I) Gene Expression, *Am. J. Physiol. 275*, G1480–1485.

97. Teupser, D., Thiery, J., and Seidel, D. (1999) α-Tocopherol Down-Regulates Scavenger Receptor Activity in Macrophages, *Atherosclerosis 144*, 109–115.

98. Ricciarelli, R., Zingg, J.-M., and Azzi, A. (2000) Downregulation of the CD36 Scavenger Receptor by α-Tocopherol, *Circulation 102*, 82–87.

Chapter 15

Vitamin E and Selenium Effects on Differential Gene Expression

Gerald Rimbach[a], Alexandra Fischer[b], Josef Pallauf[b], and Fabio Virgili[c]

[a]School of Food Bioscience, Hugh Sinclair Human Nutrition Unit, University of Reading, UK
[b]Insititute of Animal Nutrition and Nutrition Physiology, Justus Liebig University, Giessen, Germany
[c]National Institute of Food and Nutrition Research, Rome, Italy

Introduction

Vitamin E, one of the most effective lipid-soluble, chain-breaking antioxidants, protects cell membranes from peroxidative damage (1–4). Furthermore, vitamin E has been demonstrated to affect specific genes, in a manner partially independent of its antioxidant/radical-scavenging ability (5). At the post-translational level, α-tocopherol is known to inhibit protein kinase C in smooth muscle cells (6) and 5-lipoxygenase in activated human monocytes (7), whereas γ-tocopherol has been reported to inhibit cyclooxygenase activity in macrophages and epithelial cells. Several genes, including α-tocopherol transfer protein (8) and α-tropomyosin (9), are modulated by α-tocopherol at the transcriptional level. It has been demonstrated in cultured cells that vitamin E inhibits inflammation, cell adhesion, and platelet aggregation. However, the exact molecular targets through which vitamin E mediates beneficial effects have yet not been fully characterized.

Selenium (Se) is an essential trace element known to affect a wide range of physiologic processes. The biological functions of selenium known to date in mammals are mediated mainly by selenoproteins (10). Selenoprotein enzymes are glutathione peroxidases (GPx) (11), iodothyronine deiodinases (12), thioredoxin reductase (13), and selenophosphate synthetase (14). Cytosolic GPx catalyzes the two–electron reduction of hydrogen peroxide and organic hydroperoxides, thereby playing an important role in eliminating reactive oxygen species (ROS). Overproduction of ROS can cause oxidative damage to lipids, proteins and DNA and is widely believed to play a pivotal role in aging and degenerative diseases. Furthermore, selenium-dependent peroxidases [phospholipid hydroperoxide GPx (PhGPx), plasma GPx (pGPx), gastrointestinal (GI)-GPx] and thioredoxin reductase are associated with the modulation of redox-sensitive enzyme cascades, thereby regulating leucotriene synthesis, inflammatory processes, cell proliferation, and apoptosis (15).

Methods for large-scale measurement of gene expression are becoming important tools in the field of free radical research (16–18). Importantly, cDNA arrays can help to discover redox-regulated genes and potential biomarkers of oxidative stress. To obtain a more comprehensive understanding of the molecular mechanisms involved in the physiology of selenium and α-tocopherol and the pathophysiology of Se and vitamin E deficiency, a global gene expression profile in the rat liver and primary human endothelial cells was determined using cDNA microarrays.

Differential Gene Expression in Rat Liver

Four groups of male albino rats (n = 8; Wistar, Unilever, Harlan/Winkelmann) with an initial mean live weight of 35 g were randomly assigned to semisynthetic diets based on torula yeast (30%) and tocopherol-stripped corn oil (5%) for 7 wk. Control rats (+vitamin E, +Se, group IV) received diets supplemented with vitamin E (75 mg/kg as dl-α-tocopheryl acetate) and Se (200 μg/kg as sodium selenate). The −vitamin E, −Se rats (group I) were fed the basal diets deficient in both nutrients. Two other groups received the basal diets supplemented with either vitamin E (group II) or Se (group III).

Of the 465 genes spotted on the applied Atlas rat toxicology cDNA array evaluated, 22 genes in the combined Se plus vitamin E deficiency and 9 genes in the Se-deficient rats displayed a greater than twofold change in expression levels. Studies on the reproducibility and variability of array results indicate that a twofold or greater difference in the expression of a particular gene could be considered a real difference in transcript abundance (19). Interestingly, vitamin E deficiency alone did not induce any significant changes in the expression profile among the genes evaluated compared with the control rats. Possibly other genes not present on the cDNA membrane could have been differentially regulated by vitamin E. Additionally, tissues other than liver might be more susceptible to vitamin E–induced changes in gene expression.

In addition to a significant down-regulation of the cGPx gene, selenium deficiency alone was accompanied by an increase in the expression of UDP-glucuronosyltransferase 1 gene and bilirubin UDP-glucuronosyltransferase isoenzyme 2 gene (Table 15.1). These two genes encode for enzymes that have an important function in the detoxification of xenobiotics in liver. Similarly, rat liver cytochrome $P_{450}4B1$, which is also involved in xenobiotic metabolism and inducible by glucocorticoids, was induced 2.3-fold. The mRNA levels of arachidonate 12-lipoxygenase (ALOX 12) were 2.4-fold higher in selenium-deficient rats than in controls. It has been shown that ALOX 12 and PHGPx are opposing enzymes balancing the intracellular concentration of hydroperoxy lipids; an inhibition of PHGPx activity increases the enzymatic catalysis of ALOX 12 (20).

In combined selenium and vitamin E deficiency, 5% of all genes monitored were differentially expressed. The double deficiency was characterized by a significant down-regulation of genes that inhibit programmed cell death, including defender against cell death 1 protein, inhibitor of apoptosis protein 1 and Bcl2-L1 (Table 15.2). Furthermore, the expression level of early growth response protein 1, known as a suppressor of growth and transformation and an inducer of apoptosis, was increased twofold. Carbonic anhydrase III (CAIII), which was recently reported to play a role as an antioxidant preventing H_2O_2-inducible apoptosis (21), was down-regulated

TABLE 15.1
Se Deficiency–Related Changes in Gene Expression in Rat Liver

GenBank Accession	Δ –Se	(fold)	Gene	Function
X12367	↓	13.9	Cellular glutathione peroxidase I	Peroxide detoxification
M13406	↓	2.1	Histone 2A	Control of transcription
M29853	↑	2.3[a]	Cytochrome P_{450} 4B1	Xenobiotic metabolism
J02608	↑	12.3[a]	DT-diaphorase	Xenobiotic metabolism
M13506	↑	3.0	UDP-glucuronosyltransferase 1	Xenobiotic metabolism
U75903	↑	2.1	Bilirubin UDP-glucuronosyltransferase isozyme 2	Xenobiotic metabolism
L0604	↑	2.4	Arachidonate 12-lipoxygenase (ALOX12)	Balance intracellular hydroperoxide concentration
U23407	↑	4.8[a]	Cellular retinoic acid binding protein 2	Extracellular communication protein
L31883	↑	4.3[a]	Tissue inhibitor of metalloproteinase 1 precursor (TIMP1)	Protease inhibitor

[a]Gene signal at background level in one array.

TABLE 15.2
Selection of Se- and Vitamin E–Deficiency Related Changes in Gene Expression in Rat Liver[a]

GenBank accession	Δ−Se−E (fold)	Gene	Function	Se prevention	Vitamin E prevention
Apoptosis/cell cycle					
Y13336	↓2.0	Defender against cell death 1 protein (DAD1)	Protection against apoptosis	Complete	Complete
AF081503	↓2.6	Inhibitor of apoptosis protein 1	Protection against apoptosis	82%	77%
U72350	↓3.2	Bcl2-L1	Protection against apoptosis	Complete	74%
M22413	↓2.0	Carbonic anhydrase III (CAIII)	Antioxidant, protection against apoptosis	Complete	81%
D90345	↓2.2	T-complex protein 1 (CCT) α subunit	Chaperone, folding of proteins	79%	Complete
X82021	↓2.2	HSC70-interacting protein (HIP)	Stabilization of the chaperone HSC70	88%	Complete
J03969	↓2.9	Nucleophosmin (NPM)	Stimulation of normal cell growth	67%	75%
D14014	↓3.1	G1/S-specific cyclin D1 (CCND1)	Initiation of cell cycle, oncogene	76%	90%
J04154	↑2.1	Early growth response protein 1	Suppression of growth and induction of apoptosis	72%	None
U77129	↑2.0[b]	SPS1/Ste20 homolog KHS1	Transducer of signals in mitogen-activated protein kinase pathway	Complete	Complete
Antioxidant defense/stress response/inflammation					
X12367	↓18.8	Cellular glutathione peroxidase I	Peroxide detoxification	87%	None
J05181	↓3.4	γ-Glutamylcysteine synthetase (γ-GCS)	Glutathione synthesis	89%	Complete
U22424	↓2.2	11-β-Hydroxysteroid dehydrogenase 2	Conversion of corticosterone into 11-dehydrocorticosterone	Complete	Complete
L49379	↓2.3	Multispecific organic anion exporter (cMOAT)	Detoxification, export of leukotriene C_4	Complete	89%
J02608	↑15.3[b]	DT-diaphorase	Xenobiotic metabolism	Complete	None
D00753	↑2.1	SPI-3 serine protease inhibitor	Acute phase protein	None	Complete
J00696	↑2.3	α-1 Acid glycoprotein	Acute phase protein	None	81%
J00734	↑2.3	Fibrinogen γ chain	Acute phase protein	Complete	Complete
S65838	↑3.6	Metallothionein 1	Acute phase protein, antioxidant	71%	Complete

[a]The extent to which Se and vitamin E prevented alterations in gene expression is denoted as either complete (> 90%), none, or partial (% effect indicated).
[b]Gene signal at background level in one array.

twofold. A stronger tendency toward negative cell cycle progression in livers of doubly deficient rats was further suggested by the down-regulation of nucleophosmin and G1/S-specific cyclin D1, which has been characterized as an important signal in anti-apoptotic mechanisms.

Combined selenium and vitamin E deficiency also resulted in an induction of acute phase proteins (metallothionein, DT-diaphorase, α-1 acid glycoprotein) and SPI-3 serine proteinase inhibitor. A further indication of proinflammatory response in rats fed diets deficient in selenium plus vitamin E is that they exhibited higher expression of the fibrinogen γ chain, which has been shown to be significantly up-regulated in the rat liver during inflammation. The induction of proinflammatory genes was accompanied by a concerted depression of the anti-inflammatory enzyme 11-β-hydroxysteroid dehydrogenase 2, which converts the glucocorticoid corticosterone to its inactive 11-dehydro form in rats, thereby controlling glucocorticoid access to receptors.

Overall in this *in vivo* study, vitamin E deficiency alone did not induce significant changes in the expression profile among the genes evaluated, whereas Se deficiency induced genes encoding for detoxifying proteins. If Se deficiency is combined with vitamin E deficiency, much greater cellular consequences, including stress response accompanied by a proinflammatory and proapoptotic metabolic situation, can be expected. This provides experimental support for the postulated synergism between Se and vitamin E in their molecular functions as part of the so-called antioxidant network.

Differential Gene Expression in Human Endothelial Cells

Oxidation of LDL is a key event in endothelial injury and dysfunction associated with "early-stage" atherogenesis. It has been proposed that the biological action of oxidized LDL (oxLDL) may be attributed in part to its effect on gene expression in endothelial cells. *In vitro* studies demonstrate that α-tocopherol has beneficial effects on cell functions that are pivotal in atherogenesis (22). To examine the transcriptional response of oxLDL and vitamin E we applied cDNA array technology (Atlas human cardiovascular array) to cultured human umbilical vein endothelial cells (HUVEC). Endothelial cells were incubated with oxLDL for 6 h in the absence and presence of vitamin E (10 μmol/L *dl*-α-tocopherol for 24 h). Control cells were maintained in culture for the same length of time without oxLDL. Using this criterion, 78 of 588 genes were differentially expressed; 57 genes were up- and 21 genes were down-regulated in response to oxLDL (cut-off point = 2). In the same experimental model, oxLDL was also found to induce activation of DNA binding of both AP-1 and nuclear factor (NF)-κB transcription factors. Oxidized LDL significantly altered the expression of genes encoding for transcription factors (e.g., GATA-2, fos-realted antigen 2), cell receptors (e.g, AGE-related receptor precursor, thromboxane A_2 receptor), adhesion molecules (e.g., P-selectin precursor), extracellular matrix proteins (e.g., metalloproteinase 9), and enzymes involved in cholesterol metabolism (e.g., farnesyltranferase β). This effect is consistent with the proinflammatory environment associated with alterations in endothelial cell metabolism described in the early stages of atheroma formation. Interestingly, in HUVEC, some of the genes, such as fos-related antigen 2, thromboxane A_2 receptor, P-selectin precursor, and metalloproteinase 9, which were up-regulated by oxidized LDL were down-regulated by vitamin E. The experimental strategy identified several novel oxidized LDL and vitamin E–sensitive genes. Cardiovascular-specific DNA arrays are an important platform for obtaining a global genetic portrait and understanding the complex molecular events leading to atherosclerosis.

References

1. Packer, L. (1991) Protective Role of Vitamin E in Biological Systems, *Am. J. Clin. Nutr.* 53, 1050S–1055S.
2. Brigelius-Flohé, R., and Traber, M.G. (1999) Vitamin E: Function and Metabolism, *FASEB J. 13*, 1145–1155.
3. Packer, L., Weber, S.U., and Rimbach, G. (2001) Molecular Aspects of α-Tocotrienol Antioxidant Action and Cell Signaling, *J. Nutr. 131*, 369S–373S.
4. Rao, L., Puschner, B., and Prolla, T.A. (2001) Gene Expression Profiling of Low Selenium Status in the Mouse Intestine: Transcriptional Activation of Genes Linked to DNA Damage, Cell Cycle Control and Oxidative Stress, *J. Nutr. 131*, 3175–3181.
5. Azzi, A., Breyer, I., Feher, M., Pastori, M., Ricciarelli, R., Spycher, S., Staffieri, M., Stocker, A., Zimmer, S., and Zingg, J.M. (2000) Specific Cellular Responses to α-Tocopherol, *J. Nutr. 130*, 1649–1652.
6. Tasinato, A., Boscoboinik, D., Bartoli, G.M., Maroni, P., and Azzi, A. (1995) *d*-α-Tocopherol Inhibition of Vascular Smooth Muscle Cell Proliferation Occurs at Physiological Concentrations, Correlates with Protein Kinase C Inhibition, and Is Independent of Its Antioxidant Properties, *Proc. Natl. Acad. Sci. USA 92*, 12190–12194.
7. Devaraj, S., Li, D., and Jialal, I. (1996) The Effects of Alpha Tocopherol Supplementation on Monocyte Function. Decreased Lipid Oxidation, Interleukin 1 Beta Secretion, and Monocyte Adhesion to Endothelium, *J. Clin. Investig. 98*, 756–763.
8. Shaw, H.M., and Huang, C. (1998) Liver Alpha-Tocopherol Transfer Protein and Its mRNA Are Differentially Altered by Dietary Vitamin E Deficiency and Protein Insufficiency in Rats, *J. Nutr. 128*, 2348–2354.
9. Aratri, E., Spycher, S.E., Breyer, I., and Azzi, A. (1999) Modulation of α-Tropomyosin Expression by α-Tocopherol in Rat Vascular Smooth Muscle Cells, *FEBS Lett. 447*, 91–94.
10. Stadtman, T.C. (1991) Biosynthesis and Function of Selenocysteine-Containing Enzymes, *J. Biol. Chem. 266*, 16257–16260.
11. Flohé, L. (1988) Glutathione Peroxidase, *Basic Life Sci. 49*, 663–668.
12. Behne, D., Kyriakopoulos, A., Meinhold, H., and Köhrle, J. (1990) Identification of Type I Iodothyronine 5'-deiodinase as a Selenoenzyme, *Biochem. Biophys. Res. Commun. 173*, 1143–1149.
13. Tamura, T., and Stadtman, T.C. (1996) A New Selenoprotein from Human Lung Adenocarcinoma Cells: Purification, Properties, and Thioredoxin Reductase Activity, *Proc. Natl. Acad. Sci. USA 93*, 1006–1011.
14. Guimaraes, M.J., Peterson, D., Vicari, A., Cocks, B.G., Copeland, N.G., Gilbert, D.J., Jenkins, N.A., Ferrick, D.A., Kastelein, R.A., Bazan, J.F., and Zlotnik, A. (1996) Identification of a Novel selD Homolog from Eukaryotes, Bacteria, and Archaea: Is There

an Autoregulatory Mechanism in Selenocysteine Metabolism? *Proc. Nat. Acad. Sci. USA 93*, 15086–15091.

15. Flohé, L., Andreesen, J.R., Brigelius-Flohé, R., Maiorino, M., and Ursini, F. (2000) Selenium, the Element of the Moon, in Life on Earth, *IUBMB Life 49*, 411–420.

16. Watanabe, C.M.H., Wolffram, S., Ader, P., Rimbach, G., Packer, L., Maguire, J.J., Schultz, P.G., and Gohil, K. (2001) The In Vivo Neuromodulatory Effect of the Herbal Medicine *Ginkgo biloba*, *Proc. Nat. Acad. Sci. USA 98*, 6577–6580.

17. Fischer, A., Pallauf, J., Gohil, K., Weber, S.U., Packer, L., and Rimbach, G. (2001) Effect of Selenium and Vitamin E Deficiency on Differential Gene Expression in Rat Liver, *Biochem. Biophys. Res. Commun. 285*, 470–475.

18. Fischer, A., Pallauf, J., and Rimbach, G. (2002) Selenium- and Vitamin E-Dependent Gene Expression in Rats: Analysis of Differentially Expressed mRNAs, *Methods Enzymol. 347*, 267–276.

19. Hellmann, G.M., Fields, W.R., and Doolittle, D.J. (2001) Gene Expression Profiling of Cultured Human Bronchial Epithelial and Lung Carcinoma Cells, *Toxicol. Sci. 61*, 154–163.

20. Chen, C.J., Huang, H.S., Lin, S.B., and Chang, W.C. (2000) Regulation of Cyclooxygenase and 12-Lipoxygenase Catalysis by Phospholipid Hydroperoxide Glutathione Peroxidase in A431 Cells, *Prostaglandins Leukot. Essent. Fatty Acids 62*, 261–268.

21. Raisanen, S.R., Lehenkari, P., Tasanen, M., Rahkila, P., Harkonen, P.L., and Vaananen, H.K. (1999) Carbonic Anhydrase III Protects Cells from Hydrogen Peroxide-Induced Apoptosis, *FASEB J. 13*, 513–522.

22. Jialal, I., Devaraj, S., and Kaul, N. (2001) The Effect of α-Tocopherol on Monocyte Proatherogenic Activity, *J. Nutr. 131*, 389S–394S.

Chapter 16

Vitamin E and Enhancement of the Immune Response in the Aged: Cellular and Molecular Mechanisms

Sung Nim Han[a], Oskar Adolfsson[a,b], and Simin Nikbin Meydani[a,c]

[a]Jean Mayer USDA Human Nutrition Research Center on Aging at Tufts University, Boston, MA
[b]Novel Gene Evaluation/Musculoskeletal Sciences, Genetics Institute, Cambridge, MA
[c]Department of Pathology, Sackler Graduate School of Biomedical Sciences, Tufts University, Boston, MA

Introduction

Aging is associated with dysregulation of immune and inflammatory responses, which is believed to contribute to the higher morbidity and mortality from infection, neoplastic, and inflammatory diseases as well as pathologic processes associated with certain diseases in the aged. Studies indicate that a multitude of defects involving different immune cells are responsible for the dysregulation of immune and inflammatory responses observed with aging (1,2) (Fig. 16.1).

Among the immune cells, T cells are the main cells to show age-related changes. Aging is associated with reduced T-cell function, as demonstrated by decreased T-cell proliferation and interleukin (IL)-2 production, and a shift toward greater proportions of antigen (Ag)-experienced memory T cells with fewer T cells of naïve phenotype (3). IL-2 receptor expression is also decreased in cells from elderly individuals (4). In addition, functional disruption of the CD28 gene transcriptional initiator is observed in senescent T cells (2) and an age-related decline in activation of c-Jun N-terminal kinase (JNK) pathway by T-cell receptor (TCR)- and CD28-mediated signals has been reported (5). Intrinsic changes in T cells and increased production of suppressive factors such as prostaglandin (PG)E_2 from old macrophages contribute to these age-related declines in T-cell responses.

Macrophages play a key role in inflammatory responses by releasing a variety of inflammatory mediators including prostaglandins and proinflammatory cytokines (6). Prostaglandins are generated from arachidonic acid (released from the membrane phospholipids by phospholipase A_2) by the action of cyclooxygenase (COX). COX is the rate-limiting enzyme in the biosynthesis of prostaglandins with bifunctional catalytic properties, i.e., formation of PGG_2 from arachidonic acid *via* its COX activity and subsequent reduction of PGG_2 to PGH_2 *via* its peroxidase activity (7). The PGH_2 formed is isomerized enzymatically to PGE_2 or other COX products depending on the cell types. The activity of COX requires the presence of oxidant hydroperoxide as an activator (Fig. 16.2). Peroxynitrite, an inorganic hydroperoxide generated by the coupling of nitric oxide (NO) to superoxide anion ($O_2^{\bullet-}$), can activate the COX activity

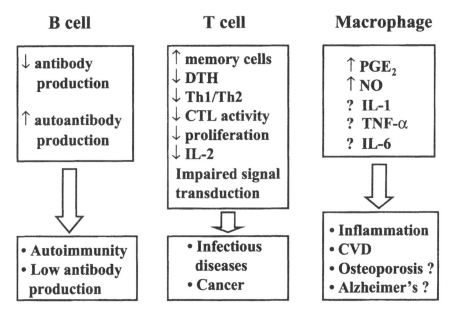

B cell	**T cell**	**Macrophage**

| ↓ antibody production

 ↑ autoantibody production | ↑ memory cells
 ↓ DTH
 ↓ Th1/Th2
 ↓ CTL activity
 ↓ proliferation
 ↓ IL-2
 Impaired signal transduction | ↑ PGE$_2$
 ↑ NO
 ? IL-1
 ? TNF-α
 ? IL-6 |

| • **Autoimmunity**
 • **Low antibody production** | • **Infectious diseases**
 • **Cancer** | • **Inflammation**
 • **CVD**
 • **Osteoporosis ?**
 • **Alzheimer's ?** |

Fig. 16.1. Effects of aging on immune cells. Abbreviations: IL, interleukin; PG prostaglandin; NO, nitric oxide; TNF, tumor necrosis factor; CVD, cardiovascular disease.

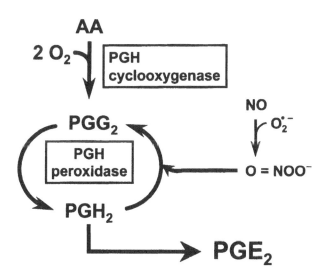

Fig. 16.2. Biochemical pathway of cyclooxygenase-dependent prostaglandin synthesis. NO, nitric oxide; O$_2^{\bullet-}$, superoxide; O = NOO, peroxynitrite; PG, prostaglandin.

by serving as a substrate for the enzymes' peroxidase activity (8). Our earlier studies showed that there is an age-associated increase in PGE_2 production, which is due to increased COX-2 activity, resulting from increased COX-2 protein and mRNA expression (9). PGE_2 has a direct inhibitory effect on the early stages of T-cell activation (10) and can modulate Th1/Th2 cytokine secretion (11). Beharka *et al.* (12) showed that increased production of PGE_2 by macrophages from the aged contributes to the age-associated decrease in T-cell function. In a co-culture study, the addition of macrophages from old mice suppressed proliferation and IL-2 production by T cells from young mice, and the addition of PGE_2 at concentrations produced by old macrophages decreased proliferation and IL-2 production by young but not old T cells.

Vitamin E and Immune Functions in the Aged

The beneficial effects of supplemental vitamin E on immune functions of the aged have been shown in animal studies and human clinical trials (13,14). Vitamin E supplementation has been shown to increase delayed-type hypersensitivity response, *in vitro* T-cell proliferation, and IL-2 production, and to decrease macrophage production of the T cell–suppressive PGE_2 (13–15). In addition, the immunostimulatory effect of vitamin E is associated with increased resistance against infectious agents (16). Vitamin E supplementation significantly decreased pulmonary influenza virus titer and increased IL-2 and interferon (IFN)-γ production by splenocytes in the old mice infected with influenza virus (16). Detailed reviews on effect of vitamin E on immune response are available (17,18).

Mechanisms of Immunoenhancing Effect of Vitamin E

Several mechanisms have been proposed to explain the immunostimulatory effects of vitamin E including its effect on membrane integrity, PGE_2 production, and signal transduction pathways that are particularly sensitive to oxidative stress such as nuclear factor (NF)-κB and AP-1. Vitamin E can exert its effect either directly on T cells or indirectly through reducing the production of suppressive factors from macrophages.

Several lines of evidence indicate that vitamin E exerts its immunostimulatory effects partly by lowering macrophage PGE_2 production (12,15). Preincubation of macrophages from old mice with vitamin E *in vitro* increased proliferation of co-cultures containing T cells from either young or old mice and increased IL-2 production of co-cultures containing T cells from old mice accompanied by decreased production of PGE_2 (12). These data suggest that the immunoenhancing effect of vitamin E is mediated in part through lowering PGE_2 production of old macrophages. Vitamin E supplementation of old mice eliminates the age-associated increase in PGE_2 production and suppresses the age-associated increase in COX-2 activity. The vitamin E–induced suppression of COX activity was not due to a decrease in expression of protein or mRNA for COX-1 or COX-2, indicating that regulation of COX activity by vitamin E is at the post-translational level (15).

To determine the mechanisms involved in the vitamin E–induced decrease in production of PGE_2 and COX-2 activity in macrophages from old mice, Beharka *et al.* (19) investigated whether vitamin E–induced inhibition of COX activity was mediated through a decrease in peroxynitrite formation. Peroxynitrite, a product of nitric oxide (NO) and superoxide ($O_2^{\bullet-}$), was shown to increase the activity of COX without affecting its expression levels (8). Macrophages from old mice had significantly higher PGE_2 levels, COX activity, and NO levels compared with young mice, whose levels were all significantly reduced by vitamin E supplementation in the old macrophages. To demonstrate that vitamin E–induced inhibition of COX activity was due to a decrease in peroxynitrite, old mice were fed diets containing 30 or 500 µg/g *dl*-α-tocopheryl acetate. Peritoneal macrophages from mice fed 500 µg/g vitamin E were separated and cultured in presence or absence of NO generators (*S*-nitroso-*N*-acetyl-penicillamine, SNAP), $O_2^{\bullet-}$ generators (xanthine/xanthine oxidase), or peroxynitrite generators (SNAP + xanthine/xanthine oxidase or 3-morpholinosydonimine *N*-ethylcarbamide, SIN-1). On the other hand, macrophages from mice fed 30 µg/g vitamin E were cultured in the presence of NO inhibitors (*N*-monoethyl-L-arginine, L-NMMA), $O_2^{\bullet-}$ inhibitors [Mn (III) tetrakis (1-methyl-4-pyridyl) porphyrin, MnTMPyP], or both. PGE_2 production and COX activity and expression levels were measured (19). When inhibitors of NO and $O_2^{\bullet-}$ were added together, COX activity was significantly reduced in the macrophages from old mice fed 30 µg/g vitamin E diet (Table 16.1). However, adding NO or $O_2^{\bullet-}$ inhibitors alone had no effect on inhibiting COX activity. When peroxynitrite levels were increased using Sin-1, which provides a continuous source of peroxynitrite by decomposing to generate NO and $O_2^{\bullet-}$, or the combination of SNAP (NO generator) and xanthine/xanthine oxidase ($O_2^{\bullet-}$ generator), COX activity in macrophages from old mice fed 500 µg/g vitamin E diet increased significantly, whereas there was no change in macrophages from old mice fed 30 µg/g vitamin E diet (Table 16.2). These results strongly suggest that peroxynitrite plays an important role in the vitamin E–induced inhibition of COX activity. Thus, we propose that vitamin E reduces COX activity through reduction in peroxy-

TABLE 16.1
Effects of $O_2^{\bullet-}$, NO, and NO_3^- Inhibitors on Cyclooxygenase (COX) Activity[a,b]

	Dietary group	
	30 µg/g vitamin E	500 µg/g vitamin E
Inhibitors	(% change)	
MnTMPyP ($O_2^{\bullet-}$ inhibitor)	−13	−15
L-NMMA (NO inhibitor)	−15	−15
MnTMPyP + L-NMMA	−40[c]	−24

[a]*Source:* Data adapted from Beharka *et al.* (19).
[b]Macrophages were cultured with 5 µg/mL of lipopolysacchride (LPS) in the presence or absence of MnTMPyP (20 µmol/L), L-NMMA (300 µmol/L), or L-NMMA + MnTMPyP. The percentage of change in COX activity was calculated as COX activity [pg PGE_2/(µg protein·10 min)] in the presence of LPS plus treatment minus COX activity in the presence of LPS alone, divided by COX activity in the presence of LPS alone, and multiplied by 100.
[c]Significant decrease in COX activity with inhibitor treatment, $P < 0.05$.

TABLE 16.2

Effects of $O_2^{\bullet-}$, NO, and NO_3^- Generators on Cyclooxygenase (COX) Activity[a,b]

	Dietary group	
	30 µg/g vitamin E	500 µg/g vitamin E
Inhibitors	(% change)	
X/XO ($O_2^{\bullet-}$ generator)	−10	−2
SNAP (NO generator)	14	40
Sin-1 ($O_2^{\bullet-}$ and NO generator)	25	78[c]
X/XO + SNAP ($O_2^{\bullet-}$ and NO generator)	24	65[c]

[a]*Source*: Data adapted from Beharka *et al.* (19).
[b]Macrophages were cultured with 5 µg/mL of lipopolysacchride (LPS) in the presence or absence of NO generator (300 µmol/L SNAP), $O_2^{\bullet-}$ generator (100 µmol/L xanthine/0.2 U/mL xanthine oxidase, X/XO), or NO_3^- generator (350 µmol/L Sin-1 or X/XO + 150 µmol/L SNAP). The percentage of change in COX activity was calculated as COX activity [(pg PGE_2/(µg protein·10 min)] in the presence of LPS plus treatment minus COX activity in the presence of LPS alone, divided by COX activity in the presence of LPS alone, and multiplied by 100.
[c]Significant increase in COX activity with generator treatment, $P < 0.05$.

nitrite formation, which is at least in part due to a decrease in NO production by vitamin E.

In addition to its effect on reduction of PGE_2 in macrophages, vitamin E might have a direct effect on T-cell function. Previous studies from our laboratory (12) suggested that the modulating effects of vitamin E on immune function in aged mice might be mediated in part through its direct effect on T cells independently of its effect on macrophage PGE_2 production. We therefore investigated the effect of vitamin E on purified T cells obtained from young and old mice (20). Purified T cells were collected from the spleens of young and old C57BL/6 mice by negative immunomagnetic selection to eliminate macrophages, the main producer of PGE_2 in spleen. Flow cytometric data demonstrated that >94% of the enriched cell population expressed the CD3 Ag, which characterizes mature T cells. Cells staining positive for macrophages, B cells, or natural killer (NK) cells represented <2% of the purified T-cell population. Radioimmunoassays revealed no PGE_2 production in the purified cell culture media, thus eliminating the possible contribution of PGE_2 from macrophages. In addition, the use of a COX inhibitor, indomethacin, did not alter any of the T-cell responses tested. These data indicate that in the absence of macrophages, PGE_2 does not contribute to age-associated T-cell functional defects.

To investigate the effects of vitamin E on proliferative responses and IL-2 production in response to anti-CD3 and anti-CD28 ligation, we supplemented T cells with 46 µmol/L vitamin E by an *in vitro* method. In agreement with our previous studies, over a 48-h activation period, T cells from old mice proliferated less and produced less IL-2 than T cells from young mice. Vitamin E supplementation significantly increased both proliferation and total IL-2 production by T cells from old mice, whereas there was no significant effect on young T cells. These data support our hypothesis that vitamin E has a direct, PGE_2-independent, immunoenhancing effect on T-cell function in the aged.

It is well accepted that aging is accompanied by a phenotypic shift in the peripheral T-cell population, from mainly T cells that have not encountered antigen (naive) to a much greater proportion of T cells that have (memory). This shift in T-cell phenotype is a major change that influences T cell–mediated immunity in the aged (3,21). This shift results, in part, from decreased output of positively selected immature naïve T cells from the aging thymus. Compared with memory T cells, naive T cells have different response kinetics to an activating challenge, with Ag-experienced memory cells responding faster and to a lower dose of Ag than naive T cells (22,23). In addition, the requirements for cell activation have been shown to differ between naïve and memory T cells. Naive T cells are more profoundly dependent on co-stimulatory signals that are provided by antigen-presenting cells *via* B7-1/CD28 interaction (22).

The age-related functional decline has been reported for both the memory (24–26) as well as the naive (1,27) T-cell subpopulations. One of the most important roles IL-2 plays in the response of T cells to Ag challenge is its ability to up-regulate

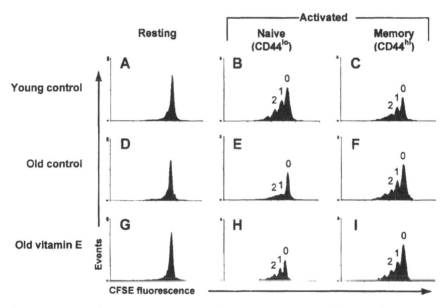

Fig. 16.3. Effect of age and vitamin E on the progression of T cells through cell cycle division. Purified T cells were preincubated with 46 μmol/L vitamin E for 4 h, labeled with carboxyfluoroscein succinimidyl ester (CFSE), and activated with immobilized anti-CD3 and soluble anti-CD28 mAb for 48 h. Cells were harvested, stained for CD44 expression, and analyzed on a flow cytometer. One representative histogram for each of young control (A, B, and C), old control (D, E, and F), and old rats preincubated with vitamin E (G, H, and I) are shown. Cell cycle division patterns are shown for unactivated T cells (A, D, and G), activated naive (CD44^lo^) T cells (B, E, and D), and activated memory (CD44^hi^) T cells (C, F, and I). Peaks representing cell division cycles 0, 1, and 2 are also indicated. *Source:* Reproduced from Adolfsson *et al.* (20) with permission.

high-affinity IL-2 receptors by itself and by neighboring T cells, thereby inducing responsiveness to secreted IL-2 (28). IL-2 is also essential for T cells to progress through the cell cycle and for the differentiation of activated naive T cells into effector T cells. Thus, diminished production of IL-2 by T cells from the aged is an important contributing factor to the decline in cell-mediated immunity in this population.

To better determine the mechanism of vitamin E–induced enhancement of T-cell function in the aged, T cells were stained with anti-CD44, a marker that identifies naïve and memory T cells on the basis of the expression level (20). The age-related defect in anti-CD3– and anti-CD28–induced IL-2 production and proliferation at 48 h was observed mainly in naive T cells. This activation protocol is particularly effective for stimulating naive T cells (22). We hypothesized that vitamin E increased the total IL-2 production by old T cells by preserving the proliferation of naive T cells, or by boosting memory T-cell function, or both.

Diminished production and secretion of IL-2 by naive T cells during the initial encounter with antigen may indeed influence the generation of functional memory T cells. This was demonstrated using responsive but IL-2–negative naive T cells that have been shown to become unresponsive memory T cells (29). Using transgenic mice with naive T cells specific for pigeon cytochrome c (PCC), Linton et al. (27) reported that naive Ag-inexperienced transgene-positive T cells isolated from old mice showed reduced IL-2 production and less proliferation compared with naive transgene-positive T cells isolated from young mice. Furthermore, Garcia and Miller

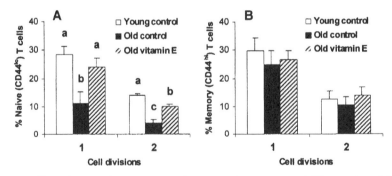

Fig. 16.4. Effect of age and vitamin E on the expression patterns of CD44 at 1 and 2 cell division cycles. Purified T cells ($n = 3$) were preincubated with 46 µmol/L vitamin E for 4 h, labeled with carboxyfluoroscein succinimidyl ester (CFSE), and activated with immobilized anti-CD3 and soluble anti-CD28 mAb for 48 h. Cells were harvested, stained for CD44 expression, and analyzed on a flow cytometer. Each 2-dimensional dot plot-region representing 0, 1, and 2 cell division cycles was evaluated for low (CD44[lo]) and high (CD44[hi]) expression patterns of the CD44 antigen. Panel A represents naive (CD44[lo]) T cells and panel B represents memory (CD44[hi]) T cells. Bars with different letters within each group are significantly different ($P < 0.05$) by an ANOVA followed by Tukey's Honestly Significant Difference post-hoc procedure. *Source*: Reproduced from Adolfsson et al. (20) with permission.

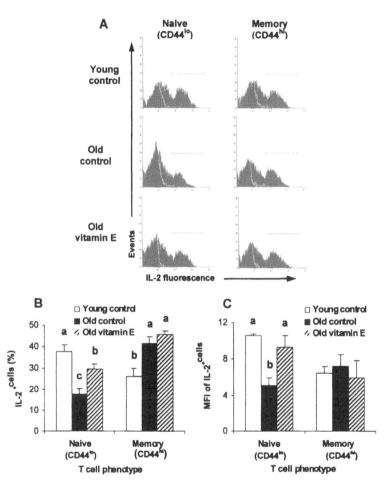

Fig. 16.5. Effect of age and vitamin E on intracellular interleukin (IL)-2 by naive and memory T-cell subsets. Purified T cells (*n* = 5) were preincubated with 46 μmol/L vitamin E for 4 h and activated with immobilized anti-CD3 and soluble anti-CD28 mAb for 48 h. Cells were treated with monensin, an inhibitor of IL-2 secretion, for the last 10 h of activation. Harvested cells were stained with fluorochrome conjugated anti-CD44 mAb, permeabilized, and stained with fluorochrome conjugated anti-IL-2. T cells were divided into naïve and memory phenotypes based on low or high expression of the CD44 antigen, respectively. Cell fluorescence was measured on a flow cytometer. Panel A shows one representative histogram for naive (CD44[lo]; left) and memory (CD44[hi]; right) T cells from each of the three groups: young control (top), old control (middle), and old vitamin E supplemented rats (bottom). Panel B shows the relative proportion (mean ± SEM) of IL-2[+] T cells, and panel C represents the linearized mean fluorescence intensity (MFI) of IL-2[+] T cells. Bars with different letters within each phenotype are significantly different (*p* < 0.05) by an ANOVA followed by Tukey's Honestly Significant Difference post-hoc procedure. *Source*: Reproduced from Adolfsson *et al.* (20) with permission.

(1) reported that naive PCC-specific T cells from old mice had at least two age-related functional defects in the early stages of T-cell activation. The cells were defective in translocating signaling proteins to the cell membrane and in forming an immunologic membrane synapse. Furthermore, T cells that did form a membrane synapse had defective nuclear translocation of NF-AT, a transcription factor critical for IL-2 expression.

The ability of individual cells to divide can be monitored by staining the cells with a nontoxic level of a fluorochrome that remains in the cytoplasm for several days and is divided equally into each daughter cell during cell division. When young and old T cells supplemented with vitamin E were examined for their ability to go through activation-induced cell division over a 48-h period, vitamin E significantly increased the ability of naive T cells from old mice to progress through one as well as two cell division cycles (Figs. 16.3 and 16.4, pp. 222 and 223) (20). This enhancing effect of vitamin E was not observed for memory T cells from old mice. Furthermore, by performing intracellular staining of IL-2, vitamin E supplementation increased IL-2 production by naive T cells from old mice, whereas there was no effect on IL-2 production by memory T cells (Fig. 16.5). Both the number of naive IL-2$^+$ T cells from old mice and the staining

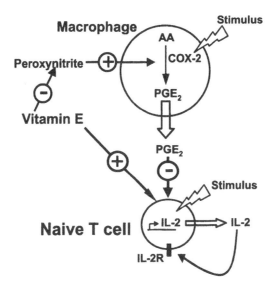

Fig. 16.6. Supplemental vitamin E increases the function of T cells from the aged by at least two different mechanisms. Vitamin E enhances T-cell function indirectly by reducing the age-related increase in the production of T-cell–suppressive prostaglandin (PG)E$_2$ by macrophages by reducing peroxynitrite formation. Additionally, a direct PGE$_2$-independent effect of vitamin E on the function of naive T cells in the aged is also shown. *Source*: Reproduced from Adolfsson *et al.* (20) with permission.

intensity, indicating the amount of IL-2 produced per cell, were increased by vitamin E. These results demonstrate that vitamin E can enhance the functions of T cells from old mice directly with the preferential effect on naive but not memory T cells. The differential effect of vitamin E on naive and memory T cells may be due to an underlying difference in the susceptibility of these cells to oxidative stress-induced damage (30). These findings will have important implications for developing strategies to reverse age-associated defects in T cell–mediated immune function.

Conclusion

Present evidence has demonstrated that vitamin E significantly increases T cell–mediated function in the aged. This effect of vitamin E is mediated both indirectly *via* the reduction of PGE_2 production by macrophages and directly by increasing proliferation and IL-2 production of T cells from old mice independent of its effect on PGE_2 levels (Fig. 16.6). The effects of vitamin E are mediated through increasing both the cell-dividing and IL-2–producing capacities of the naive T-cell subpopulation. The vitamin E–induced decrease in PGE_2 production is mediated through its reduction of peroxynitrite formation, which in turn results in a reduction of COX activity.

References

1. Garcia, G.G., and Miller, R.A. (2001) Single-Cell Analyses Reveal Two Defects in Peptide-Specific Activation of Naive T Cells from Aged Mice, *J. Immunol. 166*, 3151–3157.
2. Vallejo, A.N., Weyand, C.M., and Goronzy, J.J. (2001) Functional Disruption of the CD28 Gene Transcriptional Initiator in Senescent T Cells, *J. Biol. Chem. 276*, 2565–2570.
3. Miller, R.A. (1996) The Aging Immune System: Primer and Prospectus, *Science 273*, 70–74.
4. Nagel, J.E., Chopra, R.K., Chrest, F.J., McCoy, M.T., Schneider, E.L., Holbrook, N.J., and Adler, W.H. (1988) Decreased Proliferation, Interleukin 2 Synthesis, and Interleukin 2 Receptor Expression Are Accompanied by Decreased mRNA Expression in Phytohemagglutinin-Stimulated Cells from Elderly Donors, *J. Clin. Investig. 81*, 1096–1102.
5. Kirk, C., Freilich, A., and Miller, R. (1999) Age-Related Decline in Activation of JNK by TCR- and CD28-Mediated Signals in Murine T-Lymphocytes, *Cell. Immunol. 197*, 75–82.
6. Nathan, C.F. (1987) Secretory Products of Macrophages, *J. Clin. Investig. 79*, 319–326.
7. DeWitt, D., and Smith, W. (1988) Primary Structure of Prostaglandin G/H Synthase from Sheep Vesicular Gland Determined from the Complementary DNA Sequence, *Proc. Natl. Acad. Sci. USA 85*, 1412–1416.
8. Landino, L.M., Crews, B.C., Timmons, M.D., Morrow, J.D., and Marnett, L.J. (1996) Peroxynitrite, the Coupling Product of Nitric Oxide and Superoxide, Activates Prostaglandin Biosynthesis, *Proc. Natl. Acad. Sci. USA 93*, 15069–15074.
9. Hayek, M.G., Mura, C., Wu, D., Beharka, A.A., Han, S.N., Paulson, K.E., Hwang, D., and Meydani, S.N. (1997) Enhanced Expression of Inducible Cyclooxygenase with Age in Murine Macrophages, *J. Immunol. 159*, 2445–2451.
10. Vercammen, C., Ceuppens, J.L. (1987) Prostaglandin E_2 Inhibits T-Cell Proliferation After Crosslinking of the CD3-Ti Complex by Directly Affecting T Cells at an Early Step of the Activation Process, *Cell. Immunol. 104*, 24–36.

11. Gately, M.K., Renzetti, L.M., Magram, J., Stern, A.S., Adorini, L., Gubler, U., and Presky, D.H. (1998) The Interleukin-12/Interleukin-12-Receptor System: Role in Normal and Pathologic Immune Responses, *Annu. Rev. Immunol. 16*, 495–521.

12. Beharka, A.A., Wu, D., Han, S.N., and Meydani, S.N. (1997) Macrophage Prostaglandin Production Contributes to the Age-Associated Decrease in T Cell Function Which Is Reversed by the Dietary Antioxidant Vitamin E, *Mech. Ageing Dev. 93*, 59–77.

13. Meydani, S.N., Meydani, M., Verdon, C.P., Shapiro, A.A., Blumberg, J.B., and Hayes, K.C. (1986) Vitamin E Supplementation Suppresses Prostaglandin E_2 Synthesis and Enhances the Immune Response of Aged Mice, *Mech. Ageing Dev. 34*, 191–201.

14. Meydani, S.N., Meydani, M., Blumberg, J.B., Leka, L.S., Siber, G., Loszewski, R., Thompson, C., Pedrosa, M.C., Diamond, R.D., and Stollar, B.D. (1997) Vitamin E Supplementation and In Vivo Immune Response in Healthy Subjects, *J. Am. Med. Assoc. 277*, 1380–1386.

15. Wu, D., Mura, C., Beharka, A.A., Han, S.N., Paulson, K.E., Hwang, D., and Meydani, S.N. (1998) Age-Associated Increase in PGE_2 Synthesis and COX Activity in Murine Macrophages Is Reversed by Vitamin E, *Am. J. Physiol. 275*, C661–C668.

16. Han, S.N., Wu, D., Ha, W.K., Beharka, A., Smith, D.E., Bender, B.S., and Meydani, S.N. (2000) Vitamin E Supplementation Increases T Helper 1 Cytokine Production of Old Mice Infected with Influenza Virus, *Immunology 100*, 487–493.

17. Meydani, S.N., and Han, S.N. (2001) Nutrient Regulation of the Immune Response: The Case of Vitamin E, in *Present Knowledge in Nutrition*, 8th ed. (Bowman, B.A., and Russell, R.M., eds.) ILSI Press, Washington.

18. Bendich, A. (1990) Antioxidant Vitamins and Their Functions in Immune Responses, in *Antioxidant Nutrients and Immune Functions* (Bendich, A., Phillips, M., Tengerdy, R.P., eds.) Plenum Press, New York.

19. Beharka, A., Wu, D., Serafini, M., and Meydani, S. (2002) Mechanism of Vitamin E Inhibition of Cyclooxygenase Activity in Macrophages from Old Mice: Role of Peroxynitrite, *Free Radic. Biol. Med. 32*, 503–511.

20. Adolfsson, O., Huber, B.T., and Meydani, S.N. (2001) Vitamin E-Enhanced IL-2 Production in Old Mice: Naive but Not Memory T Cells Show Increased Cell Division Cycling and IL-2-Producing Capacity, *J. Immunol. 167*, 3809–3817.

21. Ernst, D.N., Hobbs, M.V., Torbett, B.E., Glasebrook, A.L., Rehse, M.A., Bottomly, K., Hayakawa, K., Hardy, R.R., and Weigle, W.O. (1990) Differences in the Expression Profiles of CD45RB, Pgp-1, and 3G11 Membrane Antigens and in the Patterns of Lymphokine Secretion by Splenic CD4+ T Cells from Young and Aged Mice, *J. Immunol. 145*, 1295–1302.

22. Dubey, C., Croft, M., and Swain, S.L. (1996) Naive and Effector CD4 T Cells Differ in Their Requirements for T Cell Receptor Versus Costimulatory Signals, *J. Immunol. 157*, 3280–3289.

23. Rogers, P.R., Dubey, C., and Swain, S.L. (2000) Qualitative Changes Accompany Memory T Cell Generation: Faster, More Effective Responses at Lower Doses of Antigen, *J. Immunol. 164*, 2338–2346.

24. Hobbs, M.V., Weigle, W.O., Noonan, D.J., Torbett, B.E., McEvilly, R.J., Koch, R.J., Cardenas, G.J., and Ernst, D.N. (1993) Patterns of Cytokine Gene Expression by CD4+ T Cells from Young and Old Mice, *J. Immunol. 150*, 3602–3614.

25. Lerner, A., Yamada, T., and Miller, R.A. (1989) Pgp-1hi T Lymphocytes Accumulate with Age in Mice and Respond Poorly to Concanavalin A, *Eur. J. Immunol. 19*, 977–982.

26. Philosophe, B., and Miller, R.A. (1990) Diminished Calcium Signal Generation in Subsets of T Lymphocytes That Predominate in Old Mice, *J. Gerontol. 45*, B87–B93.

27. Linton, P.-J., Haynes, L., Klinman, N.R., and Swain, S.L. (1996) Antigen-Independent Changes in Naive CD4 T Cells with Aging, *J. Exp. Med. 184*, 1891–1900.

28. Katzen, D., Chu, E., Terhorst, C., Leung, D.Y., Gesner, M., and Miller, R.A. (1985) Mechanisms of Human T Cell Response to Mitogens: IL-2 Induces IL-2 Receptor Expression and Proliferation but Not IL-2 Synthesis in PHA-Stimulated T Cells, *J. Immunol. 135*, 1840–1845.

29. Saparov, A., Wagner, F.H., Zheng, R., Oliver, J.R., Maeda, H., Hockett, R.D., and Weaver, C.T. (1999) Interleukin-2 Expression by a Subpopulation of Primary T Cells Is Linked to Enhanced Memory/Effector Function, *Immunity 11*, 271–280.

30. Lohmiller, J.J., Roellich, K.M., Toledano, A., Rabinovitch, P.S., Wolf, N.S., and Grossmann, A. (1996) Aged Murine T-Lymphocytes Are More Resistant to Oxidative Damage Due to the Predominance of the Cells Possessing the Memory Phenotype, *J. Gerontol. 51*, B132–B140.

Chapter 17

Is There a Role for Vitamin E in the Prevention of Atherosclerosis?

Roland Stocker[a,b], Andrew C. Terentis[a], Joanne M. Upston[a], and Leonard Kritharides[b,c]

[a]Biochemistry Group, the Heart Research Institute, Sydney NSW 2050, Australia
[b]Centre for Thrombosis and Vascular Research, University of New South Wales, UNSW Sydney, NSW 2052 Australia
[c]Clinical Research Group, the Heart Research Institute, Sydney NSW 2050, Department of Cardiology, Concord Hospital, Sydney, NSW 2139, Australia

Introduction

The role of lipoprotein oxidation, and particularly that of low density lipoprotein (LDL), in atherogenesis has been the subject of intense investigation. Major reviews of LDL oxidation have been published (see, e.g., Refs. 1–6). The aim of this article is to review randomized controlled studies investigating the clinical use of vitamin E supplements to prevent or treat coronary heart disease (CHD), and discuss the scientific basis for a role of vitamin E in the inhibition of lipoprotein lipid oxidation in the vessel wall.

Data Suggesting Benefit of Vitamin E Supplementation

Absolute rates of CHD vary among countries with similar mean serum cholesterol and may relate to factors such as consumption of saturated fat and antioxidants, and nondietary factors such as cigarette smoking. Many investigators have found that populations with low rates of CHD consume diets rich in antioxidants such as vitamin E (7), and this agrees with studies measuring plasma levels of some antioxidants (8). Gey and co-workers (7) found that the apparent clinical benefit was great despite only small difference, in apparent plasma concentrations of α-tocopherol (biologically the most active form of vitamin E) in high- and low-intake populations. Although different dietary antioxidants, including flavonoids (9), have been reported to confer protection against vascular disease, overall, vitamin E has received the most attention.

Prospective cohort studies. Prospective cohort studies have correlated high self-reported intakes of antioxidants with low rates of CHD. Although at first glance supportive of a role for vitamin E in protecting against CHD, significant discrepancies exist in the data of these studies. Perhaps most apparent is that the benefit attributable to small amounts of vitamin E ingested from food (7) is far greater than the dose required for benefit from supplemental vitamin E. There are also potentially important yet unexplained differences between studies in terms of dose and source of vitamin E (food vs. supplement). The 87,000 population Nurse's Health Study (10) and the 40,000 population male

Health Professional Study (11) showed benefit with 100 IU/d supplemental vitamin E but not with low dose supplemental or high dietary intake of vitamin E. Contrary to this, a report from a population of 35,000 women (12) showed that a modest increase in dietary vitamin E intake from <4.9 (lowest quartile) to 9.6 IU/d (highest quartile) to be associated with a significant reduction in cardiovascular risk. However, in this study there was no benefit attributable to vitamin E supplements (12).

Vitamin E and intima-to-media thickness (IMT). Carotid IMT is commonly used as a surrogate marker of atherosclerotic disease. IMT correlates weakly with the prevalence and extent of coronary artery disease measured by arteriography (13), but is useful for longitudinal evaluation of study populations. Results pertaining to vitamin E and IMT appear to parallel those of observational studies of vitamin E intake, in which benefits appear most marked in self-selected populations (14). The only two randomized studies to date however suggest a null effect of vitamin E (see below).

Basic Research on Role of Vitamin E in Atherosclerosis

The scientific rationale for vitamin E supplements protecting against atherosclerosis and its clinical complications is based primarily on the "LDL oxidation theory" of atherosclerosis (1). Accordingly, α-tocopherol (the most active form of vitamin E) is the most abundant radical scavenger of LDL (15) and may prevent a number of atherogenic events. These include the adhesion of leukocytes to endothelial cells (16), endothelial dysfunction (17), platelet aggregation (18), and the proliferation of vascular smooth muscle cells (19).

In vitro *LDL oxidation.* There is unequivocal evidence of lipid and protein oxidation in atherosclerotic lesions. *In vitro* studies have indicated that vitamin E can inhibit the oxidation of LDL lipid. However, depending on the experimental conditions used, vitamin E can also demonstrate prooxidant activity (20). Because the precise conditions for *in vivo* LDL oxidation remain obscure, the overall effect of vitamin E on LDL oxidation *in vivo* is not established. *In vitro* LDL oxidation studies remove LDL from its natural environment, leaving their biological relevance open to question. The arterial wall, where LDL oxidation is expected to take place, is difficult to replicate, and atherosclerotic lesions contain large quantities of antioxidants including α-tocopherol and vitamin C (21). In addition, diverse oxidative processes identified within atherosclerotic plaque (22) are likely differentially responsive to vitamin E (23).

F_2-isoprostanes and in vivo *effects of vitamin E.* F_2-isoprostanes are prostaglandin (PG)-like compounds produced during the free radical–catalyzed oxidation of arachidonate and are commonly used as noninvasive markers of *in vivo* lipid oxidation (24). Urinary levels of 8-epi-PGF$_{2\alpha}$ are elevated in people with hypercholesterolemia (25) and diabetes (26), and in smokers (27), consistent with the occurrence of lipid oxidation. Human (28) and mouse atherosclerotic blood vessels (29) contain elevated levels of F_2-isoprostanes compared with healthy vessels, and a pharmacologic dose of vita-

min E has been reported to significantly decrease F_2-isoprostanes in the aortas of apolipoprotein E gene–deficient (apoE-/-) mice (29). Supplementation with vitamin E also significantly reduced urinary 8-epi-PGF$_{2\alpha}$ in hypercholesterolemic (25) and diabetic subjects (26). In contrast, vitamin E supplements (or aspirin) did not decrease urinary 8-epi-PGF$_{2\alpha}$ in healthy adults (30) or in smokers (31,32) where vitamin C supplements were effective (31).

However, most of the studies described here were uncontrolled, and the apparent conflicting results may point to variations in assays of isoprostanes or population differences. Whether the negative results with vitamin E relate to the lack of efficacy of the supplement or to the complexity of factors affecting isoprostane production/metabolism is unclear. It is also unclear to what extent plasma or urinary isoprostanes reflect LDL oxidation.

RESULTS

Randomized Clinical Trials with Vitamin E in Cardiovascular Disease (CVD)

General issues and limitations. Variations in plasma concentrations of α-tocopherol identified in nutritional studies as relevant to the incidence of CHD incidence are very small, i.e., in the low micromolar range. They are exceeded greatly by most studies using antioxidant supplements, including the α-tocopherol, β-carotene (ATBC) study (see below) for which the 50-mg vitamin E/d dose used is often criticized for being too low. Nevertheless, the possibility that higher doses matching those used in prospective cohort studies may have provided protection cannot be excluded. It is also important to note that results obtained from a follow-up of ~5 y (as is conventional for pharmaceutical studies) used in the primary prevention studies, do not preclude benefits that may derive from a lifetime of dietary intake of vitamin E. Furthermore, study entry criteria have generally not included *in vivo* quantitative indices of oxidative stress or antioxidant deficiency, so that individual responses to vitamin E are unpredictable and not measured (30).

Primary prevention studies. In the Finnish ATBC study (33–35), there was no effect of vitamin E on the incidence of fatal or nonfatal myocardial infarction (MI) (Table 17.1). An increased risk of hemorrhagic stroke and decreased risk of prostate cancer were identified with vitamin E but are of uncertain importance, given their absence in the Gruppo Italiano per lo Studio della Supravivenza nell'Infarto Miocardico (GISSI), Cambridge Heart Antioxidant Study (CHAOS), or Heart Outcomes Prevention Evaluation Study (HOPE). The Antioxidant Supplementation in Atherosclerosis Prevention (ASAP) study (36) randomized 520 men and women to vitamin E (91 mg/d), 250 mg/d slow release vitamin C, or both for 3 y. The rate of progression of IMT was unaffected by the consumption of any antioxidant(s) in women and by the taking of either antioxidant alone in men (Table 17.1). In the collaborative Primary Prevention Project (PPP), open-label, low-dose aspirin, and vitamin E treatments were investigated in general practice (37). Stopped prematurely because of the evidence of favorable effects of aspirin shown in other studies, this study also confirmed that aspirin prevented cardiovascular events, and showed that vitamin E had no effect.

Secondary prevention. In the CHAOS study, vitamin E reduced the risk of nonfatal acute MI (AMI), but caused a nonsignificant increase in fatal AMI (38) (Table 17.2). Vitamin E also achieved a major decrease in rates of AMI [relative risk (RR) 0.3, 95% confidence interval (CI), 0.11–0.78], $P = 0.016$) in a small group of hemodialysis patients [secondary prevention with antioxidants of CVD in endstage renal disease (SPACE) study] (39) (Table 17.2). For both studies, the extent of this decrease is remarkable and unexpected given the results of other studies and the very short duration of follow-up. However, in the CHAOS study, some baseline characteristics were not balanced, and the reduction in nonfatal AMI was at odds with the trend of effect on fatal AMI, pointing at potential limitations. The result of the SPACE study may be attributable to small patient numbers or special characteristics of the renal failure population, raising the possibility that certain subjects, perhaps those with increased oxidative stress, may benefit from supplemental vitamin E. This warrants further investigation.

In the large GISSI-Prevenzione study, n-3 polyunsaturated fatty acids (PUFA, 1 g/d) reduced the relative risk of the combined end point of cardiovascular death, nonfatal MI, and stroke to 0.80 (95% CI, 0.68–0.95), whereas vitamin E showed a nonsignificant trend (RR 0.88, 95% CI, 0.75–1.04) (40) (Table 17.2). Unfortunately, substantial discontinuation rates, dietary changes, the lack of independent monitors, and the open-label treatment confound the interpretation of the GISSI-P study. That n-3 PUFA offered protection and that there was no evidence of interaction by combining vitamin E and PUFA is not immediately consistent with the LDL oxidation theory of atherosclerosis. Thus, PUFA supplements would be expected to increase rather than decrease LDL oxidation, and one might have expected vitamin E to protect PUFA from oxidation. The difference in outcome between GISSI and CHAOS is unexplained.

In the HOPE study, patients with known CVD, or diabetes plus another risk factor were treated with vitamin E or ramipril or both. Treatment with vitamin E exerted no benefit (41) (Table 17.2), whereas ramipril substantially reduced risk of the combined primary endpoint (42). The negative result is compelling; compared with CHAOS, HOPE combined a larger, multinational cohort of patients with longer (4 y) follow-up, and used 400 IU/d vitamin E, which matches the dose administered in the CHAOS study. However, the CHAOS and HOPE studies differed in baseline medication usage. In the study to evaluate carotid ultrasound changes in patients treated with ramipril and vitamin E (SECURE) substudy of the HOPE study, 732 patients with vascular disease or diabetes and at least one other risk factor for coronary disease were investigated for IMT progression. Again, vitamin E had no effect, whereas ramipril reduced IMT progression in a dose-dependent manner (43).

Vitamin E plus other antioxidants. In the ASAP study referred to above, the combination of vitamins E and C significantly decreased disease progression in men (36) (Table

TABLE 17.1
Vitamin E and CVD: Primary Prevention Studies[a]

Trial	Design	Subjects (× 10^3)	History	Years	Vitamin E (IU/d)	Outcome	RR
ATBC	DB, 2 × 2	29	Male smokers	6.1	50	IHD/stroke deaths	0.95
ASAP	DB, 2 × 2	0.12	MW, serum cholesterol ≥5 mmol/L	3.0	272	IMT progression	0.56/1.05[b]
PPP	OL, 2 × 2	4.5	Risk factor	3.6	448	CVD deaths	1.07

[a]Abbreviations: ASAP, Antioxidant Supplementation in Atherosclerosis Prevention; ATBC, α-Tocopherol, β-Carotene; CVD, cardiovascular disease; DB, double-blind; IHD, ischemic heart disease; IMT, intima-to-media thickness; M, men; OL, open label; PPP, Collaborative Primary Prevention Project; RR, relative risk; 2 × 2, 2 × 2 factorial design comparing placebo, agent A, agent B, and combination of agent A and agent B; W, women.
[b]Effects of vitamin E supplements on relative risks for men/women were not significantly different from placebo (36).

TABLE 17.2
Vitamin E and CVD: Secondary Prevention Studies[a]

Trial	Design	Subjects (× 10^3)	History	Years	Vitamin E (IU/d)	Outcome	RR
CHAOS	DB, 2 × 2	2	CAD patients	1.4	800/400	CVD deaths	1.18
						Nonfatal MI	0.23[b]
ATBC$_{sub}$	DB, 2 × 2	1.8	Male smokers	5.3	50	Coronary events	0.97
GISSI	OL, 2 × 2	11	Recent MI	3.5	448	CVD deaths and nonfatal MI	0.88
HOPE	DB, 2 × 2	9.5	CVD+	4.5	400	CVD deaths	1.05
SPACE	DB	0.16	HD + CVD	1.4	800	CVD, MI	0.46[b]
SECURE	DB, 3 × 2	0.7	CVD+	4.5	400	IMT progression	1

[a]Abbreviations: See Table 17.1. CAD, coronary artery disease; CHAOS, Cambridge Heart Antioxidant Study; GISSI, Gruppo Italiano per lo Studio della Supravivenza nell'Infarto Miocardico; HD, hemodialysis patients; HOPE, Heart Outcomes Prevention Evaluation Study; MI, myocardial infarction; SECURE, study to evaluate carotid ultrasound changes in patients treated with ramipril and vitamin E; SPACE, secondary prevention with antioxidants of cardiovascular disease in endstage renal disease.
[b]Significant effect.

17.3). The robustness of this effect appears limited, however, because it did not occur in women, and the results of other studies with antioxidant combinations also do not offer convincing benefit. In the HDL-Atherosclerosis Treatment Study (HATS) (44), 160 subjects with known coronary artery disease, low high density lipoprotein (HDL), "normal" LDL, were randomized to simvastatin plus niacin, antioxidant supplements (combination of 800 IU α-tocopherol, 1000 mg vitamin C, 25 mg β-carotene, and 100 μg selenium/d), both simvastatin-niacin and antioxidants, or neither. Supplemental antioxidants decreased plasma HDL_2 and had no significant effect on the incidence of clinical endpoints and progression of coronary stenosis. In contrast, simvastatin-niacin significantly reduced LDL and the rate of stenosis progression and clinical events, and elevated total HDL and HDL_2. Of concern, antioxidants inhibited the favorable effect of simvastatin-niacin on lesion progression, clinical events, and HDL_2.

The Medical Research Council (MRC)/British Heart Foundation (BHF) Heart Protection Study (HPS) addressed the effect of antioxidant combinations in addition to lipid-lowering therapy in subjects with known coronary disease. Over 20,000 subjects were randomized to 40 mg simvastatin, antioxidant combination (600 mg vitamin E, 250 mg vitamin C, 20 mg β-carotene), both or neither for a mean of 5.5 y. Simvastatin reduced all-cause mortality ($P < 0.001$) and cardiovascular events ($P < 0.0001$), whereas antioxidants had no effect whether alone or when combined with simvastatin (Table 17.3). In this study, antioxidants did not overcome the favorable effects of simvastatin.

Role of Vitamin E in Lipid Oxidation in the Vessel Wall

For vitamin E supplements to prevent atherosclerosis and thereby provide clinical benefit, one may expect α-tocopherol to become limited as disease progresses, vitamin E to favorably affect a biological process, and that process to be causally linked to atherogenesis. Concerning the "biological process" involved, by far most attention has been placed on the inhibition of lipoprotein (LDL) oxidation.

Vitamin E concentration in atherosclerotic lesions. Previous studies employing freshly obtained carotid endarterectomy specimens, i.e., samples representing end-stage disease, indicate that the concentrations of lipid-adjusted α-tocopherol are essentially intact (21,45). To confirm this and to rule out the possibility that α-tocopherol may decrease during the early stages of atherogenesis, we determined the concentration of the vitamin in human aortic tissue representing different stages of the disease. Tissue levels of α-tocopherol remained remarkably constant as disease progressed, independent of whether results were expressed per protein, cholesterol, or cholesteryl esters, and whether homogenate of aortas or their lipoprotein fraction were analyzed (46,47).

In addition, we determined the oxidation products of α-tocopherol, using a gas chromatography/mass spectrometry–based assay (48). As disease developed, the proportion of the vitamin present as tocopherylquinone or tocopherylquinone epoxide remained more or less unaltered, but accounted for only ~15% of the vitamin (46). Therefore, available data indicate that α-tocopherol does not become depleted as atherosclerosis develops.

TABLE 17.3

Cardiovascular Disease (CVD) and Vitamin E Plus Other Antioxidants[a]

	ASAP	HATS	HPS
Design	PP, DB, 2 × 2; 3 y	SP, DB, 2 × 2; 3.5 y	SP, DB, 2 × 2; 5.5 y
Population	113; men; serum cholesterol ≥ 5 mmol/L	160; Coronary disease	20,000; Coronary disease
Treatment	272 IU Vitamin E + 250 mg vitamin C	800 IU Vitamin E, 1 g vitamin C 25 mg β-carotene, 100 µg selenium	600 mg Vitamin E, 250 mg vitamin C, 20 mg β-carotene
Outcome	Decrease in IMT progression[b]	Decrease in HDL_2	No effect on AMI, CVD deaths and stroke
Comment	No significant effect in women	Antioxidants decreased the protection conferred by simvastatin plus niacin	Simvastatin decreased AMI, CVD deaths and stroke

[a]Abbreviations: See Tables 17.1 and 17.2. AMI, nonfatal acute myocardial infarction; HATS, HDL-Atherosclerosis Treatment Study; HDL, high density lipoprotein; HPS, Heart Protection Study; PP, primary prevention; SP, secondary prevention.
[b]Significant effect.

Does vitamin E inhibit lipoprotein lipid oxidation in the vessel wall? Atherosclerotic lesions are characterized by the presence of both apparently normal concentrations of α-tocopherol and substantial amounts of oxidized, lipoprotein-derived lipids. Such coexistence could be rationalized if lipoprotein lipid oxidation were to occur in the presence of vitamin E, as is observed *in vitro* when lipoproteins are oxidized *via* tocopherol-mediated peroxidation (20,49).

To address this question, we determined the distribution of regioisomers of cholesteryl linoleate hydroxides that are derived from cholesteryl linoleate hydroperoxides, the single major, primary oxidation product produced in lipoproteins undergoing oxidation. Free radical–induced lipid oxidation generates a complex, although not random array of products, and configurational isomer specificity is influenced by the presence of hydrogen donors, i.e., specifically, the presence of α-tocopherol in lipoproteins selectively generates *cis, trans*-isomers of cholesteryl linoleate hydro(pero)xides during *in vitro* lipid peroxidation (50,51).

We observed that free radical–mediated oxidation of LDL isolated from diseased human aortic tissue by sequential ultracentrifugation, resulted in the initial formation of *cis, trans*-cholesteryl linoleate hydro(pero)xides, as α-tocopherol was being consumed (52). After depletion of the vitamin, the *trans, trans*-isomers formed predominantly, similar to the situation with isolated plasma lipoproteins (51). Further, esterified *cis, trans*-18:2-O(O)H were the primary products found in human lesions and in a rabbit model of arterial injury, and there was no evidence for a significant contribution of lipoxygenase to lipoprotein oxidation (52).

Together, these data suggest that *in vivo* lipoprotein lipid peroxidation in the vessel wall occurs in the presence rather than the absence of α-tocopherol, indicating that the vitamin does not effectively prevent LDL oxidation in diseased vessels.

Is there a causal relationship between lipoprotein lipid oxidation and atherogenesis? Despite the very large body of literature supporting a role for lipoprotein lipid oxidation in atherogenesis, direct evidence for a causal relationship between these two processes remains scarce. Arguably the strongest supportive evidence comes from a study in apoE-/- mice in which atherogenesis was reported to result in an increase in $iPF_{2\alpha}$-VI, an F2-isoprostane, in vascular tissue, and supplemental vitamin E to significantly reduce aortic lesion areas and $iPF_{2\alpha}$-VI levels (29). We also observed inhibition of both atherosclerosis and lipoprotein lipid (per)oxidation in the vessel wall in two different mouse models of atherosclerosis, using co-antioxidants as supplements and measuring tissue concentrations of cholesteryl linoleate hydro(pero)xides (53–55). The results of these studies support although they do not prove a causal relationship between aortic lipoprotein lipid oxidation and atherogenesis.

However, such association of the two processes is not observed consistently. Thus, in LDL receptor–deficient rabbits, prevention of aortic accumulation of cholesteryl linoleate hydro(pero)xides by a synthetic co-antioxidant had no effect on atherosclerosis (56). Also, in cholesterol-fed, balloon-injured rabbits, a large dose of supplemental vitamin E increased disease, assessed by the intima-to-media ratio, without significantly affecting aortic concentrations of esterified lipid hydroperoxides (57). In addition, inhibition of atherosclerosis in the thoracic and abdominal aorta of apoE-/- mice was not associated with a decrease in aortic cholesteryl linoleate hydroperoxides (58). Perhaps most strikingly, probucol treatment of apoE-/- mice significantly increased lesion size in the aortic root (58,59), yet this is associated with a decrease in the parent lipid-standardized concentration of tissue 8-epi-PGF$_{2\alpha}$ and cholesteryl linoleate hydroperoxides (2001, unpublished data). Therefore, lipoprotein lipid oxidation in the vessel wall can be dissociated from atherogenesis, thereby questioning their causal relationship.

Accumulation of nonoxidized and oxidized lipoprotein lipids during atherogenesis. Numerous previous studies reported the presence of certain lipid or protein oxidation products in lesions, although a systematic investigation measuring several oxidation parameters and the accumulation of nonoxidized lipids and antioxidants at various stages of atherosclerosis has not been carried out in the same tissue. Using the intimal lipoprotein-containing fraction of human aortic lesions, we observed that cholesterol accumulated with lesion development and that this increase was already significant at the fatty streak stage. By comparison, cholesteryl esters increased significantly only in fibro-fatty and more complex lesions that also contained significantly increased amounts of cholesteryl ester hydro(pero)xides, the major lipid oxidation product detected (52). When standardized per parent lipid, the increase in tissue cholesteryl ester hydro(pero)xides was significant only in the most advanced lesions (Stary Class V and VI), and accounted for $\leq 2.3\%$ of the cholesteryl esters (52), an amount below that required to convert LDL into "high-uptake" LDL *in vitro* by exposure to Cu(II). Thus, in human atherosclerosis, primary lipoprotein lipid (per)oxidation products do not significantly accumulate until late in lesion development, a finding inconsistent with the LDL oxidation theory, according to which oxidation of LDL is hypothesized to occur early in the disease process and to precede foam cell accumulation and fatty streak formation.

Of the oxidized protein moieties measured, only *o,o*-dityrosine increased with disease, although chloro-tyrosines were present at relatively high levels in all lesions compared with healthy vessels (47). Consistent with a role of hypochlorite in early atherogenesis, the concentration of α-tocopherylquinone was relatively high in fatty streaks (46). Together, these findings raise the possibility that 2*e*-oxidants rather than 1*e*-oxidants (which give rise to lipoprotein lipid peroxidation) are more important in early atherogenesis. If so, one would not expect vitamin E to provide efficient protection because α-tocopherol appears to be a relatively poor scavenger of 2*e*-oxidants, as indicated by its lack of protection against hypochlorite-induced LDL oxidation (23).

DISCUSSION

Randomized Controlled Studies

The strength of the association between food antioxidant consumption and the prevention of coronary events is strongest in observational studies, which are unfortunately confounded by self-selection of patients and co-consumption of other nutrients in whole foods. In IMT studies, vitamin E alone exerted no effect in most randomized controlled studies, but an isolated observation suggests that administration of both vitamin E and vitamin C may be beneficial. The data of large and well-designed, randomized, placebo-controlled studies powered to detect clinical events (ATBC, HOPE, GISSI, PPP, HPS) overall indicate a null effect of vitamin E. Two secondary prevention studies with relatively small numbers and short follow-up, CHAOS and SPACE, suggest that certain subpopulations may benefit from vitamin E supplements.

Plasma concentration of antioxidants may serve as an indirect measure of antioxidant intake as well as a marker of consumption of healthy whole foods. Confounding nondietary lifestyle variables may also contribute to apparent favorable effects in nonrandomized studies. Also, there are many different antioxidants within single food groups. Overall, the findings strengthen the recommendation that diets rich in antioxidants, but not antioxidant supplements, be advocated (60). In support of this, The Lyon Heart Study (61), found a clear reduction in recurrence of CHD after randomizing patients to an entire diet containing increased fiber intake, n-3 PUFA intake, as well as fruit and vegetable consumption.

Revisiting Lipoprotein Lipid Oxidation and Atherogenesis

There is no doubt that lipoprotein oxidation is evident in the arterial wall, and that oxidized lipids and proteins exert many potential biological effects. What is less clear is whether these effects are causal for, rather than consequential to, atherosclerosis or for any of its complications (47). Animal studies currently dissociating antiatherogenic and antioxidant effects of pharmacologic agents (see above) may help resolve these issues. Selected targeting of those processes for which oxidation is causal may permit benefits of antioxidants to be revealed. This will also require identification of which types of oxidants are responsible for the oxidative modification of important targets.

For vitamin E, there now is a sound scientific rationale to explain why α-tocopherol alone may not be effective in preventing lipoprotein lipid oxidation, at least that inflicted by one-electron oxidants (i.e., radicals). Furthermore, vitamin E appears to provide ineffective protection against 2e-oxidants, increasingly implicated in the early stages of atherosclerosis (62). Therefore, a reevaluation of the relevant biological consequences of oxidation, identification of the oxidative processes operative in the arterial wall at different stages of atherogenesis, as well as evaluation of the consequences of their inhibition in animal models are required.

Vitamin E Supplements Against CHD?

Several issues must be addressed before vitamin E or any other antioxidant supplements can be recommended. First, criteria for the identification of patient subgroups requiring or deficient in vitamin E/antioxidant(s) must be established. For example, it would be helpful to identify the uniqueness of the SPACE study population. Similar considerations may apply to a potential interaction of vitamin E with the nitric oxide synthase genotype in the CHAOS study (63). Second, the effects of antioxidant supplementation should be monitored by *in vivo* response to antioxidant therapy, such as the measurement of plasma F_2-isoprostanes in specialized laboratories. Third, the above criteria also apply to the potential use of multiple antioxidants that may allow cooperative antioxidant activity. Given the evidence to date from HPS and HATS (Table 17.3), it appears unlikely, however, that antioxidant combinations will substantially alter the current conclusion of null effect derived from studies of single antioxidants. We note the recent report that treatment of patients after cardiac transplantation for 1 y with pravastatin plus vitamin E (800 IU/d) and vitamin C (1000 mg/d) significantly decreased transplant-associated intimal thickening compared with subjects receiving pravastatin alone (64). However, this double-blind prospective study employed only small numbers of patients (n = 40) and the relevance of transplant-associated arteriosclerosis to atherosclerosis-based CHD is unclear.

Summary

Although the absence of harm with recent large vitamin E studies is reassuring, recommendations to take vitamin E supplements for the prevention or treatment of CHD require convincing proof of positive effect. However, current controlled data do not provide this proof, and scientific evidence is now available that can explain why vitamin E supplements may not protect against atherosclerosis.

Acknowledgments

The authors acknowledge the generous support from the National Health and Medical Research Council and the National Heart Foundation of Australia.

References

1. Steinberg, D., Parthasarathy, S., Carew, T.E., Khoo, J.C., and Witztum, J.L. (1989) Beyond Cholesterol: Modifications of Low-Density Lipoprotein That Increase Its Atherogenicity, *N. Engl. J. Med. 320*, 915–924.
2. Berliner, J.A., and Heinecke, J.W. (1996) The Role of Oxidized Lipoproteins in Atherogenesis, *Free Radic. Biol. Med. 20*, 707–727.
3. Diaz, M.N., Frei, B., Vita, J.A., and Keaney, J.F., Jr. (1997) Antioxidants and Atherosclerotic Heart Disease, *N. Engl. J. Med. 337*, 408–416.
4. Heinecke, J.W. (1998) Oxidants and Antioxidants in the Pathogenesis of Atherosclerosis: Implications for the Oxidized Low Density Lipoprotein Hypothesis, *Atherosclerosis 141*, 1–15.
5. Stocker, R. (1999) Dietary and Pharmacological Antioxidants and Atherosclerosis, *Curr. Opin. Lipidol. 10*, 589–597.
6. Chisolm, G.M., and Steinberg, D. (2000) The Oxidative Modification Hypothesis of Atherogenesis: An Overview, *Free Radic. Biol. Med. 28*, 1815–1826.
7. Gey, K.F., Puska, P., Jordan, P., and Moser, U.K. (1991) Inverse Correlation Between Plasma Vitamin E and Mortality from Ischemic Heart Disease in Cross-Cultural Epidemiology, *Am. J. Clin. Nutr. 53*, 326S–334S.

8. Riemersma, R.A., Wood, D.A., Macintyre, C.C., Elton, R.A., Gey, K.F., and Oliver, M.F. (1991) Risk of Angina Pectoris and Plasma Concentrations of Vitamins A, C, and E and Carotene, *Lancet 337*, 1–5.

9. Keli, S.O., Hertog, M.G., Feskens, E.J., and Kromhout, D. (1996) Dietary Flavonoids, Antioxidant Vitamins, and Incidence of Stroke: The Zutphen Study, *Arch. Intern. Med. 156*, 637–642.

10. Stampfer, M.J., Hennekens, C.H., Manson, J.E., Colditz, G.A., Rosner, B., and Willett, W.C. (1993) Vitamin E Consumption and the Risk of Coronary Disease in Women, *N. Engl. J. Med. 328*, 1444–1449.

11. Rimm, E.B., Stampfer, M.J., Ascherio, A., Giovannucci, E., Colditz, G.A., and Willett, W.C. (1993) Vitamin E Consumption and the Risk of Coronary Heart Disease in Men, *N. Engl. J. Med. 328*, 1450–1456.

12. Kushi, L.H., Folsom, A.R., Prineas, R.J., Mink, P.J., Wu, Y., and Bostick, R.M. (1996) Dietary Antioxidant Vitamins and Death from Coronary Heart Disease in Postmenopausal Women, *N. Engl. J. Med. 334*, 1156–1162.

13. Adams, M.R., Nakagomi, A., Keech, A., Robinson, J., McCredie, R., Bailey, B.P., Freedman, S.B., and Celermajer, D.S. (1995) Carotid Intima-Media Thickness Is Only Weakly Correlated with the Extent and Severity of Coronary Artery Disease, *Circulation 92*, 2127–2134.

14. Azen, S.P., Qian, D., Mack, W.J., Sevanian, A., Selzer, R.H., Liu, C.R., Liu, C.H., and Hodis, H.N. (1996) Effect of Supplementary Antioxidant Vitamin Intake on Carotid Arterial Wall Intima-Media Thickness in a Controlled Clinical Trial of Cholesterol Lowering, *Circulation 94*, 2369–2372.

15. Esterbauer, H., Gebicki, J., Puhl, H., and Jürgens, G. (1992) The Role of Lipid Peroxidation and Antioxidants in Oxidative Modification of LDL, *Free Radic. Biol. Med. 13*, 341–390.

16. Martin, A., Foxall, T., Blumberg, J.B., and Meydani, M. (1997) Vitamin E Inhibits Low-Density Lipoprotein-Induced Adhesion of Monocytes to Human Aortic Endothelial Cells In Vitro, *Arterioscler. Thromb. Vasc. Biol. 17*, 429–436.

17. Keaney, J.F., Jr. , Gaziano, J.M., Xu, A., Frei, B., Curran-Celentano, J., Shwaery, G.T., Loscalzo, J., and Vita, J.A. (1993) Dietary Antioxidants Preserve Endothelium-Dependent Vessel Relaxation in Cholesterol-Fed Rabbits, *Proc. Natl. Acad. Sci. USA 90*, 11880–11884.

18. Freedman, J.E., Farhat, J.H., Loscalzo, J., and Keaney, J.F., Jr. (1996) α-Tocopherol Inhibits Aggregation of Human Platelets by a Protein Kinase C-Dependent Mechanism, *Circulation 94*, 2434–2440.

19. Boscoboinik, D., Szewczyk, A., and Azzi, A. (1991) α-Tocopherol (Vitamin E) Regulates Vascular Smooth Muscle Cell Proliferation and Protein Kinase C Activity, *Arch. Biochem. Biophys. 286*, 264–269.

20. Upston, J.M., Terentis, A.C., and Stocker, R. (1999) Tocopherol-Mediated Peroxidation (TMP) of Lipoproteins: Implications for Vitamin E as a Potential Antiatherogenic Supplement, *FASEB J. 13*, 977–994.

21. Suarna, C., Dean, R.T., May, J., and Stocker, R. (1995) Human Atherosclerotic Plaque Contains Both Oxidized Lipids and Relatively Large Amounts of α-Tocopherol and Ascorbate, *Arterioscler. Thromb. Vasc. Biol. 15*, 1616–1624.

22. Heinecke, J.W. (1997) Mechanisms of Oxidative Damage of Low Density Lipoprotein in Human Atherosclerosis, *Curr. Opin. Lipidol. 8*, 268–274.

23. Hazell, L.J., and Stocker, R. (1997) α-Tocopherol Does Not Inhibit Hypochlorite-Induced Oxidation of Apolipoprotein B-100 of Low-Density Lipoprotein, *FEBS Lett. 414*, 541–544.

24. Meagher, E.A., and FitzGerald, G.A. (2000) Indices of Lipid Peroxidation In Vivo: Strengths and Limitations, *Free Radic. Biol. Med. 28*, 1745–1750.

25. Davi, G., Alessandrini, P., Mezzetti, A., Minotti, G., Bucciarelli, T., Costantini, F., Cipollone, F., Bon, G.B., Ciabattoni, G., and Patrono, C. (1997) In Vivo Formation of 8-Epi-Prostaglandin $F_{2\alpha}$ Is Increased in Hypercholesterolemia, *Arterioscler. Thromb. Vasc. Biol. 17*, 3230–3235.

26. Davi, G., Ciabattoni, G., Consoli, A., Mezzetti, A., Falco, A., Santarone, S., Pennese, E., Vitacolonna, E., Bucciarelli, T., Costantini, F., Capani, F., and Patrono, C. (1999) In Vivo Formation of 8-Iso-Prostaglandin $F_{2\alpha}$ and Platelet Activation in Diabetes Mellitus: Effects of Improved Metabolic Control and Vitamin E Supplementation, *Circulation 99*, 224–229.

27. Morrow, J.D., Frei, B., Longmire, A.W., Gaziano, J.M., Lynch, S.M., Shyr, Y., Strauss, W.E., Oates, J.A., and Roberts, L.J., 2nd. (1995) Increase in Circulating Products of Lipid Peroxidation (F_2-Isoprostanes) in Smokers. Smoking as a Cause of Oxidative Damage, *N. Engl. J. Med. 332*, 1198–1203.

28. Gniwotta, C., Morrow, J.D., Roberts, L.J.I., and Kühn, H. (1997) Prostaglandin F_2-Like Compounds, F_2-Isoprostanes, Are Present in Increased Amounts in Human Atherosclerotic Lesions, *Arterioscler. Thromb. Vasc. Biol. 17*, 3236–3241.

29. Pratico, D., Tangirala, R.K., Radar, D., Rokach, J., and FitzGerald, G.A. (1998) Vitamin E Suppresses Isoprostane Generation *In Vivo* and Reduces Atherosclerosis in ApoE-Deficient Mice, *Nat. Med. 4*, 1189–1192.

30. Meagher, E.A., Barry, O.P., Lawson, J.A., Rokach, J., and FitzGerald, G. (2001) Effects of Vitamin E on Lipid Peroxidation in Healthy Persons, *J. Am. Med. Assoc. 285*, 1178–1182.

31. Reilly, M., Delanty, N., Lawson, J.A., and FitzGerald, G.A. (1996) Modulation of Oxidant Stress In Vivo in Chronic Cigarette Smokers, *Circulation 94*, 19–25.

32. Patrignani, P., Panara, M.R., Tacconelli, S., Seta, F., Bucciarelli, T., Ciabattoni, G., Alessandrini, P., Mezzetti, A., Santini, G., Sciulli, M.G., Cipollone, F., Davi, G., Gallina, P., Bon, G.B., and Patrono, C. (2000) Effects of Vitamin E Supplementation on F_2-Isoprostane and Thromboxane Biosynthesis in Healthy Cigarette Smokers, *Circulation 102*, 539–545.

33. The α-Tocopherol, β-Carotene Cancer Prevention Group (1994) The Effect of Vitamin E and Beta Carotene on the Incidence of Lung Cancer and Other Cancers in Male Smokers, *N. Engl. J. Med. 330*, 1029–1035.

34. Rapola, J.M., Virtamo, J., Ripatti, S., Huttunen, J.K., Albanes, D., Taylor, P.R., and Heinonen, O.P. (1997) Randomised Trial of α-Tocopherol and Beta-Carotene Supplements on Incidence of Major Coronary Events in Men with Previous Myocardial Infarction, *Lancet 349*, 1715–1720.

35. Virtamo, J., Rapola, J.M., Ripatti, S., Heinonen, O.P., Taylor, P.R., Albanes, D., and Huttunen, J.K. (1998) Effect of Vitamin E and Beta Carotene on the Incidence of Primary Nonfatal Myocardial Infarction and Fatal Coronary Heart Disease, *Arch. Intern. Med. 158*, 668–675.

36. Salonen, J.T., Nyyssonen, K., Salonen, R., Lakka, H.M., Kaikkonen, J., Porkkala-Sarataho, E., Voutilainen, S., Lakka, T.A., Rissanen, T., Leskinen, L., Tuomainen, T.P.,

Valkonen, V.P., Ristonmaa, U., and Poulsen, H.E. (2000) Antioxidant Supplementation in Atherosclerosis Prevention (ASAP) Study: A Randomized Trial of the Effect of Vitamins E and C on 3-Year Progression of Carotid Atherosclerosis, *J. Intern. Med. 248*, 377–386.

37. (PPP) Primary Prevention, and Project (2001) Low-Dose Aspirin and Vitamin E in People at Cardiovascular Risk: A Randomised Trial in General Practice, *Lancet 357*, 89–95.

38. Stephens, N.G., Parsons, A., Schofield, P.M., Kelly, F., Cheeseman, K., Mitchinson, M.J., and Brown, M.J. (1996) Randomised Controlled Trial of Vitamin E in Patients with Coronary Disease: Cambridge Heart Antioxidant Study (CHAOS), *Lancet 347*, 781–786.

39. Boaz, M., Smetana, S., Weinstein, T., Matas, Z., Gafter, U., Iaina, A., Knecht, A., Weissgarten, Y., Brunner, D., Fainaru, M., and Green, M.S. (2000) Secondary Prevention with Antioxidants of Cardiovascular Disease in Endstage Renal Disease (SPACE): Randomised Placebo-Controlled Trial, *Lancet 356*, 1213–1218.

40. GISSI Prevenzione (1999) Dietary Supplementation with n-3 Polyunsaturated Fatty Acids and Vitamin E After Myocardial Infarction: Results of the GISSI-Prevenzione Trial, *Lancet 354*, 447–455.

41. Yusuf, S., Dagenais, G., Pogue, J., Bosch, J., and Sleight, P. (2000) Vitamin E Supplementation and Cardiovascular Events in High-Risk Patients. The Heart Outcomes Prevention Evaluation Study Investigators, *N. Engl. J. Med. 342*, 154–160.

42. Yusuf, S., Sleight, P., Pogue, J., Bosch, J., Davies, R., and Dagenais, G. (2000) Effects of an Angiotensin-Converting Enzyme Inhibitor, Ramipril, on Cardiovascular Events in High-Risk Patients. The Heart Outcomes Prevention Evaluation Study Investigators, *N. Engl. J. Med. 342*, 145–153.

43. Lonn, E.M., Yusuf, S., Dzavik, V., Doris, C.I., Yi, Q., Smith, S., Moore-Cox, A., Bosch, J., Riley, W.A., and Teo, K.K. (2001) Effects of Ramipril and Vitamin E on Atherosclerosis: The Study to Evaluate Carotid Ultrasound Changes in Patients Treated with Ramipril and Vitamin E (SECURE), *Circulation 103*, 919–925.

44. Brown, B.G., Zhao, X.Q., Chait, A., Fisher, L.D., Cheung, M.C., Morse, J.S., Dowdy, A.A., Marino, E.K., Bolson, E.L., Alaupovic, P., Frohlich, J., and Albers, J.J. (2001) Simvastatin and Niacin, Antioxidant Vitamins, or the Combination for the Prevention of Coronary Disease, *N. Engl. J. Med. 345*, 1583–1592.

45. Niu, X., Zammit, V., Upston, J.M., Dean, R.T., and Stocker, R. (1999) Co-Existence of Oxidized Lipids and α-Tocopherol in All Lipoprotein Fractions Isolated from Advanced Human Atherosclerotic Plaques, *Arterioscler. Thromb. Vasc. Biol. 19*, 1708–1718.

46. Terentis, A.C., Thomas, S.R., Burr, J.A., Liebler, D.C., and Stocker, R. (2002) Vitamin E Oxidation in Human Atherosclerotic Lesions, *Circ. Res. 90*, 333–339.

47. Upston, J.M., Niu, X., Brown, A.J., Mashima, R., Wang, H., Senthilmohan, R., Kettle, A.J., Dean, R.T., and Stocker, R. (2002) Disease Stage-Dependent Accumulation of Lipid and Protein Oxidation Products in Human Atherosclerosis, *Am. J. Pathol. 160*, 701–710.

48. Liebler, D.C., Burr, J.A., Philips, L., and Ham, A.J. (1996) Gas Chromatography-Mass Spectrometry Analysis of Vitamin E and Its Oxidation Products, *Anal. Biochem. 236*, 27–34.

49. Stocker, R. (1999) The Ambivalence of Vitamin E in Atherogenesis, *Trends Biochem. Sci. 24*, 219–223.

50. Porter, N.A., and Wujek, D.G. (1984) Autoxidation of Polyunsaturated Fatty Acids, an Expanded Mechanistic Study, *J. Am. Chem. Soc. 106*, 2626–2629.

51. Kenar, J.A., Havrilla, C.M., Porter, N.A., Guyton, J.R., Brown, S.A., Klemp, K.F., and Selinger, E. (1996) Identification and Quantification of the Regioisomeric Cholesteryl

Linoleate Hydroperoxides in Oxidized Human Low Density Lipoprotein and High Density Lipoprotein, *Chem. Res. Toxicol. 9*, 737–744.

52. Upston, J.M., Terentis, A.C., Morris, K., Keaney, J.F., Jr., and Stocker, R. (2002) Oxidized Lipid Accumulates in the Presence of α-Tocopherol in Atherosclerosis, *Biochem. J. 363*, 753–760.

53. Witting, P.K., Pettersson, K., Östlund-Lindqvist, A.-M., Westerlund, C., Westin Eriksson, A., and Stocker, R. (1999) Inhibition by a Co-Antioxidant of Aortic Lipoprotein Lipid Peroxidation and Atherosclerosis in Apolipoprotein E and Low Density Lipoprotein Receptor Gene Double Knockout Mice, *FASEB J. 13*, 667–675.

54. Witting, P.K., Pettersson, K., Letters, J., and Stocker, R. (2000) Anti-Atherogenic Effect of Coenzyme Q_{10} in Apolipoprotein E Gene Knockout Mice, *Free Radic. Biol. Med. 29*, 295–305.

55. Thomas, S.R., Leichtweis, S.B., Pettersson, K., Croft, K.D., Mori, T.A., Brown, A.J., and Stocker, R. (2001) Dietary Co-Supplementation with Vitamin E and Coenzyme Q_{10} Inhibits Atherosclerosis in Apolipoprotein E Gene Knockout Mice, *Arterioscler. Thromb. Vasc. Biol. 21*, 585–593.

56. Witting, P.K., Pettersson, K., Östlund-Lindqvist, A.-M., Westerlund, C., Wågberg, M., and Stocker, R. (1999) Dissociation of Atherogenesis from Aortic Accumulation of Lipid Hydro(Pero)Xides in Watanabe Heritable Hyperlipidemic Rabbits, *J. Clin. Investig. 104*, 213–220.

57. Upston, J.M., Witting, P.K., Brown, A.J., Stocker, R., and Keaney, J.F., Jr. (2001) Effect of Vitamin E on Aortic Lipid Oxidation and Intimal Proliferation After Vascular Injury in Cholesterol-Fed Rabbits, *Free Radic. Biol. Med. 31*, 1245–1253.

58. Witting, P.K., Pettersson, K., Letters, J., and Stocker, R. (2000) Site-Specific Anti-Atherogenic Effect of Probucol in Apolipoprotein E Deficient Mice, *Arterioscler. Thromb. Vasc. Biol. 20*, e26-e33.

59. Zhang, S.H., Reddick, R.L., Avdievich, E., Surles, L.K., Jones, R.G., Reynolds, J.B., Quarfordt, S.H., and Maeda, N. (1997) Paradoxical Enhancement of Atherosclerosis by Probucol Treatment in Apolipoprotein E-Deficient Mice, *J. Clin. Investig. 99*, 2858–2866.

60. Kritharides, L., and Stocker, R. (2002) The Use of Antioxidant Supplements in Coronary Heart Disease, *Atherosclerosis 164*, 211–219.

61. de Lorgeril, M., Salen, P., Martin, J.L., Monjaud, I., Delaye, J., and Mamelle, N. (1999) Mediterranean Diet, Traditional Risk Factors, and the Rate of Cardiovascular Complications After Myocardial Infarction: Final Report of the Lyon Diet Heart Study, *Circulation 99*, 779–785.

62. Mashima, R., Witting, P.K., and Stocker, R. (2001) Oxidants and Antioxidants in Atherosclerosis, *Curr. Opin. Lipidol. 12*, 411–418.

63. Hingorani, A.D., Liang, C.F., Fatibene, J., Lyon, A., Monteith, S., Parsons, A., Haydock, S., Hopper, R.V., Stephens, N.G., O'Shaughnessy, K.M., and Brown, M.J. (1999) A Common Variant of the Endothelial Nitric Oxide Synthase (Glu[298]→Asp) Is a Major Risk Factor for Coronary Artery Disease in the UK, *Circulation 100*, 1515–1520.

64. Fang, J.C., Kinlay, S., Beltrame, J., Hikiti, H., Wainstein, M., Behrendt, D., Suh, J., Frei, B., Mudge, G.H., Selwyn, A.P., and Ganz, P. (2002) Effect of Vitamins C and E on Progression of Transplant-Associated Arteriosclerosis: A Randomised Trial, *Lancet 359*, 1108–1113.

Chapter 18

Prospective Vitamin E Clinical Trials

Ishwarlal Jialal and Sridevi Devaraj

Center for Human Nutrition and Division of Clinical Biochemistry and Human Metabolism, The University of Texas Southwestern Medical Center at Dallas, TX

Introduction

Cardiovascular disease (CVD) is the leading cause of morbidity and mortality in the Western world. Several lines of evidence support a role for oxidative stress and inflammation in the pathogenesis of atherosclerosis. Furthermore, epidemiologic studies appear to suggest that low levels of α-tocopherol (AT) are associated with increased risk for CVD, and increased intakes appear to be protective (1,2). Studies *in vitro* have shown that AT, in addition to functioning as an antioxidant, inhibits smooth muscle cell proliferation, platelet adhesion and aggregation, and monocyte endothelial adhesion (3,4). In addition, some studies in animal models have shown a decrease in lesion progression with supplementation. Supplementation with AT in humans has been shown by numerous groups to result in the following effects: decreased lipid peroxidation [decreased low density lipoprotein (LDL) oxidative susceptibility and decreased F_2-isoprostanes, a measure of *in vivo* lipid peroxidation], decreased platelet adhesion and aggregation, and an anti-inflammatory effect (decreased C-reactive protein, proinflammatory cytokines, and soluble cell adhesion molecules) (1–4).

Clinical Trials

In this review, we will focus on the larger prospective clinical trials that have tested the effect of AT supplementation on cardiovascular events in different populations. The seven clinical trials that will be discussed are as follows: (i) the ATBC Study, (ii) the CHAOS Study, (iii) the GISSI study, (iv) the HOPE Study, (v) the SPACE Study, (vi) the ASAP Study, and (vii) the PPP study. Although the HATS study is not a true vitamin E trial, it will be discussed because it has created much confusion.

The Alpha-Tocopherol Beta-Carotene (ATBC) Cancer Prevention Study

The ATBC Cancer Prevention Study (5) was designed to determine whether vitamin E (synthetic, all-*rac*-α-tocopherol acetate, 50 mg/d, 50 ɪᴜ/d) and β-carotene (20 mg/d), either alone or in combination, would decrease lung cancer incidence. A total of 29,133 male smokers age 50–69 y from southwestern Finland were randomly assigned to one of the three regimens or placebo and followed for 5–8 y. At entry, the mean age was 57.2 y; the men smoked 20.4 cigarettes/d and had smoked for 35.9 y. Although AT supplementation had no effect on the primary end point (lung cancer), an 18% increase in lung cancer incidence was observed in the β-carotene–supplemented group ($P = 0.01$).

Serum AT and β-carotene concentrations were measured before and after supplementation. The median values of serum AT at baseline were 26.7 μmol/L and increased to 40.2 μmol/L after 3 y of supplementation. A similar dose of 50 mg synthetic α-tocopheryl acetate supplements was administered to nonsmoking men and women (aged 21–31 y) by Princen et al. (6). They reported that plasma AT concentrations were 24 ± 3.6 μmol/L at baseline and increased 4 μmol/L with supplementation to 28.7 ± 5.1 μmol/L, in sharp contrast to the 14 μmol/L increase observed in the ATBC study. It is likely that the greater increase in serum AT observed in the ATBC study resulted from higher lipid levels in the older subjects because plasma vitamin E is confined to the lipoprotein fraction (7). It is unknown how delivery of AT to tissues is affected when plasma AT concentrations are elevated as a result of lipid levels compared with elevations resulting from higher intakes of AT. Certainly, reporting measurements of the plasma ratios of AT:cholesterol could aid in the interpretation of vitamin E status (8).

In this study in smokers without preexisting CVD, AT therapy had no significant effect on the first major coronary event (fatal and nonfatal) (9). In a further analysis of the ATBC study in male smokers with previous myocardial infarction (MI), although there were no significant effects on the number of major coronary events or fatal coronary artery disease, there was a significant reduction in the multivariate-adjusted relative risk (RR) for nonfatal coronary artery disease [0.62; confidence interval (CI), 0.41–0.96] in the AT group (10). In a subsequent report of the ATBC study, Rapola et al. (11) also showed that AT supplementation was associated with a minor decrease in the incidence of angina pectoris (RR, 0.91; CI, 0.83–0.99; $P = 0.04$).

The incidence and mortality from stroke in the ATBC trial have been examined in detail (12,13). Because AT is carried in lipoproteins, its relationship with serum lipids confounds interpretation. For example, the risk of cerebral infarction was increased in those with serum total cholesterol concentrations >7.0 mmol/L. However, pretrial high serum AT, which is dependent upon serum lipid levels, decreased the risk of intracerebral hemorrhage by half and cerebral infarction by one third (10). AT supplementation appeared to increase the risk of subarachnoid hemorrhage by 50% (95% CI, –3 to 132%; $P = 0.07$), but decreased cerebral infarction 14% (95% CI, –25 to –1%; $P = 0.03$). The increase in mortality caused by subarachnoid hemorrhage with AT supplements was 181% (95% CI, 37–479%; $P = 0.01$). The overall net effects of supplementation on the incidence and mortality from total stroke were not significant. The interpretation that AT supplements increase the incidence of hemorrhagic stroke is not uniformly accepted because this adverse effect was not observed in the other intervention trials with vitamin E (see below). In fact, Steiner et al. (14) showed in a double-blind, randomized study of 100 patients with transient ischemic attacks (TIA), minor strokes, or residual neurologic deficits that the patients who received AT and the antiplatelet agent, aspirin, (400 ɪu/d and 325 mg/d, respectively) had a significant reduction in ischemic strokes and recurrent TIA compared with patients taking aspirin alone. Moreover, no increase in hemorrhagic strokes was observed in a study that was designed to evaluate neurological function in patients with Alzheimer's disease consuming 2000 ɪu/d of supplemental all-*rac*-AT for 2 y (15). However, the number of subjects and the trial length may be insufficient to detect an effect. Nevertheless, because AT has antiplatelet effects (16) that may promote bleeding, the observation that AT supplementation increases hemorrhagic stroke incidence in smokers should be considered with caution.

The Cambridge Heart Antioxidant Study (CHAOS)

The CHAOS study was a prospective, randomized, placebo-controlled, double-blind single-center trial in the East Anglian region of England that examined the effects of AT therapy on coronary artery disease (CAD) (17). A total of 2002 subjects with overt clinical and angiographic evidence of CAD were randomly assigned to receive natural or *RRR*-AT (n = 1035) or placebo (n = 967). The first 546 subjects in the AT group were given 800 ɪu/d for a median of 731 d (range: 3–981) and the remainder were given 400 ɪu/d for 366 d (range: 8–961); however, the two groups were combined for statistical analysis (the trial was not designed to determine dose-response effects of AT on the primary end points). Participants requested 73.2% of all prescribed AT or placebo as follow-up medications. Treatment with AT was well tolerated with only 11 of the 2002 patients (0.55%) discontinuing therapy due to diarrhea, dyspepsia, or rash. There was no significant difference between the treatment groups for these side effects. Both 400 and 800 ɪu/d of AT significantly increased serum AT levels at least twofold over baseline as reported previously in the literature (18). Importantly, in the placebo group, there was no significant increase in serum AT levels during follow-up. The primary outcome variables were a combined end point of cardiovascular death and nonfatal MI, and nonfatal MI alone (17). After a median follow-up of 510 d (range: 3–981), those receiving AT experienced a significant 47% reduction (95% CI, –66 to –17%; *P* = 0.005) in CAD death and nonfatal MI, which was the primary trial end point (17). This effect was due to a significant 77% reduction (95% CI, –89 to –53%; *P* = 0.005) in the risk for nonfatal MI. There was a nonsignificant effect on CAD death alone (*P* = 0.78) or total mortality (*P* = 0.31). The nonsignificant increase in deaths due to CAD in this trial was subjected to a subsequent analysis (19). With regard to total mortality, there were now 120 deaths from all causes, 68 in the AT group of 1035 and 52 in the placebo group of 967 patients (cardiovascular deaths 53 vs. 44, *P* = 0.48). This further analysis of the deaths due to heart disease revealed that the majority of deaths (78%) occurred in patients who were noncompliant with AT therapy. This subsequent analysis lessens concern about possible dangers of the use of AT in patients with established CAD.

The salient characteristics of the CHAOS trial are as follows: the effect was examined in a homogenous and stable population with established CAD consuming an English diet; the CHAOS trial used a dose of natural AT (≥ 400 IU/d) that has been shown to decrease LDL oxidizability and platelet aggregation; compliance was assessed by both drug accountability and serum levels of AT that rose 2- to 2.5-fold in the AT group but were essentially unchanged in the placebo group; despite the larger dose used compared with the ATBC study, AT supplementation in CHAOS was not associated with an increased risk for hemorrhagic stroke even though these patients were also receiving antiplatelet therapy.

GISSI (Gruppo Italiano per lo Studio della Supravivenza nell'Infarto Miocardico) Prevenzione Trial

The GISSI trial was conducted in Italy in 11,324 patients within 3 mo of a MI; it investigated whether n-3 polyunsaturated fatty acids (PUFA) (1 g/d, n = 2836), all-*rac*-AT (300 mg daily or 330 IU/d, n = 2830), a combination of n-3 PUFA and AT (n = 2830) or control (n = 2828) over 3.5 y had an effect on the primary combined efficacy end point of death, MI, and stroke (20). This was a multicenter study with an open-label design. The patients received, in addition to the supplements, the usual preventative measures including aspirin, β-blockers, and angiotensin-converting enzyme (ACE) inhibitors. A major strength of this study, in contrast to the other clinical trials, was that dietary information was detailed for fish, fruit, vegetable, and olive oil intake. The primary combined efficacy end points were the cumulative rate of all-cause death, nonfatal MI, and nonfatal stroke and the cumulative rate of cardiovascular death, nonfatal MI, and nonfatal stroke.

In this study, n-3 PUFA resulted in a significant 10% decrease in the combined primary end point of death, nonfatal MI, and nonfatal stroke in a two-way factorial analysis ($P = 0.048$). However, the decrease in the risk of other combined end points of cardiovascular death, nonfatal MI, and nonfatal stroke tended to be significant ($P = 0.053$). The four-way analysis, which compared the n-3 PUFA group with the unsupplemented control group, provided a clearer picture of the effects of n-3 PUFA with a relative decrease in the risk of the combined end point of 15% ($P = 0.023$) and for cardiovascular death, nonfatal MI and nonfatal stroke of 20% ($P = 0.008$). For the primary end points, patients receiving AT did not differ significantly from controls when analyzed according to the two-way factorial analysis. In addition, the combination of AT and n-3 PUFA did not confer greater benefit over n-3 PUFA alone. However, when the more appropriate four-way analysis was conducted, there was a significant 20% reduction in cardiovascular death in the AT group compared with the unsupplemented control group (CI, 0.65–0.99) (21).

A weakness of this study was that it was an open-label trial with an ~25% dropout at the end of the study. Additionally, no objective assessment of compliance such as measurement of n-3 PUFA or AT was provided, even in a subgroup of participants. Furthermore, because these patients were consuming a healthy Mediterranean diet, this could have also attenuated the benefits of AT. This is clearly a diet rich in antioxidants. However, it should be emphasized that the GISSI Prevenzione trial demonstrated a significant 20% reduction in cardiovascular death, i.e., RR for cardiac death = 0.77 (0.61–0.97), coronary death = 0.75 (0.59–0.96), and sudden death = 0.65 (0.48–0.89) after AT supplementation.

The Heart Outcomes Prevention Evaluation (HOPE) Study

In the HOPE Study, 2545 women and 6996 men ≥55 y of age who were at high risk for cardiovascular events because they had CVD or diabetes in addition to one other risk factor were enrolled (22). These patients were randomly assigned according to a 2 × 2 factorial design to receive either 400 IU of vitamin E daily from natural sources or matching placebo and either an ACE inhibitor (ramipril) or matching placebo. They were followed up for a period of 4–6 y (mean 4.5 y). The primary outcome was a composite end point of MI, stroke, and death from CVD. Secondary outcomes included unstable angina, congestive heart failure, revascularization or amputation, death from any cause, complications of diabetes, and cancer.

There was no significant difference between the patients who received vitamin E (n = 4761) and those receiving placebo (n = 4780) with respect to the primary end point. In addition, there was no significant difference in number of deaths from cardiovascular causes. Furthermore, there was no significant difference in incidence of secondary cardiovascular outcomes or in death from any cause. Also, there was no increase in hemorrhagic stroke associated with vitamin E use even though 77% of the patients were receiving an antiplatelet agent. Finally, there were no significant adverse effects from vitamin E supplementation in this study. The authors concluded in this study that in patients at high risk for cardiovascular events, treatment with vitamin E for a mean of 4.5 y had no apparent effect on cardiovascular outcomes. It should be emphasized that because of the overwhelming positive findings for the ACE inhibitor, ramipril, this study was stopped by the Data and Safety Monitoring Board before their revised recommendation of a 3.5-y follow-up instead of the 5 y originally planned.

This is an important study that arrives at the negative conclusion that vitamin E is without effect in patients at high risk for CVD. However, this study suffers from certain deficiencies (23). Although the study was undertaken in many geographic areas (United States, Canada, Western Europe, and South America) with clearly different dietary intakes, data on the dietary intakes, especially antioxidants, were not reported. In addition, for no subgroup were plasma levels of vitamin E provided, as in the CHAOS Study, to confirm supplementation. Furthermore, the HOPE investigators appear to have used natural vitamin E, which is made up of tocopherols and tocotrienols. Because AT is the most potent member of the vitamin E family, this could also have a bearing on the findings due to the scanty information on the other forms of vitamin E. In addition, 400 IU/d of AT does not appear to have anti-inflammatory effects (24).

Secondary Prevention with Antioxidants of CVD in End-Stage Renal Disease (SPACE)

The SPACE study was a double-blind, placebo-controlled, randomized, secondary prevention trial performed at six hemodialysis units in Israel that examined the effect of high-dose AT supplementation on CVD outcomes in hemodialysis patients with preexisting CVD. Hemodialysis patients with preexisting CVD (n = 196) aged 40–75 y at baseline from six dialysis centers were enrolled and randomly assigned to receive 800 IU/d *RRR*-AT (n = 97) or matching placebo (n = 99). Patients were followed for a median of 519 d. The primary end point was a composite variable consisting of MI (fatal and nonfatal), ischemic stroke, peripheral vascular disease (excluding the arteriovenous fistula), and unstable angina. Secondary outcomes included each of the component outcomes, total mortality, and CVD mortality. Lipid-adjusted AT levels were monitored; they rose from 22.0 ± 7.7 μmol/L in the AT group to 27.8 ± 9.3 μmol/L on-treatment and were unchanged in the placebo group (23.3 ± 10.7 μmol/L at baseline and 20.2 ± 6.9 μmol/L on-treatment). Treatment with AT significantly decreased primary cardiovascular end points (54% reduction in primary end point risk in the AT group; $P = 0.014$). There was a 39% nonsignificant reduction in CAD mortality (RR, 0.61; 95% CI, 0.28–1.3; $P = 0.25$). Also, AT supplementation was associated with a 70% reduction in total MI rate ($P = 0.016$). Furthermore, treatment with AT was associated with a 62% reduction in peripheral vascular disease but was not significant (RR, 0.38; 95% CI, 0.1–1.4; $P = 0.13$). There were no significant differences between the number of side effects reported for the placebo and AT groups. Thus, like CHAOS, the SPACE study also reported a significant reduction in composite CVD end points and MI with AT supplementation in patients with preexisting CVD. In addition, plasma AT levels were measured in this study. Thus, it appears that higher doses of AT (800 IU/d) would be beneficial for secondary prevention of CAD because this dose exerts both antioxidant and anti-inflammatory effects.

Collaborative Group for the Primary Prevention Project (PPP)

In the Primary Prevention Project (PPP), the investigators followed 4495 people with hypertension, hypercholesterolemia, diabetes, obesity, family history of premature MI, or those who were elderly (25). The mean age of the patients was 64.4 y and 58% were women. The patients were prescribed either aspirin (100 mg/d) or all-*rac*-AT (300 mg/d). This was a 2 × 2 factorial design study with a mean follow-up period of 3.6 y. The primary end point of this study was cardiovascular death, nonfatal MI, and nonfatal stroke. In this study, aspirin lowered the frequency of all end points, and was significant for cardiovascular death with a RR of 0.56 (0.31–0.99) and total cardiovascular events 0.77 (0.62–0.95). Also, severe bleeding was more frequent in the aspirin group than in the nonaspirin group (1.1 vs. 0.3%; $P = 0.0008$). However, AT supplementation had no benefit on the primary end point, i.e., RR 1.07 (0.74–1.56), total cardiovascular events or disease 0.94 (0.77–1.16). The investigators reported a benefit of AT therapy on peripheral artery disease, i.e., RR 0.54 (0.3–0.99). Thus, this study reported a 46% reduction in the incidence of peripheral artery disease among patients taking vitamin E ($P = 0.043$). Like the other clinical trials, this study suffers from certain weaknesses. There was no objective measure of compliance, i.e., measurement of plasma AT or biomarkers of oxidative stress and inflammation. Second, as the authors themselves point out, the findings for vitamin E could be regarded as a false-negative result because of the inadequate power of a prematurely interrupted trial. Thus, this trial, although yielding a negative result on the primary end point, does yield a benefit of AT therapy that was underemphasized by the investigators in their paper. Furthermore, the authors used a synthetic form of AT that does not appear to yield a benefit for markers of inflammation in supplementation studies, although it decreased LDL oxidation when supplemented (24).

Antioxidant Supplementation in Atherosclerosis Prevention Study (ASAP Study)

The ASAP study was a randomized trial of the effect of vitamin E and C on 3-y progression of carotid atherosclerosis. This was a placebo-controlled, randomized, 2 × 2 factorial trial in hypercholesterolemic patients and consisted of an 8-wk placebo lead-in phase, followed by a 3-y double-blind treatment period (26). Subjects (n = 520) were randomly assigned to receive *RRR*-AT (136 IU twice a day), slow release ascorbate (250 mg twice a day), or both *RRR*-AT and ascorbate or placebo. Carotid atherosclerosis was assessed by quantitating common carotid intima-media thickness (IMT) over semiannual assessments. Plasma levels of vitamins E and C were measured and were significantly increased in the groups randomly assigned to the respective vitamins. The average increase in common carotid arteries (CCA)-IMT was 0.02 mm/y in men randomly assigned to placebo, 0.018/y in men who received only vitamin E, 0.017 mm/y in men who received only vitamin C, and 0.011 mm/y in men who received both vitamins E and C ($P = 0.043$ for heterogeneity). IMT progression was significantly reduced in men randomly assigned to both vitamins compared with all other men ($P = 0.009$) or men who received placebo ($P = 0.008$). No significant differences were observed in women. Covariate-adjusted IMT increase in men was 45% less with both vitamins compared with placebo ($P = 0.049$); the largest treatment effect was in smoking men (64% less) vs. nonsmoking men (30% less).

The findings of the study suggest that antioxidant vitamin (vitamins E and C) supplementation significantly retards carotid atherosclerosis progression in men, especially those who are at increased oxidative stress and have insufficient dietary antioxidant status (smokers). Both supplements were safe, and the bioavailability of the supplements was good. The authors suggested that the lack of beneficial effect in women is due to the higher baseline levels of vitamin C in the population studied and insufficient statistical power in women, who had smaller baseline carotid wall thickness and less atherosclerotic progression during the study.

In a study cohort, lipid peroxidation measurements were carried out in 100 consecutive men at entry and repeated at 12 mo. The plasma F_2-isoprostane concentration was lowered by 17.3% (95% CI, 3.9–30.8%) in the vitamin E group ($P = 0.006$ for the change, compared with the placebo group). On the contrary, vitamin C had no significant effect on plasma F_2-isoprostanes compared with the placebo group. There was also no interaction in the effect between these vitamins.

The strengths of this study were that plasma levels of the vitamins were measured and significantly increased after supplementation, and biomarkers of oxidative stress were measured in a cohort. However, biomarkers of inflammation were not studied; it is possible that this dose may not have been sufficient and that the duration of the study was too short to observe significant effects on atherosclerosis.

HDL-Atherosclerosis Treatment Study (HATS)

In this 3-y, double blind study, 160 patients with CAD and low HDL-cholesterol (women and men, values <40 and 35 mg/dL, respectively) were entered into the study (27). Coronary artery disease was defined as previous MI, coronary interventions, or confirmed angina and with at least 3 stenoses of at least 30% of the luminal diameter or 1 stenosis of at least 50%. The four groups included a placebo group, a group that received simvastatin and niacin, an antioxidant group, and a simvastatin, niacin, and antioxidant group. The antioxidant supplement that was given twice daily resulted in doses of 800 IU/d of *RRR*-AT, 1000 mg/d of vitamin C, 25 mg/d of β-carotene, and 100 μg/d of selenium. The patients were followed up for 3 y and the end point was arteriographic evidence of a change in coronary stenosis and occurrence of a first cardiovascular event (death, MI, stroke, and revascularization). The average age of the patients was 53 y, and 13% were women. The antioxidant supplementation resulted in significant increases in plasma levels of AT, ascorbate, and β-carotene, with the most profound increase in the β-carotene levels of 380%. In addition, the antioxidant cocktail resulted in a significant 35% prolongation in the lag phase of LDL oxidation ($P <$ 0.001). For the mean change in the percentage of stenosis, compared with placebo, simvastatin and niacin, and the combination of simvastatin, niacin, and antioxidants resulted in a significant decrease, i.e., placebo = 3.9% increase, simvastatin and niacin = 0.4% decrease, simvastatin, niacin, and antioxidants = 0.7% increase. Also, compared with placebo, the antioxidant cocktail resulted in ~50% reduction in the percentage of stenosis (1.9%); however, this was not significant, $P = 0.16$. It should be emphasized that this study had a small number of patients in each of the four groups: placebo, n = 34; antioxidants, n = 39; simvastatin and niacin, n = 33; simvastatin, niacin, and antioxidants, n = 40. When the end point or mean change in minimal lumen diameter was examined, there was a significant reduction with all therapies, simvastatin and niacin, simvastatin, niacin, and antioxidants, and antioxidants alone. This included the change in minimal luminal diameter of the nine proximal lesions or all lesions.

Although this is an important study that clearly emphasizes the benefit in a low HDL-cholesterol group with coronary artery disease of the combined therapy of simvastatin and niacin, the antioxidant limb of the study suffers from a small sample size (n = 39). Also, it should be emphasized that in this study, an antioxidant cocktail was given; thus, the benefit of a high-dose AT supplementation alone was not tested. In addition, it is unclear whether the high dose of vitamin C that was given to a population whose dietary intake of vitamin C (110 mg/d) exceeded the RDA could have had beneficial or deleterious effects. In addition, in the antioxidant group, there was a significant increase in weight and triglycerides, which the authors dismissed in the study. Although the authors ascribe the lesser benefit of the simvastatin, niacin, and antioxidant group compared with the simvastatin and niacin group to the lowering of HDL_2-cholesterol, it should be emphasized in this study that although the HDL_2-cholesterol levels were reduced from 3.9 to 3.1 mg/dL in the antioxidant group, there was a significant increase in apolipoprotein A-1, the predominant apolipoprotein of HDL, from 110 to 116 mg/dL ($P < 0.01$). Finally, one cannot extrapolate the results of the findings in this small study in patients with coronary disease with low HDL-cholesterol to other populations including primary

TABLE 18.1

Summary of Prospective Vitamin E Clinical Trials[a,b]

Study	Dose	Primary end point	Other end points	Adverse effect
ATBC (Primary and secondary prevention)	50 iu/d all-*rac*-AT	↔ (Cancer)	↑[c]	↓[d]
CHAOS (Secondary prevention)	400 and 800 iu/d *RRR*-AT	↑[e]	↔	↔
GISSI (Secondary prevention)	330 iu/d all-*rac*-AT	↔	↑[f]	↔
HOPE (Primary and secondary prevention)	400 iu/d Natural source vitamin E	↔	↔	↔
SPACE (Secondary prevention)	800 iu/d *RRR*-AT	↑[g]	↔	↔
PPP (Primary prevention)	330 iu/d all-*rac*-AT	↔	↑[h]	↔
ASAP (Primary prevention)	272 iu/d *RRR*-AT	↔[i]	↔	↔

[a]↑ denotes positive finding; ↔ denotes no effect; ↓ denotes negative finding.
[b]Abbreviations: AT, α-tocopherol; ATBC, α-Tocopherol β-Carotene; CHAOS, Cambridge Heart Antioxidant Study; GISSI, Gruppo Italiano per lo Studio della Supravivenza nell'Infarto Miocardico; HOPE, Heart Outcomes Prevention Evaluation; SPACE, Secondary Prevention with Antioxidants of CVD in End-Stage Renal Disease; PPP, Primary Prevention Project; ASAP, Antioxidant Supplementation in Atherosclerosis Prevention Study.
[c]Decrease in the incidence of angina, nonfatal coronary artery disease in patients with previous myocardial infarction (MI) and reduction in cerebral infarction.
[d]Increase in subarachnoid hemorrhage mortality.
[e]Cardiovascular death and nonfatal MI.
[f]Reduction in cardiovascular death.
[g]Composite cardiovascular disease end point and MI.
[h]Peripheral arterial disease.
[i]Benefit on carotid intima-media thickness seen with the combination of AT (272 iu/d) + vitamin C (500 mg/d) in men.

prevention; the benefit that other antioxidant therapies such as high-dose AT might have in other patients with CVD and also in patients with diabetes, or smokers, for example, has not been ruled out.

Conclusion

In this review, we have critically appraised the major prospective clinical trials investigating the effect of AT supplementation on cardiovascular end points. Although two studies (CHAOS and SPACE) clearly showed a reduction in both cardiovascular death and nonfatal MI, the defined primary end point, as shown in Table 18.1, the GISSI, ATBC, and PPP studies also demonstrated benefit on certain end points in spite of the lack of significance for the primary end point. It should be emphasized that the primary end point in the ATBC study was cancer and not CVD. The only study that was negative for all end points was the HOPE study. The increase in mortality from subarachnoid hemorrhage in male smokers in the ATBC study does not concur with the other studies, which in fact used higher doses, coupled with antiplatelet agents. This unexpected finding will be settled in ongoing trials (28). Thus, although the data from the prospective clinical trials are not overwhelming to date, the majority of studies appear to suggest a benefit of AT supplementation in patients with CAD.

Studies of this nature could be more informative if they included dietary intakes of antioxidants, measurement of AT levels and antioxidant status, and measurement of biomarkers of oxidative stress such as F_2-isoprostanes, nitrotyrosine, or LDL oxidizability. It is quite possible that vitamin E–replete individuals might not gain additional benefit from supplements. It should also be emphasized that the trials used different doses of AT ranging from 50–400 IU/d on different study populations (primary and secondary prevention). Also, the duration of the studies in patients without CVD must be longer and populations with increased oxidative stress, e.g., end-stage renal disease, CAD, diabetes, or smoking, should be the focus of future studies. In spite of these deficiencies, these studies concur with the general body of evidence (epidemiologic, *in vitro*, and animal models); the totality of evidence would support that AT supplementation is beneficial in patients with preexisting CVD (28).

Acknowledgments

Studies cited were supported by NIH grant K-24AT00596.

References

1. Devaraj, S., and Jialal, I. (2000) Antioxidants and Vitamins to Reduce Cardiovascular Disease, *Curr. Atheroscler. Rept. 2*, 342–351.
2. Kaul, N., Devaraj, S., and Jialal, I. (2001) Alpha Tocopherol and Atherosclerosis, *Exp. Biol. Med. 226*: 5–12.
3. Jialal, I., Devaraj, S., and Kaul, N. (2001) The Effect of Alpha Tocopherol on Monocyte Pro-Atherogenic Activity, Symposium: Molecular Mechanisms of Protective Effects of Vitamin E in Atherosclerosis, *J. Nutr. 131*: 389S–394S.

4. Devaraj, S., and Jialal, I. (1998) The Effects of α-Tocopherol on Critical Cells in Atherogenesis, *Curr. Opin. Lipidol. 9*, 11–15.

5. The Alpha-Tocopherol, Beta-Carotene Cancer Prevention Study Group (1994) The Effect of Vitamin E and Beta Carotene on the Incidence of Lung Cancer and Other Cancers in Male Smokers, *N. Engl. J. Med. 330*, 1029–1035.

6. Princen, H. M., van Duyvenvoorde, W., Buytenhek, R., van der Laarse, A., van Poppel, G., Leuven, J.A.G., and van Hinsbergh, V.W.M. (1995) Supplementation with Low Doses of Vitamin E Protects LDL from Lipid Peroxidation in Men and Women, *Arterioscler. Thromb. Vasc. Biol. 15*, 325–333.

7. Traber, M.G. (1999) Vitamin E, in *Modern Nutrition in Health and Disease* (Shils, M.E., Olson, J.A., Shike, M., and Ross, A.C., eds.) pp. 347–362, Williams & Wilkins, Baltimore.

8. Traber, M.G., and Jialal, I. (2000) Measurement of Lipid-Soluble Vitamins—Further Adjustment Needed? *Lancet 355*, 2013–2014.

9. Virtamo, J., Rapola, J.M., Ripatti, S., Heinonen, O.P., Taylor, P.R., Albanes, D., and Huttunen, J.K. (1998) Effect of Vitamin E and Beta Carotene on the Incidence of Primary Nonfatal Myocardial Infarction and Fatal Coronary Heart Disease, *Arch. Intern. Med. 158*, 668–675.

10. Rapola, J.M., Virtamo, J., Ripatti, S., Huttunen, J.K., Albanes, D., Taylor, P.R., and Heinonen, O.P. (1997) Randomised Trial of Alpha-Tocopherol and Beta-Carotene Supplements on Incidence of Major Coronary Events in Men with Previous Myocardial Infarction, *Lancet 349*, 1715–1720.

11. Rapola, J.M., Virtamo, J., Haukka, J.K., Heinonen, O.P., Albanes, D., Taylor, P.R., Huttunen, J.K. (1996) Effect of Vitamin E and Beta Carotene on the Incidence of Angina Pectoris: a Randomized, Double-Blind, Controlled Trial [see comments], *J. Am. Med. Assoc. 275*, 693–698.

12. Leppala, J.M., Virtamo, J., Fogelholm, R., Albanes, D., and Heinonen, O.P. (1999) Different Risk Factors for Different Stroke Subtypes: Association of Blood Pressure, Cholesterol, and Antioxidants, *Stroke 30*, 2535–2540.

13. Leppala, J.M., Virtamo, J., Fogelholm, R., Huttunen, J.K., Albanes, D., Taylor, P.R., and Heinonen, O.P. (2000) Controlled Trial of Alpha-Tocopherol and Beta-Carotene Supplements on Stroke Incidence and Mortality in Male Smokers, *Arterioscler. Thromb. Vasc. Biol. 20*, 230–235.

14. Steiner, M., Glantz, M., and Lekos, A. (1995) Vitamin E Plus Aspirin Compared with Aspirin Alone in Patients with Transient Ischemic Attacks, *Am. J. Clin. Nutr. 62*, 1381S–1384S.

15. Sano, M., Ernesto, C., Thomas, R.G., Klauber, M.R., Schafer, K., Grundman, M., Woodbury, P., Growdon, J., Cotman, C.W., Pfeiffer, E., Schneider, L.S., and Thal, L.J. (1997) A Controlled Trial of Selegiline, Alpha-Tocopherol, or Both as Treatment for Alzheimer's Disease: the Alzheimer's Disease Cooperative Study, *N. Engl. J. Med. 336*, 1216–1222.

16. Freedman, J.E., Farhat, J.H., Loscalzo, J., Keaney, J.F. (1996) AT Inhibits Aggregation of Human Platelets by a Protein Kinase C Dependent Mechanism, *Circulation 94*, 2434–2440.

17. Stephens, N.G., Parsons, A., Schofield, P.M., Kelly, F., Cheeseman, K., Mitchinson, M.J., and Brown, M.J. (1996) Randomised Controlled Trial of Vitamin E in Patients with Coronary Disease: Cambridge Heart Antioxidant Study (CHAOS), *Lancet 347*, 781–786.

18. Jialal, I., Fuller, C.J., and Huet, B.A. (1995) The Effect of Alpha Tocopherol Supplementation on LDL Oxidation: A Dose Response Study, *Arterioscler. Thromb. 15*, 190–198.

19. Mitchinson, M.J., Stephens, N.G., Parsons, A., Bligh, E., Schofield, P.M., and Brown, M.J. (1999) Mortality in the CHAOS Trial, *Lancet 353*, 381–382.

20. GISSI-Prevenzione Investigators (1999) Dietary Supplementation with n-3 Polyunsaturated Fatty Acids and Vitamin E After Myocardial Infarction: Results of the GISSI-Prevenzione Trial, *Lancet 354*, 447–455.

21. Jialal, I., Devaraj, S., Huet, B.A., and Traber, M. (1999) GISSI-Prevenzione Trial, *Lancet 354*, 1554.

22. Yusuf, S., Dagenais, G., Pogue, J., Bosch, J., and Sleight, P. (2000) Vitamin E Supplementation and Cardiovascular Events in High-Risk Patients, The Heart Outcomes Prevention Evaluation Study Investigators, *N. Engl. J. Med. 342*, 154–160.

23. Jialal, I., and Devaraj, S. (2000) HOPE Study, *N. Engl. J. Med. 342*, 1917.

24. Kaul, N., Devaraj, S., Grundy, S. and Jialal, I. (2001) Failure to Demonstrate a Major Anti-Inflammatory Effect with Alpha Tocopherol Supplementation (400 IU/day), *Am. J. Cardiol. 87*, 1320–1323.

25. Collaborative Group of the Primary Prevention Project (PPP) (2001) Low-Dose Aspirin and Vitamin E in People at Cardiovascular Risk: A Randomised Trial in General Practice, *Lancet 357*, 89–95.

26. Salonen, J.T., Nyyssönen, K., Salonen, R., Lakka, H.-M., Kaikkonen, J., Porkkala-Sarataho, E., Voutilainen, S., Lakka, T.A., Rissanen, T., Leskinen, L., Tuomainen, T.-P., Valkonen, V.-P., Ristonmaa, U., and Poulsen, H.E. (2000) Antioxidant Supplementation in Atherosclerosis Prevention (ASAP) Study: A Randomized Trial of the Effect of Vitamins E and C on 3-Year Progression of Carotid Atherosclerosis, *J. Intern. Med. 248*, 377–386.

27. Brown, B.G., Zhao, X., Chait, A., Fisher, L.D., Cheung, M.C., Morse, J.S., Dowdy, A.A., Marino, E.K., Bolson, E.L., Alaupovic, P., Frohlich, J., and Albers, J.J. (2001) Simvastatin and Niacin, Antioxidant Vitamins, or the Combination for the Prevention of Coronary Disease, *N. Engl. J. Med. 345*, 1583–1592.

28. Pryor, W. (2000) Vitamin E and Heart Disease: Basic Science to Clinical Intervention Trials, *Free Radic. Biol. Med. 28*, 141–164.

Chapter 19

Vitamin E in Disease Prevention and Therapy: Future Perspectives

Lester Packer[a] and Ute C. Obermüller-Jevic[b]

[a]University of Southern California, School of Pharmacy, Department of Molecular Pharmacology and Toxicology, Los Angeles, CA 90089–9121
[b]University of California, Davis, Department of Internal Medicine, Division of Pulmonary and Critical Care Medicine, Davis, CA 95616

Introduction

In numerous epidemiologic studies, it has been observed that a high intake of vitamin E is associated with a lower risk of age-related and chronic diseases, and experimental studies have suggested substantial health benefits from vitamin E in disease prevention and therapy. Most large-scale human intervention trials on the prevention of cardiovascular disease (CVD), the main cause of morbidity and mortality in the Western world, however, have failed to show convincingly that (relatively short-term) supplementation with vitamin E lowers the incidence of cardiovascular events or the rate of mortality from heart disease or stroke. Interestingly, a plethora of small clinical trials report a significant improvement of health status and/or retardation of disease progression by vitamin E supplementation in patients suffering from CVD or other chronic and age-related diseases. This chapter will summarize the role of vitamin E in the prevention and therapy of common chronic diseases, focusing on some newly emerging areas in which α-tocopherol, the major form of vitamin E in the human body, appears to have important health benefits; these include reproductive diseases (preeclampsia), age-related eye diseases [cataract, age-related macular degeneration (AMD)], metabolic disorders (diabetes mellitus), neurodegenerative disorders (Alzheimer's and Parkinson's disease), skin aging (photoaging), and healthy aging. In conclusion, future directions for research will be proposed.

Reproduction (Preeclampsia)

Preeclampsia is a disease with unknown etiology affecting pregnant women; worldwide, it is one of the leading causes of infant and maternal death (1). Preeclampsia manifests itself as gestational hypertension with proteinuria, edema, and activation of the coagulation system due to endothelial and placental dysfunction. Oxidative stress seems to play an important role in preeclampsia (2). Elevated levels of lipid peroxidation products were found in plasma and placental tissue of preeclamptic women along with low blood vitamin E levels (3–9). Although a few contradicting results exist, i.e., normal or increased vitamin E levels in preeclamptic women (10–12) and no increase in lipid peroxidation (13), the totality of evidence favors a role of vitamin E in preeclampsia.

In two initial clinical trials, pregnant women with severe preeclampsia did not benefit from supplementation with vitamin E (100–800 IU/d) alone or in combination with vitamin C (1000 mg/d) (14,15). However, pregnant women at risk of preeclampsia (i.e., abnormal uterine artery Doppler or previous history of the disease) had a substantial health benefit from prophylactic supplementation with vitamin E (400 IU/d) and vitamin C (1000 mg/d) given during the second half of pregnancy, as demonstrated by a significant 61% lower incidence of preeclampsia and improved endothelial and placental dysfunction (PAI-1/PAI-2 ratio) (16). A further trial in high-risk women showed that the clinical benefit obtained from intervention with these antioxidants was associated with an improvement of antioxidant status and protection against oxidative stress (12). Plasma levels of 8-epi-prostaglandin $F_{2\alpha}$ (8-epi-$PGF_{2\alpha}$, a marker for lipid peroxidation), which were elevated in these high-risk women, could be lowered effectively with vitamin supplementation, reaching levels similar to those of women at low risk of disease. In contrast, 8-epi-$PGF_{2\alpha}$ levels of untreated women (placebo group) remained elevated throughout the progression of pregnancy. The decrease of lipid peroxidation in the treated group was particularly associated with an increase of vitamin E levels in plasma but was not related to vitamin C levels. It has been suggested that early supplementation with vitamin E may improve vascular endothelial function and ameliorate, or even prevent preeclampsia. Other biological functions of vitamin E on the vascular system apart from its antioxidant activity (e.g., regulation of vascular homeostasis, inhibition of platelet aggregation) have not yet been investigated with respect to their relationship to the pathophysiology of preeclampsia.

The few *in vitro* studies that have been conducted on the effects of vitamin E on preeclampsia showed antioxidant activity of vitamin E in human placental mitochondria (17), and an inhibition of nuclear factor (NF)-κB activation and intercellular adhesion molecule-1 (ICAM-1) expression in endothelial cells cultured with plasma from preeclamptic women (3). These findings led to the suggestion that vitamin E might inhibit endothelial cell activation *via* inhibition of lipid peroxide formation, NF-κB activation, and ICAM-1 expression. This may be the way in which vitamin E exerts beneficial effects on endothelial dysfunction. Another experimental study showed that vitamin E prevented magnesium-induced apoptosis in placental explants (18), an important factor in preeclampsia. Animal data on effects of vitamin E on preeclampsia are lacking.

To date, the only effective strategy for managing preeclampsia is considered elective delivery, and apparently no therapeutic regimen has proven adequate for prevention or delay of the onset of this disease (19). Also, there is at present no accepted marker for predicting preeclampsia. A prophylactic intake of vitamin E and C supplements may prove useful for pregnant women with low vitamin E plasma levels and particularly for those women at known risk of preeclampsia.

Age-Related Eye Diseases (Cataract and AMD)

Cataract and AMD are common eye diseases and the leading causes of vision impairment and blindness among older individuals worldwide. In the United States, >1.6 million individuals have AMD (age > 60 y) and >20 million have cataract (age > 65 y) (20). In less developed countries, the prevalence of cataract and AMD is even higher and occurs earlier in life than in developed countries (21). To date, no medical treatment against these eye diseases is available and although cataract surgery has become a frequent and successful intervention in elderly people, the only available treatment for AMD, i.e., laser photocoagulation, is not applicable to most patients and is of limited benefit. As an antioxidant, vitamin E is thought to protect eye tissues such as the lens and retina against oxidative damage from sunlight, cigarette smoke, or by-products of metabolism, and thus to prevent age-related eye diseases.

Cataract

Opacification of the ocular lens, or cataract, occurs when lens proteins (crystallins) precipitate. This may result from an accumulation of modified (e.g., glycated) or oxidatively damaged lens proteins. The whole lens contains high amounts of vitamin E with a gradient found from the (outer) epithelium and cortex decreasing to lower amounts in the (inner) nucleus (22). A protective effect of vitamin E on cataractogenesis was found in several animal studies in which vitamin E successfully delayed the progression of chemically induced cataracts (23–26).

Numerous epidemiologic studies have examined the associations between dietary vitamin E intake, supplement use, and blood vitamin E levels in relation to risk of different forms of cataract. As reviewed extensively by Taylor and Hobbs (27), most recent studies, whether retrospective or prospective, found an inverse association between vitamin E intake and risk of cataract, primarily on cortical rather than nuclear cataract. A recently published retrospective trial, the Nutrition and Vision Project, compared long-term nutrient intake and risk of early cataract among 478 participants of the Nurses' Health Study cohort (age 53–73 y). Intake and blood levels of vitamin E and C were inversely correlated with nuclear cataract; however, only the inverse correlation between blood vitamin E levels and risk of disease remained significant after adjustment for other nutrients (28).

A prospective study, the Longitudinal Study of Cataract (n = 764 participants), reported a 31% lower risk of newly diagnosed nuclear cataract by regular intake of multivitamins, and an ~50% risk reduction with regular use of vitamin E supplements or with higher blood vitamin E levels (29). In the prospective Beaver Dam Eye Study cohort (n = 1354 participants), no association between vitamin E intake and risk of cataract was observed, but blood vitamin E levels were inversely correlated with nuclear cataract (30), and a lower risk of cataract among long-term users of multivitamins or any dietary supplement containing vitamin C or E was observed (31).

Age-Related Macular Degeneration (AMD)

AMD is characterized by a progressive loss of photoreceptors and degradation of the macula, leading to a progressive loss of central and sharp vision. As in most ocular tissues, the retina and the macular area in particular are severely exposed to oxidative stress due to the high consumption of oxygen and cumulative exposure to sunlight. The presence of photosensitizers and a high content of polyunsaturated fatty acids (PUFA) contribute to the susceptibility to photooxidative damage (32). The retina also contains high amounts of vitamin E, protecting against retinal oxidative damage, and vitamin E deficiency results in retinal degeneration, accumulation of lipofuscin in the retinal pigment epithelium, and loss of PUFA (33). Early experimental studies demonstrated protective (antioxidative) effects of vitamin E in eye tissues (34), leading to the suggestion that vitamin E may counteract degenerative processes in the retina. Data from animal studies, however, are lacking.

In epidemiologic studies, it was observed that AMD patients have significantly lower blood vitamin E levels than control subjects; moreover, vitamin E levels are inversely correlated with the severity of AMD (35). The Pathologies Oculaires Liées à l'Age (POLA) survey, a prospective, population-based study on risk factors for AMD and cataract (n = 2584 participants) reported an 82% risk reduction of AMD in the highest vs. the lowest quintile of lipid-standardized blood vitamin E levels, and an inverse association with early signs of AMD (36). In the Beaver Dam Eye Study, an increased risk of AMD with low intake of vitamin E was observed (37) .

Clinical Trials

The role of vitamin E in the prevention of age-related eye diseases has yet to be investigated in clinical trials. A first large-scale intervention trial on the effects of vitamin E supplementation (500 IU/d) on the risk of AMD or cataract, the Vitamin E and Cataract Prevention Study (VECAT) with 1205 participants, has just been completed. However, the results of the VECAT study have not yet been reported (38).

A few clinical trials have been published providing data on the effects of mixtures of antioxidants on age-related eye diseases. The Alpha-Tocopherol Beta Carotene (ATBC) study with middle-aged male Finnish smokers reported that daily supplementation with vitamin E (50 mg) and β-carotene (20 mg) for 5–8 y did not affect incidences of cataract surgeries (39) or AMD (40). In contrast, two recently completed clinical trials reported a benefit from antioxidant supplements on either cataract or AMD. The Roche European American Cataract Trial (REACT) with 297 adults from the United States and the United Kingdom was a 3-y controlled trial on the effects of antioxidant supplementation (600 IU vitamin E, 750 mg vitamin C, and 18 mg β-carotene per day) on cataract progression in patients with early cataract (41). The results show a significantly reduced progression of cataract in participants from the United States; however, in the UK cohort, the effect was not significant. The AREDS study on antioxidant vitamins and AMD (trial part one) or cataract (trial part two) is an ongoing multicenter trial (n = 4757 participants) with elderly AMD patients. Subjects are assigned to receive antioxidants (400 IU vitamin E, 500 mg vitamin C, and 15 mg β-carotene per day), zinc (80 mg/d), antioxidants plus zinc, or placebo. To date, after an average 6.3 y of treatment, antioxidant supplementation had no apparent effect on the risk of development or progression of lens opacities or vision loss (42). However, a significant reduction of AMD progression in subjects supplemented with a mixture of antioxidants plus zinc was observed (43). This suggests that even well-nourished people can benefit from the intake of high-dose antioxidant supplements. Long-term intake of antioxidants plus zinc was recommended for elderly people at high risk for AMD, i.e., patients with early stage drusen or advanced AMD, and without contraindications such as smoking. Further studies in patients with a familial history of AMD or early development of drusen are required. In conclusion, the data from epidemiologic and experimental studies clearly suggest a protective role of vitamin E in age-related eye diseases. Results from clinical studies using vitamin E supplements in patients with AMD or cataract are now warranted.

Metabolic Disorders (Diabetes Mellitus)

Diabetes comprises a group of metabolic disorders characterized by a defect in insulin secretion and/or insulin insufficiency leading to a disturbed glucose homeostasis and to hyperglycemia. The current classification of diabetes distinguishes among (i) *type 1 diabetes,* which is the result of β-cell destruction leading to absolute insulin deficiency (formerly insulin-dependent diabetes); (ii) *type 2 diabetes*, characterized by insulin resistance with relative insulin deficiency (formerly noninsulin-dependent diabetes); (iii) *gestational diabetes*, which has its time of first recognition during pregnancy, and (iv) other types of diabetes (44).

In subjects with undiagnosed or poorly controlled diabetes, chronic hyperglycemia causes micro- and macroangiopathies with subsequent dysfunction and multiple organ failure of primarily the visual, kidney, nerve, and cardiovascular systems. A number of hypotheses exist on the origin of these complications apart from hyperglycemia, such as oxidative damage resulting from autoxidation of glucose, glycated proteins, inflammatory processes, and impaired antioxidative defense (45), altered lipoprotein metabolism (46,47), and increased protein kinase C (PKC) activity due to glucose-induced diacylglycerol synthesis (48), all of which may be potential targets for vitamin E action. Studies of experimentally induced diabetic animals demonstrate that vitamin E prevents diabetic complications, thus providing support for the use of vitamin E supplements in patients (49–52). A large number of small clinical trials, mainly randomized and placebo-controlled, reported amelioration of disease parameters and diabetic complications by short-term supplementation with vitamin E.

Diabetes Mellitus, Type 1

Low vitamin E status is regarded as a potential risk factor for the development of type 1 diabetes (53); this is likely because oxidative stress plays a role in the autoimmune processes leading to destruction of the β cells in the pancreas (54). Vitamin E has proven beneficial for patients with type 1 diabetes, ameliorating several disease parameters. In diabetic children, supplementation with vitamin E (100 IU/d for 3 mo) significantly lowered malondialdehyde (MDA) and increased glutathione (GSH) levels in erythrocytes; the intervention also reduced hemoglobin A_{1c} (HbA_{1c}) levels in erythrocytes, which is a marker for protein glycation and severity of disease (55). In another trial with type 1 diabetic patients, vitamin E (750 IU/d for 1 y) did not significantly change antioxidant capacity and blood viscosity or lipid composition of erythrocytes; however, an improvement in the susceptibility of lipoproteins to oxidation was observed (56).

Vitamin E supplements also improve vascular function in diabetic patients. This was shown in individuals with a short disease duration (<10 y) who were given 1800 IU/d vitamin E for 4 mo resulting in an improvement in retinal blood flow to values comparable to those of nondiabetic controls (57). In the same study, vitamin E had no effect on blood HbA_{1c} levels but it normalized renal hyperfiltration (elevated creatinine clearance levels). Another clinical trial with young type 1 diabetes patients showed that vitamin E (1000 IU/d for 3 mo) improves endothelial vasodilator function, which is an early sign of diabetic vascular disease (58). Endothelial-dependent vasodilation is apparently directly related to the LDL vitamin E content and LDL particle size (59). Long-term supplementation with vitamin E for type 1 diabetics may be advisable and an early start with vitamin E intervention may be of particular importance to prevent diabetic complications.

Diabetes Mellitus, Type 2

Type 2 diabetes is increasingly common throughout the world with the highest incidence in developed countries. In the United States, it is likely that ~10% of adults are affected and ~50% of the people with the disease are undiagnosed (60). In several small, clinical trials with type 2 diabetic patients, vitamin E supplementation (600–1200 IU/d for 1–6 mo) improved conditions of oxidative stress *via* decreasing parameters of lipid peroxidation in blood (61–64) or low density lipoprotein (LDL) oxidizability (62,65–67). Supplementation with vitamin E (900 mg/d for 3 mo) also protected against oxidative DNA damage as determined by the comet assay (68). However, other studies using a lower dose of vitamin E (400 IU/d for 8 wk) did not find improvement of oxidative susceptibility of LDL (69) or oxidative DNA damage (70).

In addition to its effects on parameters of oxidative stress, it was suggested that vitamin E may lower peripheral insulin resistance in diabetic patients because high-dose supplementation with vitamin E (800 IU/d for 1 mo) improved plasma lipid parameters, lowered fasting glucose and fructosamine levels, and increased plasma levels of C-reactive protein (C-RP) and insulin (61).

Vitamin E may also have anti-inflammatory properties in diabetes. In patients with or without macrovascular disease, a 3-mo supplementation with high-dose vitamin E (1200 IU/d) decreased monocyte activity as determined by superoxide anion release, monocyte adhesion, and release of interleukin (IL)-6, IL-1β, and tumor necrosis factor (TNF)-α (65). In the same study population, vitamin E therapy also significantly lowered levels of C-RP. A similar effect was reported from patients with well-controlled type 2 diabetes receiving 800 IU/d vitamin E for 4 wk, which resulted in a 49% reduction of plasma C-RP levels [for comparison, tomato juice consumption (500 mL/d) and vitamin C supplementation (500 mg/d) had no such effect] (66). An earlier report on anti-inflammatory effects of vitamin E in diabetic patients using a lower dose of vitamin E (600 IU/d for 4 wk) found no inhibition of cytokine production and thus no anti-inflammatory properties (64). A dose-response study of vitamin E supplementation in diabetic patients may be useful to determine the threshold level of anti-inflammatory action of vitamin E.

Several other clinical trials have been conducted with type 2 diabetic patients, indicating a benefit from vitamin E supplementation on diabetic complications. In one study, vitamin E supplementation (500 IU/d for 10 wk) lowered plasminogen activator inhibitor type 1 (PAI-1) levels, a marker for impaired fibrinolytic activity and vascular dysfunction, but no improvement of glycometabolic parameters or antioxidant defense (as determined by ferric reducing ability of plasma) was observed (71). In another trial, vitamin E (800 IU/d for 6 mo) lowered platelet expression of adhesion molecules and fibrinogen; however, co-aggregation of platelets and leukocytes increased (67). Chronic administration of vitamin E to type 2 diabetic patients also improved brachial reactivity, an index of endothelial dysfunction and early atherosclerosis, in a trial with patients receiving 600 mg/d vitamin E for 8 wk. Although insulin resistance was not improved, the intracellular content of magnesium was increased by vitamin E, suggesting a beneficial role in smooth muscle cell relaxation and thus vascular function (63).

Oxidative stress in diabetes is associated with an imbalance in the activity of the cardiac autonomous nervous system (72) that is often responsible for sudden death (73). Vitamin E (600 mg/d for 4 mo) improved plasma indices of oxidative stress, plasma catecholamine levels, and cardiac sympathovagal balance in diabetic patients with cardiac autonomic neuropathy (74). This is the only study reported in humans on the effects of vitamin E on the autonomic nervous system.

Vitamin E supplements are also beneficial for diabetic patients with renal complications. Short-term supplementation with vitamin E (680 IU/d) plus vitamin C (1250 mg/d) for 1 mo improved the urinary excretion rate of albumin in diabetic patients with persistent albuminuria (75). Elevated levels of urinary albumin excretion predict a high risk for progression to end-stage renal disease, and its reduction is an important treatment goal (76). In patients with end-stage renal disease undergoing hemodialysis, particularly high levels of parameters of oxidative stress occur, resulting from bioincompatibility as well as "oversaturation" of transferrin induced by intravenous treatment with hemoglobin. These patients exhibit altered antioxidative defenses and altered levels of antioxidative vitamins, including vitamin E (77,78). Many patients also have increased levels of atherogenic oxidized LDL (oxLDL); in that group of patients, the incidence of atherosclerotic cardiac disease is extremely high (79,80). A single oral dose of 1200 IU vitamin E taken 6 h before the hemodialysis session lowered lipid peroxidation [analysis by thiobarbituric acid-reactive substances (TBARS)] in plasma of patients given intravenous iron (81). Similar beneficial effects were also achieved when vitamin E–coated cellulose was used in dialysis membranes (82,83). In a clinical trial using vitamin E supplements, the Secondary Prevention with Antioxidants of Cardiovascular Disease in Endstage Renal Disease (SPACE) study, a long-term intake of vitamin E (800 IU/d for 2 y) lowered the risk of composite cardio-vascular end points and myocardial infarction in hemodialysis patients with a history of CVD (84). These data suggest that vitamin E may be a useful adjunct therapy for the prevention of atherosclerosis in diabetic patients, or may even support established treatment of persisting vascular complications (85).

To date, no large-scale human intervention study has been carried out on the effects of vitamin E supplementation and risk of diabetes as a primary end point in either healthy or high-risk (e.g., obese) individuals. Large-scale studies with patients with type 1 or 2 diabetes are warranted to evaluate the effects of vitamin E in prevent-ing diabetic complications. A subgroup analysis of the Heart Outcomes Prevention Evaluation (HOPE) study on 3654 primarily type 2 diabetic patients showed no bene-ficial effects of vitamin E (400 IU/d for 4.5 y) on the end points "diabetic microvascu-lar complications" including nephropathy, dialysis, and laser therapy (86). Overall, a plethora of small clinical trials have shown considerable improvement of disease para-meters by vitamin E. At present, the amount of vitamin E required to prevent increased oxidative stress and thus to prevent complications in these patients is unclear. Obviously, the Recommended Dietary Allowance (RDA) for vitamin E of the Food and Nutrition Board is for healthy individuals and, hence, the use of supple-ments seems advisable.

The question remains whether dietary vitamin E intake is related to the risk of diabetes in healthy individuals. Insulin resistance precedes the development of type 2 diabetes and it is associated with an increase in oxLDL and circulating lipid peroxida-tion products in healthy, nondiabetic individuals at an early, preclinical stage as well as in patients with glucose intolerance (87–89). In a clinical trial, insulin-mediated glucose disposal was inversely related to self-reported dietary intake of vitamin E (90). In another study with apparently healthy volunteers, blood vitamin E levels were higher in insulin-sensitive subjects than in insulin-resistant subjects, and a significant negative association between plasma vitamin E and lipid hydroperoxides was observed (91). Thus, vitamin E likely plays a beneficial role in the prevention of type 2 diabetes. Furthermore, a beneficial role for vitamin E (alone or in combination with other antioxidants) was suggested in the prevention of postprandial insults to endothe-lial function, which may contribute to the development of CVD because in healthy volunteers, co-administration of vitamins E (800 IU) and C (2 g) together with an oral glucose load could significantly prevent the negative postprandial effects of hyper-glycemia on arterial dilation (92).

Vitamin E and Pregnancy

Vitamin E may also be important in gestational diabetes. Pregnancies in women with type 1 diabetes, despite glycemic control, are at risk for spontaneous abortions, preterm delivery, and/or birth defects such as fetal malformations and growth retardation, respiratory distress, and vascular complications (93–96). This teratogenicity results from fetal hyperglycemia, hyperlipidemia, hyperinsulinemia, and chronic fetal hypoxia due to reduced maternal blood flow. Oxidative stress during diabetic pregnancy is also reflected in fetuses and is an important determinant of fetal injury (97,98). A study in mothers with gestational diabetes and their newborns (fetal age 34–39 wk) revealed increased lipid peroxidation and protein oxidative damage in erythrocytes of both mother and fetus compared with controls (99).

Animal studies have repeatedly demonstrated that embryonic damage can be markedly decreased by antioxidants (100–104), strongly suggesting a rationale for using vitamin E supplements for pregnant women with diabetes to protect the fetus against damage. In diabetic pregnancies, mothers are also at increased risk for preeclampsia and progression of microvascular disease (105–107). Supplementation with vitamin E may be useful for diabetic women before and during pregnancy as part of preconception care of individuals who are at high risk for gestational diabetes, thus decreasing the risk of complications. Because human studies are lacking, clinical trials with vitamin E and C supplementation in diabetic pregnancy are urgently required.

Neurodegenerative Disorders (Alzheimer's and Parkinson's Disease)

In the elderly, age-related brain dysfunction and progressive occurrence of dementia are common causes of disability. Moreover, an increasing number of individuals are affected by Alzheimer's (AD) and Parkinson's Disease (PD). In both diseases, the underlying processes are a progressive pathologic degeneration of neurons, apoptotic cell death, and a substantial neuronal loss. This leads to dementia with cognitive and functional impairments and ultimately early death (108,109). Progressive formation of protein aggregates such as senile plaques (amyloid β-peptide protein) and neurofibrillary tangles (τ protein) in AD patients and Lewy bodies (α-synuclein protein) in PD patients is concomitant with severe oxidative stress associated with altered metabolic pathways, inflammatory processes, and neurodegeneration (110,111).

The brain tissue of AD or PD patients shows increased oxidative damage to lipids (112,113), proteins (114) and DNA (115,116), higher levels of advanced glycation end products (117), and transition metals (118,119) compared with tissue from nondemented elderly persons. In several animal studies using models of aging and neurogeneration, vitamin E effectively prevented oxidative damage (120–122). Protective effects of vitamin E have also been found in experiments using cell culture systems in which vitamin E exerted anti-inflammatory properties through suppression of microglial activation (123) or prevention of 24-hydroxycholesterol or amyloid β-peptide neurotoxicity (124,125).

These data suggest that supplementation with antioxidants, and vitamin E in particular, might delay or even prevent development of these neuronal disorders in humans. Although vitamin E cannot easily cross the blood-brain barrier, its concentrations in the brain can be increased 50–100% *via* dietary supplementation (126). Thus antioxidant intervention may be a useful tool for the management of neurodegenerative diseases (127,128).

A unique role of vitamin E in brain function is evident because inadequate tissue concentrations lead to brain dysfunction and peripheral neuropathy in humans (129–131). For example, patients with vitamin E deficiency manifest neurologic symptoms such as dementia, hyporeflexia, and ataxia, and these conditions can be improved considerably by administration of high-dose vitamin E (132–134). A newly developed experimental model of vitamin E deficiency in mice lacking α-tocopherol binding protein (α-TTP) showed, in comparison to wild-type mice, an increase in oxidative brain damage and progressive neurodegeneration accompanying symptoms of ataxia (135). In the same report, supplementation with vitamin E diminished oxidative damage and prevented neurodegeneration.

Epidemiologic evidence from human observational studies supports a protective role of vitamin E against AD. In a prospective study with 633 healthy older individuals (age > 65 y) who were users of vitamin E and/or C supplements, none of the subjects was diagnosed with AD after 4 y of follow-up in contrast to nonusers, suggesting that these antioxidant vitamins may lower the incidence of disease (136). Moreover, in several studies, AD patients had lower vitamin E levels in blood or cerebrospinal fluid compared with control subjects (137–141). The effects of vitamin E on PD are unclear because this association was not found in PD patients (138, 142–146). However, The Rotterdam Study, a cross-sectional study with 5342 individuals without dementia, including 31 patients with PD, showed that high intake of vitamin E was associated with lowered risk of PD (147).

Only a few clinical trials have investigated the effects of vitamin E supplementation in patients with AD or PD, and the results were diverse (148–151). In a small clinical trial with patients with moderate AD, the effect of vitamin E alone (400 IU/d for 1 mo) or in combination with vitamin C (1000 mg/d) on lipid peroxidation was studied. Although supplementation with vitamin E alone was unable to decrease the oxidizability of lipoproteins in plasma and cerebrospinal fluid *in vitro*, a combination of vitamin E plus C decreased lipoprotein susceptibility, suggesting an antioxidant function of both vitamins in human brain (148).

The Alzheimer's Disease Cooperative Study, the first, large, randomized and placebo-controlled clinical trial, found a beneficial effect of vitamin E in patients with moderately severe AD. In this study, long-term supplementation with vitamin E (2000 IU/d for 2 y) alone or in combination with selegiline (a monoamino oxidase inhibitor, 10 mg/d) slowed the progression of disease as determined by delay in institutionalization and the onset of severe dementia (149). An improvement of cognitive test parameters was not observed, which may have been due to the advanced stage of the disease in these patients. It has been questioned whether intervention with vitamin E may regenerate neurons that are already irreversibly damaged as found in advanced stages of AD. No such studies exist to date. Moreover, the effect of vitamin E on cognitive function requires further study in patients suffering from milder forms of AD. It would also be of interest to establish whether early intervention with vitamin E prevents or improves dementia in high-risk individuals with no or minimal cognitive impairment.

The Deprenyl and Tocopherol Antioxidative Therapy of Parkinsonism (DATATOP) study was initiated in 1987 to examine the benefits of vitamin E (2000 IU/d) and selegiline (10 mg/d) in slowing the progression of PD in 800 untreated patients with PD. After 8 y of intervention, vitamin E had no benefits on the progression of PD as determined by the time to reach a stage of disease at which levodopa therapy was required (150). Initial beneficial effects of deprenyl could not be sustained after 8 y of follow-up either. A substudy of 18 patients in the DATATOP cohort showed that the levels of vitamin E in cerebrospinal fluid were increased by ~76% after supplementation (151). Intervention at preclinical stages to slow down disease progression seems important, but at present, no tests for preclinical diagnosis of PD are available. Clinical studies are lacking on the health benefits of vitamin E in people at high risk of developing cognitive impairment such as adults with signs of mild cognitive impairment, e.g., memory deficits, and in elderly persons.

Other Forms of Dementia

Dementia as a consequence of cerebrovascular disease, ischemia-reperfusion, and stroke is another important disorder in elderly persons. Results from observational studies showed that a low dietary intake of vitamin E was associated with an increased risk of dementia (152–154), whereas other investigations found no association (155,156). Vitamin E was strongly related to cognitive performance in the Euronut-SENECA study of elderly Europeans from different countries (157). Moreover, in a sample of 4809 elderly subjects participating in the Third National Health and Nutrition Examination Survey (NHANES III, 1988–1994), serum levels of vitamin E were inversely correlated with memory function, whereas other antioxidant nutrients including vitamin C, β-carotene, and selenium did not show such an association (153).

It may be suggested that prevention of vascular dementia by vitamin E *via* protecting neurons against oxidative damage as well as *via* inhibition of platelet aggregation or modulation of immune responses may reduce the risk of disease. In different animal models of stroke, vitamin E alone or in combination with other antioxidants was found to successfully protect against cerebral ischemic damage (158–160), e.g., vitamin E plus α-lipoic acid successfully protected against lipid peroxidation, reduced glial reactivity, and enhanced neuronal remodeling in brain tissues (161).

In humans, a clinical trial on the effects of antioxidants on dementia in persons who sustained a stroke was conducted as a substudy of The Honolulu-Asia Aging Study with Japanese-American men from Hawaii. In that study with 3385 men (age 71 to 93 y), a reduced risk of vascular dementia was associated with intake of vitamin E and/or C supplements (162). In these elderly people without dementia, an improvement of cognitive function was found but no association with AD was observed.

Skin Aging (Photoaging)

The aging of skin is an intrinsic, degenerative process over time, comparable to what occurs in other organs, that results in a thinner but smooth skin with reduced elasticity. This process is superimposed to varying degrees by premature (extrinsic) aging caused by exposure to environmental stress, e.g., ultraviolet (UV) radiation. Photoaged skin occurs at unshielded body sites such as the face and the backside of hands as a consequence of chronic overexposure to natural or artificial sunlight (163). Clinical characteristics of photoaged skin are wrinkle formation, laxity, and a leathery appearance.

Ultraviolet radiation, and mainly long-wavelength ultraviolet A (UVA) light (320–400 nm), is a strong oxidant. In UV-irradiated skin, photosensitizing reactions lead to the formation of reactive oxygen species (ROS) such as singlet oxygen, super-oxide anions, hydroxyl radicals, and hydrogen peroxide (164). ROS play a pivotal role in photoaging because they damage proteins directly (e.g., collagen) and activate matrix-degrading enzymes, leading to the destruction of the dermal layer in skin (165–167). Protection against photooxidative stress and consequently photoaging should be accomplished primarily *via* diminishing exposure to sunlight and by application of topical sunscreens. In addition, the use of antioxidants, whether topical or systemic, is a promising and common strategy to limit or prevent oxidative damage induced by UV light (168–170).

Sunscreen effect of vitamin E. Vitamin E has an absorbance maximum at 292 nm which is in the UVB range of light (290–320 nm). Like other sunscreen compounds, although with lesser activity, vitamin E may provide skin photoprotection *via* absorption of hazardous UVB radiation (169). In particular, topically applied vitamin E that does not effectively penetrate into the epidermis might act as an external sunscreen. UV irradiation of vitamin E results in formation of dimers and trimers that could contribute to photoprotection through their action as UV-absorbing compounds, could contribute to photoprotection (171).

Antioxidant action of vitamin E. Studies in humans on the antioxidant potential of vitamin E in skin are scarce. In a small clinical trial, vitamin E supplementation for 2 wk before sun exposure protected against TBARS formation in blood (172). Other studies indicate antioxidant activity of vitamin E in skin because vitamin E is readily depleted after a single suberythemal dose of UV light in human skin (173). In experimental studies with mice, repeated daily exposures to UVB light resulted in an increase in skin vitamin E levels and formation of oxidation products of vitamin E (174). Other studies in animals reported prevention of ROS formation in skin as detected by *in vivo* chemiluminescence (175), prevention of lipid peroxidation (176), and up-regulation of a network of enzymatic and nonenzymatic antioxidants by vitamin E (177).

Antioxidant action of vitamin E has also been shown *in vitro* in cultured skin fibroblasts *via* diminution of ROS generation (178) and prevention of UVA-induced up-regulation of the stress responsive gene for heme oxygenase-1 (179). Vitamin E also inhibited, to a certain degree, photoinactivation of the iron regulatory protein-1 in UVA-irradiated skin cells, and thus formation of oxidative stress by free intracellular iron (180). In primary keratinocytes, vitamin E (Trolox) diminished the UVB-induced stress signaling response as determined by a modulation of mitogen activated protein kinase (MAPK) activation (181).

In addition to direct antioxidant activity, vitamin E modulates cell signaling and gene expression in skin. Aging of skin, whether intrinsic or due to environmental stress exposure, is associated with increased PKC activity, leading to induction of collagenase (MMP-1, a matrix metalloproteinase) and thus tissue degradation. Vitamin E is able to inhibit collagenase overexpression in aging skin fibroblasts *via* PKC inhibition (182). Moreover, Trolox inhibited UVA-induced collagenase expression in skin fibroblasts *in vitro* (183). These findings strongly suggest a beneficial role of vitamin E in the prevention of skin photoaging.

Vitamin E seems to have a major protective role in human skin because it accumulates on the skin outer surface, the stratum corneum (SC) (184). Delivery of vitamin E onto the skin surface appears to be specific *via* continuous secretion by sebaceous glands (185). In sun-exposed body areas such as the face, vitamin E levels in the SC are 20-fold higher compared with unexposed areas (185). Upon exposure to environmental oxidative stress such as UV light (173) or ozone (186), a destruction of vitamin E occurs in the outer layers of the SC, indicating that vitamin E has an important antioxidant function in skin protection.

Topical application of vitamin E. In an experimental model of photoaging using hairless mice chronically exposed to UV radiation, topical application of vitamin E (5%) reduced visible skin changes and histologic alterations caused by UVB radiation (e.g., collagen damage, epidermal thickening, dermal infiltration) (187). Another study in mice revealed that vitamin E prevents UV-induced immunosuppression (188).

To date, no clinical trials have evaluated the effect of topical administration of vitamin E on parameters of photoaging in humans. One clinical study reported improved hydration of the SC and enhanced water-binding capacity of the skin after topical application of vitamin E (189). No photoprotection against erythema occurred with topical application of vitamin E (190). The efficacy of topical vitamin E depends greatly on the formulation used. In human skin *ex vivo*, the penetration of vitamin E was best after the use of encapsulated "nanotopes" followed by liposomal encapsulation and aqueous solubilization of vitamin E acetate. When applied in oil, no penetration of vitamin E into deeper layers was observed; it remained at the surface and in the SC (191). In the same study, up to 50% deacetylation of vitamin E acetate occurred by nonspecific esterases, generating the antioxidant active form of vitamin E. This effect did not occur on the skin surface or in the SC but only in the deeper skin layers where metabolically active cells are present.

Systemic application of vitamin E. In several studies vitamin E alone or in combination with other antioxidants (vitamin C, carotenoids) prevented biological end points of photodamage such as erythema in mice (192) and humans (193–196). For example, in a clinical trial using megadoses of Trolox (2 g/d), protection from erythema formation was observed. This effect was more pronounced in combination with vitamin C (3 g/d), suggesting that vitamin E and C act synergistically in an antioxidant network suppressing the sunburn reaction (194). Clinical data are not yet available on the effects of vitamin E supplementation on photoaging.

It may be hypothesized that a systemic application of vitamin E provides an additional benefit to topical application of sun protection products *via* constant accumulation in skin and, moreover, in tissues affected by UV radiation to which topical products cannot be administered (e.g., buccal mucosa cells in the oral cavity). Studies in this regard would be of interest.

Healthy Aging

The process of aging is associated with numerous biochemical and physiologic changes that cause a progressive impairment of organ functions, thus increasing the risk of morbidity and mortality. Genetic, environmental, and lifestyle factors play important roles in aging and disease development. "Successful aging," i.e., reaching old age with minimal physiologic loss, is a concept proposed by Rowe and Kahn (197) who point to diet and exercise as predominant age-extrinsic factors positively influencing healthy aging.

As an antioxidant, vitamin E is of great interest for its antiaging properties. Aging is associated with the accumulation of oxidative damage, reduced antioxidant and repair capacity involving degeneration of organ function, and the occurrence of age-related and chronic diseases (198,199). It may be assumed that individuals with a high intake of micronutrients and antioxidants will have a stronger antioxidant defense system and may lower their risk of chronic disease during aging. Consequently, it has been suggested that prevention of chronic diseases by vitamin E may contribute to "successful" aging and improve the quality of (long) life.

Elderly people are at particularly high risk for malnutrition due to decreased food intake (e.g., reduced appetite, loss of teeth), reduced energy demand, less activity and physical exercise, lower energy expenditure, and the poverty of institutionalization (200). Furthermore, increased micronutrient requirements may arise due to impaired intestinal function, nutrient absorption, and metabolism (201). Epidemiologic data consistently confirm that elderly people are a high-risk population for suboptimal intake of vitamin E and other micronutrients. An international overview on the vitamin status of the elderly in apparently well-nourished populations was reported by Haller (202) who concluded that plasma concentrations of vitamin E, among other vitamins, are not optimal in the elderly from several populations including the continental United States, Hawaii, and several European countries. This was confirmed in studies with apparently healthy elderly individuals (203,204).

A number of recent observational studies show a particular role for vitamin E in healthy aging. For example, centenarians are a group of people who have aged successfully who appear to have increased vitamin E levels compared with other elderly subjects. A study with healthy centenarians revealed a characteristic profile of nonenzymatic antioxidants with significantly higher plasma levels of vitamin E (55 μmol/L) compared with elderly controls at age 81–99 y (42 μmol/L) and 61–80 y (47 μmol/L). Although levels of vitamin A were also increased, other plasma antioxidants such as vitamin C, uric acid, thiols, and most carotenoids were decreased, and enzymatic activity of superoxide dismutase was also lower (205).

Another study showed less oxidative stress in healthy centenarians compared with "younger" elderly persons, e.g., lower levels of lipid peroxidation parameters in blood and elevated levels of vitamin E and C as well as an increased GSH/oxidized glutathione (GSSG) ratio (206). Higher levels of vitamin E and antioxidant enzymes or lower levels of oxidative stress compared with younger persons were also found in other cohorts of centenarians (207) and in subjects of an Indian tribal population, Kurichias, who are viewed as an example of successful aging and longevity (208). In contrast, no difference in vitamin E plasma values was observed in a cohort of healthy Northern Italian oldest-old (age 90–107 y) compared with younger controls (209).

A prospective study with a 7-y follow-up in 638 noninstitutionalized Dutch elderly people (age 65–85 y) revealed no relationship of plasma vitamin E levels in normal physiologic ranges (22–35 μmol/L) with mortality risk (210). However, in a study with apparently healthy Italians (≥80 y old), plasma vitamin E levels in the upper quartile were associated with a lower risk for cardiovascular events including myocardial infarction, ischemic stroke, and congestive heart failure compared with those participants with vitamin E levels in the lowest quartile (211). In this study, no association was found for vitamin C, β-carotene, or total cholesterol. Similar results were observed in a study of octogenarians in which plasma levels of vitamin E and cholesterol in relation to early signs of atherosclerosis in coronary and extracoronary blood vessels were primary study end points. It was concluded that appropriate levels of vitamin E in plasma and low levels of oxLDL might be important for healthy aging without development of atherosclerosis (212).

Aging is associated with a decline in immune status (213). Supplementation with vitamin E improves immune function, particularly age-related dysfunction in T cell–mediated immunity, which has been implicated in the development of several chronic diseases in elderly people. As reviewed by Serafini (214), vitamin E supplementation in elderly humans improves T-cell function directly and indirectly (through macrophages), enhances proliferation of lymphocytes and cytokine production, and decreases production of PGE$_2$ *via* inhibition of elevated cyclooxygenase (COX) activity. In a recent study in mice, it was demonstrated that vitamin E reduces COX activity by post-translational regulation *via* inhibition of peroxynitrite formation (215). These properties of vitamin E may contribute to the prevention of vascular disease during aging.

In healthy Dutch elderly subjects (age 65–80 y), a 6-mo supplementation with vitamin E (100 mg/d) showed a trend toward increased cellular immune function as determined by the delayed type hypersensitivity (DTH) test (216). A recent dose-response trial among healthy older adults (60, 200, 800 IU vitamin E/d over a period of 235 d) showed optimal increase of immunity at a dose of 200 IU/d and an improvement of parameters such as DTH responses and several antibody titers (217). This may be accomplished *via* its antioxidant function and/or modulation of prostaglandin synthesis. However, in an earlier study, the same dose given for 3 mo had no effect on overall immune responsiveness (218).

Physical Exercise

Regular physical exercise and training at moderate levels are important factors for disease prevention and healthy aging. Nonetheless, strenuous exercise by untrained individuals or overexertion in trained individuals greatly increases oxidative stress in the whole human body and in skeletal muscle in particular, leading to extensive tissue damage (219,220). Vitamin E may prevent exercise-induced oxidative damage (221,222). As a consequence of greater oxygen consumption during and postexercise, free radicals may result from mitochondrial leakage, from hypoxia/reoxygenation processes due to changes in blood flow in the capillary endothelium, and also from inflammatory cells infiltrating damaged tissues. An additional increase in the production of free radicals or earlier onset of oxidative stress during submaximal exercise may occur due to age, lack of training, or impaired antioxidative defense. Several markers of oxidative stress and resulting damage have been measured after exhausting exercise, e.g., increased lipid peroxidation, oxidative DNA damage, an increase in GSH peroxidase activity, a decrease in the cellular GSH/GSSG ratio, and an oxidative burst resulting from neutrophil activation (223–227).

Several studies showed that antioxidant enzymes increase along with oxidative stress and training as a physiologic response to exercise, and that vitamin E levels in plasma decrease (228–231). The strength of antioxidant defense depends on the dietary supply with vitamin E and is related to age and health status of an individual. An additional need for supplemental vitamin E and other antioxidant micronutrients seems likely in persons engaging in physical activity (221) such as cyclists, runners, and mountain climbers.

Small clinical trials showed that supplementation with vitamin E (1200 IU/d for 2 wk) lowers exercise-induced pentane production (232) and DNA damage (233). In a placebo-controlled study with cyclists, supplementation with vitamin E (100 mg/d) plus vitamin C (500 mg/d) for 15 wk during training significantly protected against acute effects of ozone on lung function (234). A cocktail of vitamin E (500 mg/d) and β-carotene (30 mg/d) for 90 d plus vitamin C (1 g/d) during the last 15 d in athletes increased the GSH/GSSG ratio and enhanced antioxidant enzyme activity (superoxide dismutase and catalase) in neutrophils (235). In contrast, the combination of a low-dose vitamin E supplement (100 mg/d) with coenzyme Q10 (90 mg/d) for 3 wk before a marathon did not affect levels of lipid peroxidation in plasma (236).

An effect of vitamin E on performance has not yet been proven, and it seems that supplementation with vitamin E has no major ergogenic effect on the end points studied (e.g., performance, muscle strength, aerobic capacity). In addition, the extent of the importance of vitamin E on exercise-induced damage has not been resolved and only a few human studies have examined the interaction of vitamin E and exercise to date. Clinical data on the effects of vitamin E on exercise-induced muscular damage are controversial and differ with biological end points. No beneficial effect of vitamin E (1200 IU/d for 30 d) was found on disruption of the band structures in muscle fibers and inflammatory response (macrophage infiltration) after a single bout of muscle contractions (237). In contrast, vitamin E supplementation (1200 IU/d) for 4 weeks before and during 6 consecutive days of endurance running lowered muscle injury as determined by leakage of creatine kinase and lactate dehydrogenase (238). In basketball players, supplementation with a cocktail of vitamin E (600 mg/d), vitamin C (1 g/d), and β-carotene (32 mg/d) for 35 d during habitual training decreased muscle leakage of lactate dehydrogenase, lowered lipid peroxides in plasma, and improved the anabolic/catabolic balance (239).

Vitamin E alone or in combination with other antioxidants seems to prevent exercise-induced oxidative damage. This affects not only skeletal muscle but the whole human body. The efficacy of vitamin E in strenuous exercise seems to depend on the dose, the amount of exercise, and on the investigated biological endpoints.

Synopsis

In conclusion, an adequate nutritional supply of vitamin E seems of crucial importance throughout life to guarantee normal function of physiologic processes in the body as well as adequate defense against oxidants generated by endogenous sources or from exposure to environmental stress. In particular, vitamin E plays a unique role in the prevention of chronic and degenerative diseases, and thus in healthy aging. A well-balanced diet rich in vitamin E as part of a healthy lifestyle seems a basic requirement for human health. Moreover, intake of supplementary vitamin E may be advisable for individuals who are unable to sustain adequate levels, or who are at high risk for, or are already suffering from chronic and degenerative diseases.

As discussed extensively above, the vast majority of small clinical trials convincingly demonstrate a beneficial role of vitamin E when given as a dietary supplement to women with preeclampsia or patients suffering from chronic diseases including diabetes and age-related degenerative diseases.

Future Perspectives

Several areas for future research on the biological effects of vitamin E and its role in disease prevention have emerged. Experimental studies are required to further describe the physiologic role of vitamin E *per se*. For example, the relative contribution of vitamin E to the antioxidant defense system in the human body is not known. Moreover, no reliable method exists to date to measure the antioxidant activity and biological efficacy of vitamin E in humans.

The effects of vitamin E and related binding proteins [e.g. tocopherol-associated protein, (TAP)] on cell signaling and gene expression must be explored further. At low levels, free radicals and oxidants are known to alter activity of some of the vital pathways for cell signaling and gene expression. Therefore, all antioxidants including all stereoisomers of vitamin E if bioavailable can be expected to exhibit cell regulatory effects.

In humans, new functional tests for the biological activity and action of vitamin E in prevention and treatment of disease must be established. This will allow the scientific community to define guidelines on how much vitamin E is required to avoid suboptimal supply and deficiency, and how much vitamin E is required to provide additional beneficial effects for chronic and degenerative diseases, and (healthy) aging. New and sensitive biomarkers are required for the determination of the functional status of vitamin E in both individuals and human populations. Only the availability of scrutinized biomarkers for vitamin E function will allow us to address these questions.

A further challenge is to identify the origin of the health effects of vitamin E in people who have a high vitamin E intake and/or supplement their diet. In this regard, it should be clarified whether any selective advantage is afforded by natural-source vitamin E or whether racemic mixtures of α-tocopherol are equally effective.

New studies to assess vitamin E intake may be useful because dietary patterns have changed and the consumption of fortified foods and supplements has increased steadily. Subpopulations at risk for vitamin E deficiency among well-nourished populations must be defined to identify those individuals who may benefit from regular intake of vitamin E supplements, e.g., elderly persons at high risk for disease, or individuals with chronic and/or age-related diseases.

Large-scale human intervention studies are required to clarify the usefulness of vitamin E supplementation in these individuals at high risk for chronic disease or age-related degeneration, and potentially to establish guidelines for vitamin E supplementation as an adjunct to standard therapy. Some disorders with highly promising results include the following:

- Preeclampsia: Clinical studies are required in women at high risk for preeclampsia. The effects of vitamin E on vascular and placental homeostasis, pathophysiology, and prevention of preeclampsia must be established. Animal models for evaluating the effects of vitamin E on preeclampsia, which are lacking at present, could provide valuable insights.
- Age-related eye diseases: Long-term studies on the prevention of cataract and age-related macular degeneration are lacking, particularly in subjects at risk for disease or early stages of disease.
- Diabetes mellitus: Long-term studies on the efficacy of vitamin E in prevention of diabetic complications and large-scale studies in individuals at risk of diabetes are required. Studies of the effects of vitamin E on prevention of postprandial insults to the vascular system and thus prevention of CVD

could be useful. Also, human data are lacking on the efficacy of vitamin E supplementation in diabetic pregnancy.

- Neurodegeneration: Studies are required on the effect of vitamin E on cognitive function in persons at risk for disease or individuals with mild forms of Alzheimer's and Parkinson's disease.
- Skin aging: Clinical trials on the role of vitamin E in prevention or amelioration of photoaging and photocarcinogenesis are warranted. In particular, studies on the health benefit from supplementation with vitamin E for protection of both skin and other tissues for which topical products cannot be administered would be of interest.
- Physical exercise: Clinical trials with new biological end points are required for prevention and repair of exercise-induced tissue damage by vitamin E and possible beneficial effects on performance.

In the future, new guidelines will be important for determining the optimal vitamin E intake to maintain health throughout life and to ameliorate disease. A particular benefit may be expected from early supplementation and in subjects at high risk for and/or in early stages of disease.

References

1. Walker, J.J. (2000) Preeclampsia, *Lancet 356*, 1260–1265.
2. Hubel, C.A. (1999) Oxidative Stress in the Pathogenesis of Preeclampsia, *Proc. Soc. Exp. Biol. Med. 222*, 222–235.
3. Takacs, P., Kauma, S.W., Sholley, M.M., Walsh, S.W., Dinsmoor, M.J., and Green, K. (2001) Increased Circulating Lipid Peroxides in Severe Preeclampsia Activate NF-κB and Upregulate ICAM-1 in Vascular Endothelial Cells, *FASEB J. 15*, 279–281.
4. El-Salahy, E.M., Ahmed, M.I., El-Gharieb, A., and Tawfik, H. (2001) New Scope in Angiogenesis: Role of Vascular Endothelial Growth Factor (VEGF), NO, Lipid Peroxidation, and Vitamin E in the Pathophysiology of Preeclampsia Among Egyptian Females, *Clin. Biochem. 34*, 323–329.
5. Panburana, P., Phuapradit, W., and Puchaiwatananon, O. (2000) Antioxidant Nutrients and Lipid Peroxide Levels in Thai Preeclamptic Pregnant Women, *J. Obstet. Gynaecol. Res. 26*, 377–381.
6. Kharb, S. (2000) Vitamin E and C in Preeclampsia, *Eur. J. Obstet. Gynecol. Reprod. Biol. 93*, 37–39.
7. Akyol, D., Mungan, T., Gorkemli, H., and Nuhoglu, G. (2000) Maternal Levels of Vitamin E in Normal and Preeclamptic Pregnancy, *Arch. Gynecol. Obstet. 263*, 151–155.
8. Kharb, S. (2000) Lipid Peroxidation in Pregnancy with Preeclampsia and Diabetes, *Gynecol. Obstet. Investig. 50*, 113–116.
9. Gratacos, E., Casals, E., Deulofeu, R., Gomez, O., Cararach, V., Alonso, P.L., and Fortuny, A. (1999) Serum and Placental Lipid Peroxides in Chronic Hypertension During Pregnancy with and Without Superimposed Preeclampsia, *Hypertens. Pregnancy 18*, 139–146.
10. Zhang, C., Williams, M.A., Sanchez, S.E., King, I.B., Ware-Jauregui, S., Larrabure, G., Bazul, V., and Leisenring, W.M. (2001) Plasma Concentrations of Carotenoids, Retinol,

and Tocopherols in Preeclamptic and Normotensive Pregnant Women, *Am. J. Epidemiol. 153,* 572–580.

11. Valsecchi, L., Cairone, R., Castiglioni, M.T., Almirante, G.M., and Ferrari, A. (1999) Serum Levels of Alpha-Tocopherol in Hypertensive Pregnancies, *Hypertens. Pregnancy 18,* 189–195.

12. Chappell, L., Seed, P., Kelly, F., Briley, A., Hunt, B., Stock, M., Charnock-Jones, S., Mallet, A., and Poston, L. (2002) Vitamin C and E Supplementation in Women at Risk of Preeclampsia Is Associated with Changes in Indices of Oxidative Stress and Placental Function, *Am. J. Obstet. Gyn. 187,* 777–784.

13. Bowen, R.S., Moodley, J., Dutton, M.F., and Theron, A.J. (2001) Oxidative Stress in Preeclampsia, *Acta Obstet. Gynecol. Scand. 80,* 719–725.

14. Gulmezoglu, A.M., Hofmeyr, G.J., and Oosthuisen, M.M. (1997) Antioxidants in the Treatment of Severe Pre-Eclampsia: An Explanatory Randomised Controlled Trial, *Br. J. Obstet. Gynaecol. 104,* 689–696.

15. Stratta, P., Canavese, C., Porcu, M., Dogliani, M., Todros, T., Garbo, E., Belliardo, F., Maina, A., Marozio, L., and Zonca, M. (1994) Vitamin E Supplementation in Preeclampsia, *Gynecol. Obstet. Investig. 37,* 246–249.

16. Chappell, L.C., Seed, P.T., Briley, A.L., Kelly, F.J., Lee, R., Hunt, B.J., Parmar, K., Bewley, S.J., Shennan, A.H., Steer, P.J., and Poston, L. (1999) Effect of Antioxidants on the Occurrence of Preeclampsia in Women at Increased Risk: A Randomised Trial, *Lancet 354,* 810–816.

17. Milczarek, R., Klimek, J., and Zelewski, L. (2000) The Effects of Ascorbate and Alpha-Tocopherol on the NADPH-Dependent Lipid Peroxidation in Human Placental Mitochondria, *Mol. Cell Biochem. 210,* 65–73.

18. Black, S., Yu, H., Lee, J., Sachchithananthan, M., and Medcalf, R.L. (2001) Physiologic Concentrations of Magnesium and Placental Apoptosis: Prevention by Antioxidants, *Obstet. Gynecol. 98,* 319–324.

19. Sibai, B.M. (1998) Prevention of Preeclampsia: a Big Disappointment, *Am. J. Obstet. Gynecol. 179,* 1275–1278.

20. Anonymous (2002) *Vision Problems in the US,* National Eye Institute, Bethesda, MD.

21. Thylefors, B., Negrel, A.D., Pararajasegaram, R., and Dadzie, K.Y. (1995) Global Data on Blindness, *Bull. WHO 73,* 115–121.

22. Yeum, K.J., Shang, F.M., Schalch, W.M., Russell, R.M., and Taylor, A. (1999) Fat-Soluble Nutrient Concentrations in Different Layers of Human Cataractous Lens, *Curr. Eye Res. 19,* 502–505.

23. Kojima, M., Shui, Y., Murano, H., Nagata, M., Hockwin, O., Sasaki, K., and Takahashi, N. (2002) Low Vitamin E as a Subliminal Risk Factor in a Rat Model of Prednisolone-Induced Cataract, *Investig. Ophthalmol. Vis. Sci. 43,* 1116–1120.

24. Ohta, Y., Yamasaki, T., Niwa, T., and Majima, Y. (2000) Preventive Effect of Vitamin E-Containing Liposome Instillation on Cataract Progression in 12-Month-Old Rats Fed a 25% Galactose Diet, *J. Ocul. Pharmacol. Ther. 16,* 323–335.

25. Ohta, Y., Yamasaki, T., Niwa, T., Majima, Y., and Ishiguro, I. (1999) Preventive Effect of Topical Vitamin E-Containing Liposome Instillation on the Progression of Galactose Cataract. Comparison Between 5-Week- and 12-Week-Old Rats Fed a 25% Galactose Diet, *Exp. Eye Res. 68,* 747–755.

26. Nagata, M., Kojima, M., and Sasaki, K. (1999) Effect of Vitamin E Eye Drops on Naphthalene-Induced Cataract in Rats, *J. Ocul. Pharmacol. Ther. 15,* 345–350.

27. Taylor, A., and Hobbs, M. (2001) Assessment of Nutritional Influences on Risk for Cataract, *Nutrition 17*, 845–857.
28. Jacques, P.F., Chylack, L.T., Jr., Hankinson, S.E., Khu, P.M., Rogers, G., Friend, J., Tung, W., Wolfe, J.K., Padhye, N., Willett, W.C., and Taylor, A. (2001) Long-Term Nutrient Intake and Early Age-Related Nuclear Lens Opacities, *Arch. Ophthalmol. 119*, 1009–1019.
29. Leske, M.C., Chylack, L.T., Jr., He, Q., Wu, S.Y., Schoenfeld, E., Friend, J., and Wolfe, J. (1998) Antioxidant Vitamins and Nuclear Opacities: The Longitudinal Study of Cataract, *Ophthalmology 105*, 831–836.
30. Lyle, B.J., Mares-Perlman, J.A., Klein, B.E., Klein, R., Palta, M., Bowen, P.E., and Greger, J.L. (1999) Serum Carotenoids and Tocopherols and Incidence of Age-Related Nuclear Cataract, *Am. J. Clin. Nutr. 69*, 272–277.
31. Mares-Perlman, J.A., Lyle, B.J., Klein, R., Fisher, A.I., Brady, W.E., VandenLangenberg, G.M., Trabulsi, J.N., and Palta, M. (2000) Vitamin Supplement Use and Incident Cataracts in a Population-Based Study, *Arch. Ophthalmol. 118*, 1556–1563.
32. Beatty, S., Koh, H., Phil, M., Henson, D., and Boulton, M. (2000) The Role of Oxidative Stress in the Pathogenesis of Age-Related Macular Degeneration, *Surv. Ophthalmol. 45*, 115–134.
33. Friedrichson, T., Kalbach, H.L., Buck, P., and van Kuijk, F.J. (1995) Vitamin E in Macular and Peripheral Tissues of the Human Eye, *Curr. Eye Res. 14*, 693–701.
34. Farnsworth, C., and Dratz, E. (1976) Oxidative Damage of Retinal Rod Outer Segment Membranes and the Role of Vitamin E, *Biochim. Biophys. Acta 443*, 556–570.
35. Belda, J., Roma, J., Vilela, C., Puertas, F., Diaz-Llopis, M., Bosch-Morell, F., and Romero, F. (1999) Serum Vitamin E Levels Negatively Correlate with Severity of Age-Related Macular Degeneration, *Mech. Ageing Dev. 107*, 159–164.
36. Delcourt, C., Cristol, J.P., Tessier, F., Leger, C.L., Descomps, B., and Papoz, L. (1999) Age-Related Macular Degeneration and Antioxidant Status in the POLA Study. POLA Study Group. Pathologies Oculaires Liées à l'Age, *Arch. Ophthalmol. 117*, 1384–1390.
37. VandenLangenberg, G.M., Mares-Perlman, J.A., Klein, R., Klein, B.E., Brady, W.E., and Palta, M. (1998) Associations Between Antioxidant and Zinc Intake and the 5-Year Incidence of Early Age-Related Maculopathy in the Beaver Dam Eye Study, *Am. J. Epidemiol. 148*, 204–214.
38. Garrett, S.K., McNeil, J.J., Silagy, C., Sinclair, M., Thomas, A.P., Robman, L.P., McCarty, C.A., Tikellis, G., and Taylor, H.R. (1999) Methodology of the VECAT Study: Vitamin E Intervention in Cataract and Age-Related Maculopathy, *Ophthalmic Epidemiol. 6*, 195–208.
39. Teikari, J.M., Rautalahti, M., Haukka, J., Jarvinen, P., Hartman, A.M., Virtamo, J., Albanes, D., and Heinonen, O. (1998) Incidence of Cataract Operations in Finnish Male Smokers Unaffected by Alpha Tocopherol or Beta Carotene Supplements, *J. Epidemiol. Community Health 52*, 468–472.
40. Teikari, J.M., Laatikainen, L., Virtamo, J., Haukka, J., Rautalahti, M., Liesto, K., Albanes, D., Taylor, P., and Heinonen, O.P. (1998) Six-Year Supplementation with Alpha-Tocopherol and Beta-Carotene and Age-Related Maculopathy, *Acta Ophthalmol. Scand. 76*, 224–229.
41. Chylack, L.T., Brown, N.P., Bron, A., Hurst, M., Kopcke, W., Thien, U., and Schalch, W. (2002) The Roche European American Cataract Trial (REACT): A Randomized Clinical Trial to Investigate the Efficacy of an Oral Antioxidant Micronutrient Mixture to Slow Progression of Age-Related Cataract, *Ophthalmic Epidemiol. 9*, 49–80.

42. Eye Disease Case Control Study Group (2001) A Randomized, Placebo-Controlled, Clinical Trial of High-Dose Supplementation with Vitamins C and E, Beta Carotene, and Zinc for Age-Related Macular Degeneration and Vision Loss: AREDS Report No. 8, *Arch. Ophthalmol. 119*, 1417–1436.

43. Eye Disease Case Control Study Group (2001) A Randomized, Placebo-Controlled, Clinical Trial of High-Dose Supplementation with Vitamins C and E and Beta Carotene for Age-Related Cataract and Vision Loss: AREDS Report No. 9, *Arch. Ophthalmol. 119*, 1439–1452.

44. The Expert Committee on the Diagnosis and Classification of Diabetes Mellitus Report, *Diabetes Care 25*, S5–S20 (2002).

45. Lipinski, B. (2001) Pathophysiology of Oxidative Stress in Diabetes Mellitus, *J. Diabet. Complicat. 15*, 203–210.

46. Ginsberg, H.N., and Illingworth, D.R. (2001) Postprandial Dyslipidemia: An Atherogenic Disorder Common in Patients with Diabetes Mellitus, *Am. J. Cardiol. 88*, 9H–15H.

47. Sniderman, A.D., Scantlebury, T., and Cianflone, K. (2001) Hypertriglyceridemic HyperapoB: The Unappreciated Atherogenic Dyslipoproteinemia in Type 2 Diabetes Mellitus, *Ann. Intern. Med. 135*, 447–459.

48. Idris, I., Gray, S., and Donnelly, R. (2001) Protein Kinase C Activation: Isozyme-Specific Effects on Metabolism and Cardiovascular Complications in Diabetes, *Diabetologia 44*, 659–673.

49. Kwag, O.G., Kim, S.O., Choi, J.H., Rhee, I.K., Choi, M.S., and Rhee, S.J. (2001) Vitamin E Improves Microsomal Phospholipase A2 Activity and the Arachidonic Acid Cascade in Kidney of Diabetic Rats, *J. Nutr. 131*, 1297–1301.

50. Sharma, A.K., Ponery, A.S., Lawrence, P.A., Ahmed, I., Bastaki, S.M., Dhanasekaran, S., Sheen, R.S., and Adeghate, E. (2001) Effect of Alpha-Tocopherol Supplementation on the Ultrastructural Abnormalities of Peripheral Nerves in Experimental Diabetes, *J. Peripher. Nerv. Syst. 6*, 33–39.

51. Kim, S.S., Gallaher, D.D., and Csallany, A.S. (2000) Vitamin E and Probucol Reduce Urinary Lipophilic Aldehydes and Renal Enlargement in Streptozotocin-Induced Diabetic Rats, *Lipids 35*, 1225–1237.

52. Ihara, Y., Yamada, Y., Toyokuni, S., Miyawaki, K., Ban, N., Adachi, T., Kuroe, A., Iwakura, T., Kubota, A., Hiai, H., and Seino, Y. (2000) Antioxidant Alpha-Tocopherol Ameliorates Glycemic Control of GK Rats, a Model of Type 2 Diabetes, *FEBS Lett. 473*, 24–26.

53. Knekt, P., Reunanen, A., Marniemi, J., Leino, A., and Aromaa, A. (1999) Low Vitamin E Status Is a Potential Risk Factor for Insulin-Dependent Diabetes Mellitus, *J. Intern. Med. 245*, 99–102.

54. Olejnicka, B.T., Andersson, A., Tyrberg, B., Dalen, H., and Brunk, U.T. (1999) Beta-Cells, Oxidative Stress, Lysosomal Stability, and Apoptotic/Necrotic Cell Death, *Antioxid. Redox. Signal. 1*, 305–315.

55. Jain, S.K., McVie, R., and Smith, T. (2000) Vitamin E Supplementation Restores Glutathione and Malondialdehyde to Normal Concentrations in Erythrocytes of Type 1 Diabetic Children, *Diabetes Care 23*, 1389–1394.

56. Engelen, W., Keenoy, B.M., Vertommen, J., and De Leeuw, I. (2000) Effects of Long-Term Supplementation with Moderate Pharmacologic Doses of Vitamin E Are Saturable and Reversible in Patients with Type 1 Diabetes, *Am. J. Clin. Nutr. 72*, 1142–1149.

57. Bursell, S.E., Clermont, A.C., Aiello, L.P., Aiello, L.M., Schlossman, D.K., Feener, E.P., Laffel, L., and King, G.L. (1999) High-Dose Vitamin E Supplementation Normalizes Retinal Blood Flow and Creatinine Clearance in Patients with Type 1 Diabetes, *Diabetes Care 22*, 1245–1251.

58. Skyrme-Jones, R.A., O'Brien, R.C., Berry, K.L., and Meredith, I.T. (2000) Vitamin E Supplementation Improves Endothelial Function in Type I Diabetes Mellitus: A Randomized, Placebo-Controlled Study, *J. Am. Coll. Cardiol. 36*, 94–102.

59. Skyrme-Jones, R.A., O'Brien, R.C., Luo, M., and Meredith, I.T. (2000) Endothelial Vasodilator Function Is Related to Low-Density Lipoprotein Particle Size and Low-Density Lipoprotein Vitamin E Content in Type 1 Diabetes, *J. Am. Coll. Cardiol. 35*, 292–299.

60. Mokdad, A.H., Ford, E.S., Bowman, B.A., Nelson, D.E., Engelgau, M.M., Vinicor, F., and Marks, J.S. (2001) The Continuing Increase of Diabetes in the US, *Diabetes Care 24*, 412.

61. Gokkusu, C., Palanduz, S., Ademoglu, E., and Tamer, S. (2001) Oxidant and Antioxidant Systems in NIDDM Patients: Influence of Vitamin E Supplementation, *Endocr. Res. 27*, 377–386.

62. Sharma, A., Kharb, S., Chugh, S.N., Kakkar, R., and Singh, G.P. (2000) Evaluation of Oxidative Stress Before and After Control of Glycemia and After Vitamin E Supplementation in Diabetic Patients, *Metabolism 49*, 160–162.

63. Paolisso, G., Tagliamonte, M.R., Barbieri, M., Zito, G.A., Gambardella, A., Varricchio, G., Ragno, E., and Varricchio, M. (2000) Chronic Vitamin E Administration Improves Brachial Reactivity and Increases Intracellular Magnesium Concentration in Type II Diabetic Patients, *J. Clin. Endocrinol. Metab. 85*, 109–115.

64. Mol, M.J., de Rijke, Y.B., Demacker, P.N., and Stalenhoef, A.F. (1997) Plasma Levels of Lipid and Cholesterol Oxidation Products and Cytokines in Diabetes Mellitus and Cigarette Smoking: Effects of Vitamin E Treatment, *Atherosclerosis 129*, 169–176.

65. Devaraj, S., and Jialal, I. (2000) Alpha Tocopherol Supplementation Decreases Serum C-Reactive Protein and Monocyte Interleukin-6 Levels in Normal Volunteers and Type 2 Diabetic Patients, *Free Radic. Biol. Med. 29*, 790–792.

66. Upritchard, J.E., Sutherland, W.H., and Mann, J.I. (2000) Effect of Supplementation with Tomato Juice, Vitamin E, and Vitamin C on LDL Oxidation and Products of Inflammatory Activity in Type 2 Diabetes, *Diabetes Care 23*, 733–738.

67. Ferber, P., Moll, K., Koschinsky, T., Rosen, P., Susanto, F., Schwippert, B., and Tschope, D. (1999) High Dose Supplementation of *RRR*-α-Tocopherol Decreases Cellular Hemostasis but Accelerates Plasmatic Coagulation in Type 2 Diabetes Mellitus, *Horm. Metab. Res. 31*, 665–671.

68. Sardas, S., Yilmaz, M., Oztok, U., Cakir, N., and Karakaya, A.E. (2001) Assessment of DNA Strand Breakage by Comet Assay in Diabetic Patients and the Role of Antioxidant Supplementation, *Mutat. Res. 490*, 123–129.

69. Sampson, M.J., Astley, S., Richardson, T., Willis, G., Davies, I.R., Hughes, D.A., and Southon, S. (2001) Increased DNA Oxidative Susceptibility Without Increased Plasma LDL Oxidizability in Type II Diabetes: Effects of Alpha-Tocopherol Supplementation, *Clin. Sci. 101*, 235–241.

70. Astley, S., Langrish-Smith, A., Southon, S., and Sampson, M. (1999) Vitamin E Supplementation and Oxidative Damage to DNA and Plasma LDL in Type 1 Diabetes, *Diabetes Care 22*, 1626–1631.

71. Bonfigli, A.R., Pieri, C., Manfrini, S., Testa, I., Sirolla, C., Ricciotti, R., Marra, M., Compagnucci, P., and Testa, R. (2001) Vitamin E Intake Reduces Plasminogen Activator Inhibitor Type 1 in T2DM Patients, *Diabetes Nutr. Metab. 14*, 71–77.

72. Dhalla, N.S., Liu, X., Panagia, V., and Takeda, N. (1998) Subcellular Remodeling and Heart Dysfunction in Chronic Diabetes, *Cardiovasc Res. 40*, 239–247.

73. Ponikowski, P., Anker, S.D., Chua, T.P., Szelemej, R., Piepoli, M., Adamopoulos, S., Webb-Peploe, K., Harrington, D., Banasiak, W., Wrabec, K., and Coats, A.J. (1997) Depressed Heart Rate Variability as an Independent Predictor of Death in Chronic Congestive Heart Failure Secondary to Ischemic or Idiopathic Dilated Cardiomyopathy, *Am. J. Cardiol. 79*, 1645–1650.

74. Manzella, D., Barbieri, M., Ragno, E., and Paolisso, G. (2001) Chronic Administration of Pharmacologic Doses of Vitamin E Improves the Cardiac Autonomic Nervous System in Patients with Type 2 Diabetes, *Am. J. Clin. Nutr. 73*, 1052–1057.

75. Gaede, P., Poulsen, H.E., Parving, H.H., and Pedersen, O. (2001) Double-Blind, Randomised Study of the Effect of Combined Treatment with Vitamin C and E on Albuminuria in Type 2 Diabetic Patients, *Diabet. Med. 18*, 756–760.

76. Parving, H.-H., Osterby, R., Anderson, P., and Hsueh, W. (1996) Diabetic Nephropathy, in *The Kidney* (Brenner, B., ed.) pp. 1864–1892, Saunders, Philadelphia.

77. Yukawa, S., Hibino, A., Maeda, T., Mimura, K., Yukawa, A., Maeda, A., Kishino, M., Sonobe, M., Mune, M., Yamada, Y., and et al. (1995) Effect of Alpha-Tocopherol on in Vitro and In Vivo Metabolism of Low-Density Lipoproteins in Haemodialysis Patients, *Nephrol. Dial. Transplant. 10*, 1–3.

78. Jackson, P., Loughrey, C.M., Lightbody, J.H., McNamee, P.T., and Young, I.S. (1995) Effect of Hemodialysis on Total Antioxidant Capacity and Serum Antioxidants in Patients with Chronic Renal Failure, *Clin. Chem. 41*, 1135–1138.

79. Tsumura, M., Kinouchi, T., Ono, S., Nakajima, T., and Komoda, T. (2001) Serum Lipid Metabolism Abnormalities and Change in Lipoprotein Contents in Patients with Advanced-Stage Renal Disease, *Clin. Chim. Acta 314*, 27–37.

80. Bommer, J., Strohbeck, E., Goerich, J., Bahner, M., and Zuna, I. (1996) Arteriosclerosis in Dialysis Patients, *Int. J. Artif. Organs 19*, 638–644.

81. Roob, J.M., Khoschsorur, G., Tiran, A., Horina, J.H., Holzer, H., and Winklhofer-Roob, B.M. (2000) Vitamin E Attenuates Oxidative Stress Induced by Intravenous Iron in Patients on Hemodialysis, *J. Am. Soc. Nephrol. 11*, 539–549.

82. Mune, M., Yukawa, S., Kishino, M., Otani, H., Kimura, K., Nishikawa, O., Takahashi, T., Kodama, N., Saika, Y., and Yamada, Y. (1999) Effect of Vitamin E on Lipid Metabolism and Atherosclerosis in ESRD Patients, *Kidney Int. Suppl. 71*, S126–S129.

83. Tarng, D.C., Huang, T.P., Liu, T.Y., Chen, H.W., Sung, Y.J., and Wei, Y.H. (2000) Effect of Vitamin E-Bonded Membrane on the 8-Hydroxy 2'-Deoxyguanosine Level in Leukocyte DNA of Hemodialysis Patients, *Kidney Int. 58*, 790–799.

84. Boaz, M., Smetana, S., Weinstein, T., Matas, Z., Gafter, U., Iaina, A., Knecht, A., Weissgarten, Y., Brunner, D., Fainaru, M., and Green, M.S. (2000) Secondary Prevention with Antioxidants of Cardiovascular Disease in Endstage Renal Disease (SPACE): Randomised Placebo-Controlled Trial, *Lancet 356*, 1213–1218.

85. Gazis, A., Page, S., and Cockcroft, J. (1997) Vitamin E and Cardiovascular Protection in Diabetes, *Br. Med. J. 314*, 1845–1846.

86. Yusuf, S., Dagenais, G., Pogue, J., Bosch, J., and Sleight, P. (2000) Vitamin E Supplementation and Cardiovascular Events in High-Risk Patients. The Heart Outcomes Prevention Evaluation Study Investigators, *N. Engl. J. Med. 342*, 154–160.

87. Facchini, F.S., Hua, N.W., Reaven, G.M., and Stoohs, R.A. (2000) Hyperinsulinemia: The Missing Link Among Oxidative Stress and Age-Related Diseases, *Free Radic. Biol. Med. 29*, 1302–1306.

88. Carantoni, M., Abbasi, F., Warmerdam, F., Klebanov, M., Wang, P.W., Chen, Y.D., Azhar, S., and Reaven, G.M. (1998) Relationship Between Insulin Resistance and Partially Oxidized LDL Particles in Healthy, Nondiabetic Volunteers, *Arterioscler. Thromb. Vasc. Biol. 18*, 762–767.

89. Nourooz-Zadeh, J., Gopaul, N.K., Barrow, S., Mallet, A.I., and Anggard, E.E. (1995) Analysis of F_2-Isoprostanes as Indicators of Non-Enzymatic Lipid Peroxidation In Vivo by Gas Chromatography-Mass Spectrometry: Development of a Solid-Phase Extraction Procedure, *J. Chromatogr. B Biomed. Appl. 667*, 199–208.

90. Facchini, F., Coulston, A.M., and Reaven, G.M. (1996) Relation Between Dietary Vitamin Intake and Resistance to Insulin-Mediated Glucose Disposal in Healthy Volunteers, *Am. J. Clin. Nutr. 63*, 946–949.

91. Facchini, F.S., Humphreys, M.H., DoNascimento, C.A., Abbasi, F., and Reaven, G.M. (2000) Relation Between Insulin Resistance and Plasma Concentrations of Lipid Hydroperoxides, Carotenoids, and Tocopherols, *Am. J. Clin. Nutr. 72*, 776–779.

92. Title, L.M., Cummings, P.M., Giddens, K., and Nassar, B.A. (2000) Oral Glucose Loading Acutely Attenuates Endothelium-Dependent Vasodilation in Healthy Adults Without Diabetes: An Effect Prevented by Vitamins C and E, *J. Am. Coll. Cardiol. 36*, 2185–2191.

93. Lepercq, J., Taupin, P., Dubois-Laforgue, D., Duranteau, L., Lahlou, N., Boitard, C., Landais, P., Hauguel-De Mouzon, S., and Timsit, J. (2001) Heterogeneity of Fetal Growth in Type 1 Diabetic Pregnancy, *Diabetes Metab. 27*, 339–344.

94. Nordstrom, L., Spetz, E., Wallstrom, K., and Walinder, O. (1998) Metabolic Control and Pregnancy Outcome Among Women with Insulin-Dependent Diabetes Mellitus. A Twelve-Year Follow-Up in the Country of Jamtland, Sweden, *Acta Obstet. Gynecol. Scand. 77*, 284–289.

95. Casson, I.F., Clarke, C.A., Howard, C.V., McKendrick, O., Pennycook, S., Pharoah, P.O., Platt, M.J., Stanisstreet, M., van Velszen, D., and Walkinshaw, S. (1997) Outcomes of Pregnancy in Insulin Dependent Diabetic Women: Results of a Five Year Population Cohort Study, *Br. Med. J. 315*, 275–278.

96. Kucera, J. (1971) Rate and Type of Congenital Anomalies Among Offspring of Diabetic Women, *J. Reprod. Med. 7*, 73–82.

97. Kinalski, M., Sledziewski, A., Telejko, B., Kowalska, I., Kretowski, A., Zarzycki, W., and Kinalska, I. (2001) Lipid Peroxidation, Antioxidant Defence and Acid-Base Status in Cord Blood at Birth: The Influence of Diabetes, *Horm. Metab. Res. 33*, 227–231.

98. Eriksson, U.J. (1999) Oxidative DNA Damage and Embryo Development, *Nat. Med. 5*, 715.

99. Kamath, U., Rao, G., Raghothama, C., Rai, L., and Rao, P. (1998) Erythrocyte Indicators of Oxidative Stress in Gestational Diabetes, *Acta Paediatr. 87*, 676–679.

100. Persson, B. (2001) Prevention of Fetal Malformation with Antioxidants in Diabetic Pregnancy, *Pediatr. Res. 49*, 742–743.

101. Cederberg, J., Siman, C.M., and Eriksson, U.J. (2001) Combined Treatment with Vitamin E and Vitamin C Decreases Oxidative Stress and Improves Fetal Outcome in Experimental Diabetic Pregnancy, *Pediatr. Res. 49*, 755–762.

102. Siman, C.M., Gittenberger-De Groot, A.C., Wisse, B., and Eriksson, U.J. (2000) Malformations in Offspring of Diabetic Rats: Morphometric Analysis of Neural Crest-Derived Organs and Effects of Maternal Vitamin E Treatment, *Teratology 61*, 355–367.

103. Kinalski, M., Sledziewski, A., Telejko, B., Zarzycki, W., and Kinalska, I.I. (1999) Antioxidant Therapy and Streptozotocin-Induced Diabetes in Pregnant Rats, *Acta Diabetol. 36*, 113–117.

104. Yang, X., Borg, L.A., Siman, C.M., and Eriksson, U.J. (1998) Maternal Antioxidant Treatments Prevent Diabetes-Induced Alterations of Mitochondrial Morphology in Rat Embryos, *Anat. Rec. 251*, 303–315.

105. Garner, P.R., D'Alton, M.E., Dudley, D.K., Huard, P., and Hardie, M. (1990) Preeclampsia in Diabetic Pregnancies, *Am. J. Obstet. Gynecol. 163*, 505–508.

106. Temple, R.C., Aldridge, V.A., Sampson, M.J., Greenwood, R.H., Heyburn, P.J., and Glenn, A. (2001) Impact of Pregnancy on the Progression of Diabetic Retinopathy in Type 1 Diabetes, *Diabetes Med. 18*, 573–577.

107. Oguz, H. (1999) Diabetic Retinopathy in Pregnancy: Effects on the Natural Course, *Semin. Ophthalmol. 14*, 249–257.

108. Sherer, T.B., Betarbet, R., and Greenamyre, J.T. (2001) Pathogenesis of Parkinson's Disease, *Curr. Opin. Investig. Drugs 2*, 657–662.

109. Maccioni, R.B., Munoz, J.P., and Barbeito, L. (2001) The Molecular Bases of Alzheimer's Disease and Other Neurodegenerative Disorders, *Arch. Med. Res. 32*, 367–381.

110. Friedman, A., and Galazka-Friedman, J. (2001) The Current State of Free Radicals in Parkinson's Disease. Nigral Iron as a Trigger of Oxidative Stress, *Adv. Neurol. 86*, 137–142.

111. Butterfield, D.A., and Kanski, J. (2001) Brain Protein Oxidation in Age-Related Neurodegenerative Disorders That Are Associated with Aggregated Proteins, *Mech. Ageing Dev. 122*, 945–962.

112. Schippling, S., Kontush, A., Arlt, S., Buhmann, C., Sturenburg, H.J., Mann, U., Muller-Thomsen, T., and Beisiegel, U. (2000) Increased Lipoprotein Oxidation in Alzheimer's Disease, *Free Radic. Biol. Med. 28*, 351–360.

113. Sayre, L.M., Zelasko, D.A., Harris, P.L., Perry, G., Salomon, R.G., and Smith, M.A. (1997) 4-Hydroxynonenal-Derived Advanced Lipid Peroxidation End Products Are Increased in Alzheimer's Disease, *J. Neurochem. 68*, 2092–2097.

114. Aksenova, M.V., Aksenov, M.Y., Payne, R.M., Trojanowski, J.Q., Schmidt, M.L., Carney, J.M., Butterfield, D.A., and Markesbery, W.R. (1999) Oxidation of Cytosolic Proteins and Expression of Creatine Kinase BB in Frontal Lobe in Different Neurodegenerative Disorders, *Dement. Geriatr. Cogn. Disord. 10*, 158–165.

115. Iida, T., Furuta, A., Nishioka, K., Nakabeppu, Y., and Iwaki, T. (2002) Expression of 8-Oxoguanine DNA Glycosylase Is Reduced and Associated with Neurofibrillary Tangles in Alzheimer's Disease Brain, *Acta Neuropathol. 103*, 20–25.

116. Zhang, J., Perry, G., Smith, M.A., Robertson, D., Olson, S.J., Graham, D.G., and Montine, T.J. (1999) Parkinson's Disease Is Associated with Oxidative Damage to Cytoplasmic DNA and RNA in Substantia Nigra Neurons, *Am. J. Pathol. 154*, 1423–1429.

117. Sasaki, N., Fukatsu, R., Tsuzuki, K., Hayashi, Y., Yoshida, T., Fujii, N., Koike, T., Wakayama, I., Yanagihara, R., Garruto, R., Amano, N., and Makita, Z. (1998) Advanced Glycation End Products in Alzheimer's Disease and Other Neurodegenerative Diseases, *Am. J. Pathol. 153*, 1149–1155.

118. Bartzokis, G., Cummings, J.L., Markham, C.H., Marmarelis, P.Z., Treciokas, L.J., Tishler, T.A., Marder, S.R., and Mintz, J. (1999) MRI Evaluation of Brain Iron in Earlier-

and Later-Onset Parkinson's Disease and Normal Subjects, *Magn. Reson. Imaging 17*, 213–222.

119. Smith, M.A., Harris, P.L., Sayre, L.M., and Perry, G. (1997) Iron Accumulation in Alzheimer Disease Is a Source of Redox-Generated Free Radicals, *Proc. Natl. Acad. Sci. USA 94*, 9866–9868.

120. Yatin, S., Yatin, M., Aulick, T., Ain, K., and Butterfiled, D. (1999) Alzheimer's Amyloid Beta-Peptide Associated Free Radicals Increase Rat Embryonic Neuronal Polyamine Uptake and Ornithine Decarboxylase Activity: Protective Effect of Vitamin E, *Neurosci. Lett. 263*, 17–20.

121. Yamada, K., Tanaka, T., Han, D., Senzaki, K., Kameyama, T., and Nabeshima, T. (1999) Protective Effects of Idebenone and Alpha-Tocopherol on Beta-Amyloid-(1-42)-Induced Learning and Memory Deficits in Rats: Implication of Oxidative Stress in Beta-Amyloid-Induced Neurotoxicity In Vivo, *Eur. J. Neurosci. 11*, 83–90.

122. Roghani, M., and Behzadi, G. (2001) Neuroprotective Effect of Vitamin E on the Early Model of Parkinson's Disease in Rat: Behavioral and Histochemical Evidence, *Brain Res. 892*, 211–217.

123. Li, Y., Liu, L., Barger, S.W., Mrak, R.E., and Griffin, W.S. (2001) Vitamin E Suppression of Microglial Activation Is Neuroprotective, *J. Neurosci. Res. 66*, 163–170.

124. Kolsch, H., Ludwig, M., Lutjohann, D., and Rao, M.L. (2001) Neurotoxicity of 24-Hydroxycholesterol, an Important Cholesterol Elimination Product of the Brain, May Be Prevented by Vitamin E and Estradiol-17β, *J. Neural. Transm. 108*, 475–488.

125. Butterfield, D.A., Koppal, T., Subramaniam, R., and Yatin, S. (1999) Vitamin E as an Antioxidant/Free Radical Scavenger Against Amyloid Beta-Peptide-Induced Oxidative Stress in Neocortical Synaptosomal Membranes and Hippocampal Neurons in Culture: Insights into Alzheimer's Disease, *Rev. Neurosci. 10*, 141–149.

126. Burton, G.W., Traber, M.G., Acuff, R.V., Walters, D.N., Kayden, H., Hughes, L., and Ingold, K.U. (1998) Human Plasma and Tissue Alpha-Tocopherol Concentrations in Response to Supplementation with Deuterated Natural and Synthetic Vitamin E, *Am. J. Clin. Nutr. 67*, 669–684.

127. Grundman, M. (2000) Vitamin E and Alzheimer Disease: The Basis for Additional Clinical Trials, *Am. J. Clin. Nutr. 71*, 630S–636S.

128. Vatassery, G., Bauer, T., and Dysken, M. (1999) High Doses of Vitamin E in the Treatment of Disorders of the Central Nervous System in the Aged, *Am. J. Clin. Nutr. 70*, 793–801.

129. McCarron, M., Russell, A., Metcalfe, R., and Deysilva, R. (1999) Chronic Vitamin E Deficiency Causing Spinocerebellar Degeneration, Peripheral Neuropathy, and Centro-Cecal Scotomata, *Nutrition 15*, 217–219.

130. Ko, H., and Park-Ko, I. (1999) Electrophysiologic Recovery After Vitamin E-Deficient Neuropathy, *Arch. Phys. Med. Rehabil. 80*, 964–967.

131. Zouari, M., Feki, M., Ben Hamida, C., Larnaout, A., Turki, I., Belal, S., Mebazaa, A., Ben Hamida, M., and Hentati, F. (1998) Electrophysiology and Nerve Biopsy: Comparative Study in Friedreich's Ataxia and Friedreich's Ataxia Phenotype with Vitamin E Deficiency, *Neuromuscul. Disord. 8*, 416–425.

132. Martinello, F., Fardin, P., Ottina, M., Ricchieri, G., Koenig, M., Cavalier, L., and Trevisan, C. (1998) Supplemental Therapy in Isolated Vitamin E Deficiency Improves the Peripheral Neuropathy and Prevents the Progression of Ataxia, *J. Neurol. Sci. 156*, 177–179.

133. Aparicio, J.M., Belanger-Quintana, A., Suarez, L., Mayo, D., Benitez, J., Diaz, M., and Escobar, H. (2001) Ataxia with Isolated Vitamin E Deficiency: Case Report and Review of the Literature, *J. Pediatr. Gastroenterol. Nutr. 33*, 206–210.

134. Sokol, R.J. (1990) Vitamin E and Neurologic Deficits, *Adv. Pediatr. 37*, 119–148.

135. Yokota, T., Igarashi, K., Uchihara, T., Jishage, K., Tomita, H., Inaba, A., Li, Y., Arita, M., Suzuki, H., Mizusawa, H., and Arai, H. (2001) Delayed-Onset Ataxia in Mice Lacking Alpha-Tocopherol Transfer Protein: Model for Neuronal Degeneration Caused by Chronic Oxidative Stress, *Proc. Natl. Acad. Sci. USA 98*, 15185–15190.

136. Morris, M.C., Beckett, L.A., Scherr, P.A., Hebert, L.E., Bennett, D.A., Field, T.S., and Evans, D.A. (1998) Vitamin E and Vitamin C Supplement Use and Risk of Incident Alzheimer Disease, *Alzheimer Dis. Assoc. Disord. 12*, 121–126.

137. Bourdel-Marchasson, I., Delmas-Beauvieux, M.C., Peuchant, E., Richard-Harston, S., Decamps, A., Reignier, B., Emeriau, J.P., and Rainfray, M. (2001) Antioxidant Defences and Oxidative Stress Markers in Erythrocytes and Plasma from Normally Nourished Elderly Alzheimer Patients, *Age Ageing 30*, 235–241.

138. Foy, C.J., Passmore, A.P., Vahidassr, M.D., Young, I.S., and Lawson, J.T. (1999) Plasma Chain-Breaking Antioxidants in Alzheimer's Disease, Vascular Dementia and Parkinson's Disease, *Q. J. Med. 92*, 39–45.

139. Sinclair, A.J., Bayer, A.J., Johnston, J., Warner, C., and Maxwell, S.R. (1998) Altered Plasma Antioxidant Status in Subjects with Alzheimer's Disease and Vascular Dementia, *Int. J. Geriatr. Psychiatry 13*, 840–845.

140. Jimenez-Jimenez, F.J., de Bustos, F., Molina, J.A., Benito-Leon, J., Tallon-Barranco, A., Gasalla, T., Orti-Pareja, M., Guillamon, F., Rubio, J.C., Arenas, J., and Enriquez-de-Salamanca, R. (1997) Cerebrospinal Fluid Levels of Alpha-Tocopherol (Vitamin E) in Alzheimer's Disease, *J. Neural Transm. 104*, 703–710.

141. Zaman, Z., Roche, S., Fielden, P., Frost, P.G., Niriella, D.C., and Cayley, A.C. (1992) Plasma Concentrations of Vitamins A and E and Carotenoids in Alzheimer's Disease, *Age Ageing 21*, 91–94.

142. Nicoletti, G., Crescibene, L., Scornaienchi, M., Bastone, L., Bagala, A., Napoli, I.D., Caracciolo, M., and Quattrone, A. (2001) Plasma Levels of Vitamin E in Parkinson's Disease, *Arch. Gerontol. Geriatr. 33*, 7–12.

143. Molina, J.A., de Bustos, F., Jimenez-Jimenez, F.J., Benito-Leon, J., Orti-Pareja, M., Gasalla, T., Tallon-Barranco, A., Navarro, J.A., Arenas, J., and Enriquez-de-Salamanca, R. (1997) Cerebrospinal Fluid Levels of Alpha-Tocopherol (Vitamin E) in Parkinson's Disease, *J. Neural Transm. 104*, 1287–1293.

144. Tohgi, H., Abe, T., Saheki, M., Hamato, F., Sasaki, K., and Takahashi, S. (1995) Reduced and Oxidized Forms of Glutathione and Alpha-Tocopherol in the Cerebrospinal Fluid of Parkinsonian Patients: Comparison Between Before and After L-Dopa Treatment, *Neurosci. Lett. 184*, 21–24.

145. Abbott, R.A., Cox, M., Markus, H., and Tomkins, A. (1992) Diet, Body Size and Micronutrient Status in Parkinson's Disease, *Eur. J. Clin. Nutr. 46*, 879–884.

146. Logroscino, G., Marder, K., Cote, L., Tang, M., Shea, S., and Mayeux, R. (1996) Dietary Lipids and Antioxidants in Parkinson's Disease: A Population-Based, Case-Control Study, *Ann. Neurol. 39*, 89–94.

147. De Rijk, M., Breteler, M., Den Breeijen, J., Launer, L., Grobbee, D., van der Meche, F., and Hofman, A. (1997) Dietary Antioxidants and Parkinson Disease. The Rotterdam Study, *Arch. Neurol. 54*, 762–765.

148. Kontush, A., Mann, U., Arlt, S., Ujeyl, A., Luhrs, C., Muller-Thomsen, T., and Beisiegel, U. (2001) Influence of Vitamin E and C Supplementation on Lipoprotein Oxidation in Patients with Alzheimer's Disease, *Free Radic. Biol. Med. 31*, 345–354.

149. Sano, M., Ernesto, C., Thomas, R.G., Klauber, M.R., Schafer, K., Grundman, M., Woodbury, P., Growdon, J., Cotman, C.W., Pfeiffer, E., Schneider, L.S., and Thal, L.J. (1997) A Controlled Trial of Selegiline, Alpha-Tocopherol, or Both as Treatment for Alzheimer's Disease. The Alzheimer's Disease Cooperative Study, *N. Engl. J. Med. 336*, 1216–1222.

150. Shoulson, I. (1998) DATATOP: A Decade of Neuroprotective Inquiry. Parkinson Study Group. Deprenyl and Tocopherol Antioxidative Therapy of Parkinsonism, *Ann. Neurol. 44*, S160–S166.

151. Vatassery, G.T., Fahn, S., and Kuskowski, M.A. (1998) Alpha Tocopherol in CSF of Subjects Taking High-Dose Vitamin E in the DATATOP Study. Parkinson Study Group, *Neurology 50*, 1900–1902.

152. Schmidt, R., Hayn, M., Reinhart, B., Roob, G., Schmidt, H., Schumacher, M., Watzinger, N., and Launer, L. (1998) Plasma Antioxidants and Cognitive Performance in Middle-Aged and Older Adults: Results of the Austrian Stroke Prevention Study, *J. Am. Geriatr. Soc. 46*, 1407–1410.

153. Perkins, A.J., Hendrie, H.C., Callahan, C.M., Gao, S., Unverzagt, F.W., Xu, Y., Hall, K.S., and Hui, S.L. (1999) Association of Antioxidants with Memory in a Multiethnic Elderly Sample Using the Third National Health and Nutrition Examination Survey, *Am. J. Epidemiol. 150*, 37–44.

154. La Rue, A., Koehler, K., Wayne, S., Chiulli, S., Haaland, K., and Garry, P. (1997) Nutritional Status and Cognitive Functioning in a Normally Aging Sample: a 6-y Reassessment [See Comments], *Am. J. Clin. Nutr. 65*, 20–29.

155. Jama, J., Launer, L., Witteman, J., den Breeijen, J., Breteler, M., Grobbee, D., and Hofman, A. (1996) Dietary Antioxidants and Cognitive Function in a Population-Based Sample of Older Persons. The Rotterdam Study, *Am. J. Epidemiol. 144*, 275–280.

156. Goodwin, J., Goodwin, J., and Garry, P. (1983) Association Between Nutritional Status and Cognitive Functioning in a Healthy Elderly Population, *J. Am. Med. Assoc. 249*, 2917–2921.

157. Haller, J., Weggemans, R.M., Lammi-Keefe, C.J., and Ferry, M. (1996) Changes in the Vitamin Status of Elderly Europeans: Plasma Vitamins A, E, B-6, B-12, Folic Acid and Carotenoids. SENECA Investigators, *Eur. J. Clin. Nutr. 50 (Suppl. 2)*, S32–S46.

158. Tagami, M., Ikeda, K., Yamagata, K., Nara, Y., Fujino, H., Kubota, A., Numano, F., and Yamori, Y. (1999) Vitamin E Prevents Apoptosis in Hippocampal Neurons Caused by Cerebral Ischemia and Reperfusion in Stroke-Prone Spontaneously Hypertensive Rats, *Lab. Investig. 79*, 609–615.

159. van der Worp, H.B., Bar, P.R., Kappelle, L.J., and de Wildt, D.J. (1998) Dietary Vitamin E Levels Affect Outcome of Permanent Focal Cerebral Ischemia in Rats, *Stroke 29*, 1002–1005; discussion 1005–1006.

160. Stohrer, M., Eichinger, A., Schlachter, M., and Stangassinger, M. (1998) Protective Effect of Vitamin E in a Rat Model of Focal Cerebral Ischemia, *Z. Naturforsch. [C] 53*, 273–278.

161. Gonzalez-Perez, O., Gonzalez-Castaneda, R.E., Huerta, M., Luquin, S., Gomez-Pinedo, U., Sanchez-Almaraz, E., Navarro-Ruiz, A., and Garcia-Estrada, J. (2002) Beneficial Effects of Alpha-Lipoic Acid Plus Vitamin E on Neurological Deficit, Reactive Gliosis and Neuronal Remodeling in the Penumbra of the Ischemic Rat Brain, *Neurosci. Lett. 321*, 100–104.

162. Masaki, K.H., Losonczy, K.G., Izmirlian, G., Foley, D.J., Ross, G.W., Petrovitch, H., Havlik, R., and White, L.R. (2000) Association of Vitamin E and C Supplement Use with Cognitive Function and Dementia in Elderly Men, *Neurology 54*, 1265–1272.

163. Yaar, M., and Gilchrest, B. (1998) Aging Versus Photoaging: Postulated Mechanisms and Effectors, *J. Investig. Dermatol. Symp. Proc. 3*, 47–51.

164. Darr, D., and Fridovich, I. (1994) Free Radicals in Cutaneous Biology, *J. Investig. Dermatol. 102*, 671–675.

165. Sander, C., Chang, H., Salzmann, S., Mueller, C., Ekanayake-Mudiyanselage, S., Elsner, P., and Thiele, J. (2002) Photoaging Is Associated with Protein Oxidation in Human Skin In Vivo, *J. Investig. Dermatol. 118*, 618–625.

166. Yamamoto, Y. (2001) Role of Active Oxygen Species and Antioxidants in Photoaging, *J. Investig. Dermatol. 27*, S1–S4.

167. Scharffetter-Kochanek, K., Brenneisen, P., Wenk, J., Herrmann, G., Ma, W., Kuhr, L., Meewes, C., and Wlaschek, M. (2000) Photoaging of the Skin from Phenotype to Mechanisms, *Exp. Gerontol. 35*, 307–316.

168. Shapiro, S., and Saliou, C. (2001) Role of Vitamins in Skin Care, *Nutrition 17*, 839–844.

169. Krol, E.S., Kramer-Stickland, K.A., and Liebler, D.C. (2000) Photoprotective Actions of Topically Applied Vitamin E, *Drug Metab. Rev. 32*, 413–420.

170. Fuchs, J. (1998) Potentials and Limitations of the Natural Antioxidants *RRR*-α-Tocopherol, L-Ascorbic Acid and β-Carotene in Cutaneous Photoprotection, *Free Radic. Biol. Med. 25*, 848–873.

171. Krol, E.S., Escalante, D.D., and Liebler, D.C. (2001) Mechanisms of Dimer and Trimer Formation from Ultraviolet-Irradiated α-Tocopherol, *Lipids 36*, 49–55.

172. Pietschmann, A., Kuklinski, B., and Otterstein, A. (1992) Protection from UV-Light-Induced Oxidative Stress by Nutritional Radical Scavengers [Article in German], *Z. Gesamte. Inn. Med. 47*, 518–522.

173. Thiele, J.J., Traber, M.G., and Packer, L. (1998) Depletion of Human Stratum Corneum Vitamin E: an Early and Sensitive In Vivo Marker of UV Induced Photo-Oxidation, *J. Investig. Dermatol. 110*, 756–761.

174. Liebler, D.C., and Burr, J.A. (2000) Effects of UV Light and Tumor Promoters on Endogenous Vitamin E Status in Mouse Skin, *Carcinogenesis 21*, 221–225.

175. Evelson, P., Ordonez, C.P., Llesuy, S., and Boveris, A. (1997) Oxidative Stress and In Vivo Chemiluminescence in Mouse Skin Exposed to UVA Radiation, *J. Photochem. Photobiol. B 38*, 215–219.

176. Moison, R., and Beijersbergen van Henegouwen, G. (2002) Topical Antioxidant Vitamins C and E Prevent UVB-Radiation-Induced Preoxidation of Eicosapentaenoic Acid in Pig Skin, *Radiat. Res. 157*, 402–409.

177. Lopez-Torres, M., Thiele, J.J., Shindo, Y., Han, D., and Packer, L. (1998) Topical Application of Alpha-Tocopherol Modulates the Antioxidant Network and Diminishes Ultraviolet-Induced Oxidative Damage in Murine Skin, *Br. J. Dermatol. 138*, 207–215.

178. Jones, S.A., McArdle, F., Jack, C.I., and Jackson, M.J. (1999) Effect of Antioxidant Supplementation on the Adaptive Response of Human Skin Fibroblasts to UV-Induced Oxidative Stress, *Redox Rep. 4*, 291–299.

179. Obermüller-Jevic, U.C., Francz, P.I., Frank, J., Flaccus, A., and Biesalski, H.K. (1999) Enhancement of the UVA Induction of Haem Oxygenase-1 Expression by Beta-Carotene in Human Skin Fibroblasts, *FEBS Lett. 460*, 212–216.

180. Giordani, A., Martin, M.E., Beaumont, C., Santus, R., and Morliere, P. (2000) Inactivation of Iron Responsive Element-Binding Capacity and Aconitase Function of Iron Regulatory Protein-1 of Skin Cells by Ultraviolet A, *Photochem. Photobiol. 72*, 746–752.

181. Peus, D., Meves, A., Pott, M., Beyerle, A., and Pittelkow, M. (2001) Vitamin E Analog Modulates UVB-Induced Signaling Pathway Activation and Enhances Cell Survival, *Free Radic. Biol. Med. 30*, 425–432.

182. Ricciarelli, R., Maroni, P., Ozer, N., Zingg, J., and Azzi, A. (1999) Age-Dependent Increase of Collagenase Expression Can Be Reduced by Alpha-Tocopherol via Protein Kinase C Inhibition, *Free Radic. Biol. Med. 27*, 729–737.

183. Yin, L., Morita, A., and Tsuji, T. (2000) Alterations of Extracellular Matrix Induced by Tobacco Smoke Extract, *Arch. Dermatol. Res. 292*, 188–194.

184. Thiele, J.J., and Packer, L. (1999) Noninvasive Measurement of Alpha-Tocopherol Gradients in Human Stratum Corneum by High-Performance Liquid Chromatography Analysis of Sequential Tape Strippings, *Methods Enzymol. 300*, 413–419.

185. Thiele, J.J., Weber, S.U., and Packer, L. (1999) Sebaceous Gland Secretion Is a Major Physiologic Route of Vitamin E Delivery to Skin, *J. Investig. Dermatol. 113*, 1006–1010.

186. Thiele, J.J., Traber, M.G., Polefka, T.G., Cross, C.E., and Packer, L. (1997) Ozone-Exposure Depletes Vitamin E and Induces Lipid Peroxidation in Murine Stratum Corneum, *J. Investig. Dermatol. 108*, 753–757.

187. Bissett, D., Chatterjee, R., and Hannon, D. (1990) Photoprotective Effect of Superoxide-Scavenging Antioxidants Against Ultraviolet Radiation-Induced Chronic Skin Damage in the Hairless Mouse, *Photodermatol. Photoimmunol. Photomed. 7*, 56–62.

188. Steenvoorden, D., and Beijersbergen van Henegouwen, G. (1999) Protection Against UV-Induced Systemic Immunosuppression in Mice by a Single Topical Application of the Antioxidant Vitamins C and E, *Int. J. Radiat. Biol. 75*, 747–755.

189. Gehring, W., Fluhr, J., and Gloor, M. (1998) Influence of Vitamin E Acetate on Stratum Corneum Hydration, *Arzneimittelforschung 48*, 772–775.

190. Dreher, F., Denig, N., Gabard, B., Schwindt, D.A., and Maibach, H.I. (1999) Effect of Topical Antioxidants on UV-Induced Erythema Formation When Administered After Exposure, *Dermatology 198*, 52–55.

191. Baschong, W., Artmann, C., Hueglin, D., and Roeding, J. (2001) Direct Evidence for Bioconversion of Vitamin E Acetate into Vitamin E: An Ex Vivo Study in Viable Human Skin, *J. Cosmet. Sci. 52*, 155–161.

192. Quevedo, W.C., Jr., Holstein, T.J., Dyckman, J., McDonald, C.J., and Isaacson, E.L. (2000) Inhibition of UVR-Induced Tanning and Immunosuppression by Topical Applications of Vitamins C and E to the Skin of Hairless (*hr/hr*) Mice, *Pigment Cell Res. 13*, 89–98.

193. Stahl, W., Heinrich, U., Jungmann, H., Sies, H., and Tronnier, H. (2000) Carotenoids and Carotenoids Plus Vitamin E Protect Against Ultraviolet Light-Induced Erythema in Humans, *Am. J. Clin. Nutr. 71*, 795–798.

194. Fuchs, J., and Kern, H. (1998) Modulation of UV-Light-Induced Skin Inflammation by D-α-Tocopherol and L-Ascorbic Acid: A Clinical Study Using Solar Simulated Radiation, *Free Radic. Biol. Med. 25*, 1006–1012.

195. Eberlein-Konig, B., Placzek, M., and Przybilla, B. (1998) Protective Effect Against Sunburn of Combined Systemic Ascorbic Acid (Vitamin C) and d-Alpha-Tocopherol (Vitamin E), *J. Am. Acad. Dermatol. 38*, 45–48.

196. Dreher, F., Gabard, B., Schwindt, D., and Maibach, H. (1998) Topical Melatonin in Combination with Vitamins E and C Protects Skin from Ultraviolet-Induced Erythema: A Human Study In Vivo, *Br. J. Dermatol. 139*, 332–339.

197. Rowe, J.W., and Kahn, R.L. (1987) Human Aging: Usual and Successful, *Science 237*, 143–149.

198. Stahelin, H.B. (1999) The Impact of Antioxidants on Chronic Disease in Ageing and in Old Age, *Int. J. Vitam. Nutr. Res. 69*, 146–149.

199. Schmidt, K. (1999) Physiology and Pathophysiology of Senescence, *Int. J. Vitam. Nutr. Res. 69*, 150–153.

200. Drewnowski, A., and Shultz, J.M. (2001) Impact of Aging on Eating Behaviors, Food Choices, Nutrition, and Health Status, *J. Nutr. Health Aging 5*, 75–79.

201. Kasper, H. (1999) Vitamin Absorption in the Elderly, *Int. J. Vitam. Nutr. Res. 69*, 169–172.

202. Haller, J. (1999) The Vitamin Status and Its Adequacy in the Elderly: An International Overview, *Int. J. Vitam. Nutr. Res. 69*, 160–168.

203. Foote, J.A., Giuliano, A.R., and Harris, R.B. (2000) Older Adults Need Guidance to Meet Nutritional Recommendations, *J. Am. Coll. Nutr. 19*, 628–640.

204. Ravaglia, G., Forti, P., Maioli, F., Bastagli, L., Facchini, A., Mariani, E., Savarino, L., Sassi, S., Cucinotta, D., and Lenaz, G. (2000) Effect of Micronutrient Status on Natural Killer Cell Immune Function in Healthy Free-Living Subjects Aged >90 y, *Am. J. Clin. Nutr. 71*, 590–598.

205. Mecocci, P.,. Polidori, M.C., Troiano, L., Cherubini, A., Cecchetti, R., Pini, G., Straatman, M., Monti, D., Stahl, W., Sies, H., Franceschi, C., and Senin, U. (2000) Plasma Antioxidants and Longevity: A Study on Healthy Centenarians, *Free Radic. Biol. Med. 28*, 1243–1248.

206. Paolisso, G., Tagliamonte, M.R., Rizzo, M.R., Manzella, D., Gambardella, A., and Varricchio, M. (1998) Oxidative Stress and Advancing Age: Results in Healthy Centenarians, *J. Am. Geriatr. Soc. 46*, 833–838.

207. Klapcinska, B., Derejczyk, J., Wieczorowska-Tobis, K., Sobczak, A., Sadowska-Krepa, E., and Danch, A. (2000) Antioxidant Defense in Centenarians (a Preliminary Study), *Acta Biochim. Pol. 47*, 281–292.

208. Reddy, K.K., Rao, A.P., and Reddy, T.P. (1999) Serum Vitamins E, A and Lipid Peroxidation Levels in Kurichias, an Indian Tribal Population, *Indian J. Biochem. Biophys. 36*, 44–50.

209. Ravaglia, G., Forti, P., Maioli, F., Nesi, B., Pratelli, L., Savarino, L., Cucinotta, D., and Cavalli, G. (2000) Blood Micronutrient and Thyroid Hormone Concentrations in the Oldest-Old, *J. Clin. Endocrinol. Metab. 85*, 2260–2265.

210. De Waart, F.G., Schouten, E.G., Stalenhoef, A.F., and Kok, F.J. (2001) Serum Carotenoids, Alpha-Tocopherol and Mortality Risk in a Prospective Study Among Dutch Elderly, *Int. J. Epidemiol. 30*, 136–143.

211. Mezzetti, A., Zuliani, G., Romano, F., Costantini, F., Pierdomenico, S.D., Cuccurullo, F., and Fellin, R. (2001) Vitamin E and Lipid Peroxide Plasma Levels Predict the Risk of Cardiovascular Events in a Group of Healthy Very Old People, *J. Am. Geriatr. Soc. 49*, 533–537.

212. Cherubini, A., Zuliani, G., Costantini, F., Pierdomenico, S.D., Volpato, S., Mezzetti, A., Mecocci, P., Pezzuto, S., Bregnocchi, M., Fellin, R., and Senin, U. (2001) High Vitamin E Plasma Levels and Low Low-Density Lipoprotein Oxidation Are Associated with the Absence of Atherosclerosis in Octogenarians, *J. Am. Geriatr. Soc. 49*, 651–654.

213. Miller, R.A. (1997) The Aging Immune System: Subsets, Signals, and Survival, *Aging 9*, 23–24.

214. Serafini, M. (2000) Dietary Vitamin E and T Cell-Mediated Function in the Elderly: Effectiveness and Mechanism of Action, *Int. J. Dev. Neurosci. 18*, 401–410.

215. Wu, D., Hayek, M.G., and Meydani, S. (2001) Vitamin E and Macrophage Cyclooxygenase Regulation in the Aged, *J. Nutr. 131*, 382S–388S.

216. Pallast, E.G., Schouten, E.G., de Waart, F.G., Fonk, H.C., Doekes, G., von Blomberg, B.M., and Kok, F.J. (1999) Effect of 50- and 100-mg Vitamin E Supplements on Cellular Immune Function in Noninstitutionalized Elderly Persons, *Am. J. Clin. Nutr. 69*, 1273–1281.

217. Meydani, S.N., Meydani, M., Blumberg, J.B., Leka, L.S., Siber, G., Loszewski, R., Thompson, C., Pedrosa, M.C., Diamond, R.D., and Stollar, B.D. (1997) Vitamin E Supplementation and In Vivo Immune Response in Healthy Elderly Subjects. A Randomized Controlled Trial, *J. Am. Med. Assoc. 277*, 1380–1386.

218. De Waart, F.G., Portengen, L., Doekes, G., Verwaal, C.J., and Kok, F.J. (1997) Effect of 3 Months Vitamin E Supplementation on Indices of the Cellular and Humoral Immune Response in Elderly Subjects, *Br. J. Nutr. 78*, 761–774.

219. Singh, V.N. (1992) A Current Perspective on Nutrition and Exercise, *J. Nutr. 122*, 760–765.

220. Quintanilha, A.T., and Packer, L. (1983) Vitamin E, Physical Exercise and Tissue Oxidative Damage, *Ciba Found. Symp. 101*, 56–69.

221. Sacheck, J.M., and Blumberg, J.B. (2001) Role of Vitamin E and Oxidative Stress in Exercise, *Nutrition 17*, 809–814.

222. Evans, W.J. (2000) Vitamin E, Vitamin C, and Exercise, *Am. J. Clin. Nutr. 72*, 647S–652S.

223. Moller, P., Loft, S., Lundby, C., and Olsen, N.V. (2001) Acute Hypoxia and Hypoxic Exercise Induce DNA Strand Breaks and Oxidative DNA Damage in Humans, *FASEB J. 15*, 1181–1186.

224. Alessio, H.M., Hagerman, A.E., Fulkerson, B.K., Ambrose, J., Rice, R.E., and Wiley, R.L. (2000) Generation of Reactive Oxygen Species After Exhaustive Aerobic and Isometric Exercise, *Med. Sci. Sports Exerc. 32*, 1576–1581.

225. Suzuki, K., Totsuka, M., Nakaji, S., Yamada, M., Kudoh, S., Liu, Q., Sugawara, K., Yamaya, K., and Sato, K. (1999) Endurance Exercise Causes Interaction Among Stress Hormones, Cytokines, Neutrophil Dynamics, and Muscle Damage, *J. Appl. Physiol. 87*, 1360–1367.

226. Gohil, K., Viguie, C., Stanley, W.C., Brooks, G.A., and Packer, L. (1988) Blood Glutathione Oxidation During Human Exercise, *J. Appl. Physiol. 64*, 115–119.

227. Davies, K.J., Quintanilha, A.T., Brooks, G.A., and Packer, L. (1982) Free Radicals and Tissue Damage Produced by Exercise, *Biochem. Biophys. Res. Commun. 107*, 1198–1205.

228. Child, R.B., Wilkinson, D.M., and Fallowfield, J.L. (1999) Resting Serum Antioxidant Status Is Positively Correlated with Peak Oxygen Uptake in Endurance Trained Runners, *J. Sports Med. Phys. Fitness 39*, 282–284.

229. Liu, M.L., Bergholm, R., Makimattila, S., Lahdenpera, S., Valkonen, M., Hilden, H., Yki-Jarvinen, H., and Taskinen, M.R. (1999) A Marathon Run Increases the Susceptibility of LDL to Oxidation In Vitro and Modifies Plasma Antioxidants, *Am. J. Physiol. 276*, E1083–E1091.

230. Child, R.B., Wilkinson, D.M., Fallowfield, J.L., and Donnelly, A.E. (1998) Elevated Serum Antioxidant Capacity and Plasma Malondialdehyde Concentration in Response to a Simulated Half-Marathon Run, *Med. Sci. Sports Exerc. 30*, 1603–1607.
231. Pincemail, J., Deby, C., Camus, G., Pirnay, F., Bouchez, R., Massaux, L., and Goutier, R. (1988) Tocopherol Mobilization During Intensive Exercise, *Eur. J. Appl. Physiol. Occup. Physiol. 57*, 189–191.
232. Dillard, C.J., Litov, R.E., Savin, W.M., Dumelin, E.E., and Tappel, A.L. (1978) Effects of Exercise, Vitamin E, and Ozone on Pulmonary Function and Lipid Peroxidation, *J. Appl. Physiol. 45*, 927–932.
233. Hartmann, A., Niess, A.M., Grunert-Fuchs, M., Poch, B., and Speit, G. (1995) Vitamin E Prevents Exercise-Induced DNA Damage, *Mutat. Res. 346*, 195–202.
234. Grievink, L., Zijlstra, A.G., Ke, X., and Brunekreef, B. (1999) Double-Blind Intervention Trial on Modulation of Ozone Effects on Pulmonary Function by Antioxidant Supplements, *Am. J. Epidemiol. 149*, 306–314.
235. Tauler, P., Aguilo, A., Fuentespoina, E., Tur, J.A., and Pons, A. (2002) Diet Supplementation with Vitamin E, Vitamin C and Beta-Carotene Cocktail Enhances Basal Neutrophil Antioxidant Enzymes in Athletes, *Pflugers Arch. 443*, 791–797.
236. Kaikkonen, J., Kosonen, L., Nyyssonen, K., Porkkala-Sarataho, E., Salonen, R., Korpela, H., and Salonen, J.T. (1998) Effect of Combined Coenzyme Q10 and d-α-Tocopheryl Acetate Supplementation on Exercise-Induced Lipid Peroxidation and Muscular Damage: A Placebo-Controlled Double-Blind Study in Marathon Runners, *Free Radic. Res. 29*. 85–92.
237. Beaton, L., Allan, D., Tarnopolsky, M., Tiidus, P., and Phillips, S. (2002) Contraction-Induced Muscle Damage Is Unaffected by Vitamin E Supplementation, *Med. Sci. Sports Exerc. 34*, 798–805.
238. Itoh, H., Ohkuwa, T., Yamazaki, Y., Shimoda, T., Wakayama, A., Tamura, S., Yamamoto, T., Sato, Y., and Miyamura, M. (2000) Vitamin E Supplementation Attenuates Leakage of Enzymes Following Six Successive Days of Running Training, *Int. J. Sports Med. 21*, 369–374.
239. Schroeder, H., Navarro, E., Mora, J., Galiano, D., and Tramullas, A. (2001) Effects of Alpha-Tocopherol, Beta-Carotene and Ascorbic Acid on Oxidative, Hormonal and Enzymatic Exercise Stress Markers in Habitual Training Activity of Professional Basketball Players, *Eur. J. Nutr. 40*, 178–184.

Index